CALCULUS
Simplified

Other Books by Oscar Fernandez

Everyday Calculus: Discovering the Hidden Math All Around Us

The Calculus of Happiness: How a Mathematical Approach to Life Adds Up to Health, Wealth, and Love

CALCULUS *Simplified*

Oscar E. Fernandez

PRINCETON UNIVERSITY PRESS
PRINCETON AND OXFORD

Copyright © 2019 by Princeton University Press
Published by Princeton University Press
41 William Street, Princeton, New Jersey 08540
6 Oxford Street, Woodstock, Oxfordshire OX20 1TR

press.princeton.edu

All Rights Reserved

LCCN 2019936027
ISBN 978-0-691-17539-3

British Library Cataloging-in-Publication Data is available

Editorial: Vickie Kearn, Susannah Shoemaker, and Lauren Bucca
Production Editorial: Kathleen Cioffi
Text Design: Lorraine Doneker
Cover design: Layla Mac Rory
Production: Erin Suydam
Publicity: Matthew Taylor and Kathryn Stevens
Copyeditor: Theresa Kornak

This book has been composed in MinionPro

Printed on acid-free paper. ∞

Printed in the United States of America

1 3 5 7 9 10 8 6 4 2

To Emilia and Alicia

Many years from now when you confront calculus,
Come back to this book and give it a read.
Come back to me, too, and give me a hug.
As much as I love math, I love you both much, much more.

Contents

Preface

Hi. Welcome to *Calculus Simplified*. My name is Oscar Fernandez, Associate Professor of Mathematics and Faculty Director of the Pforzheimer Learning and Teaching Center at Wellesley College, and I will be your instructor.

Who Is This Book Intended For?

Here are three questions that will help you determine if this book is for *you*.

- **Do you have a background in algebra, geometry, and some exposure to functions (exposure to transcendental functions—exponentials, logs, trigonometry—is not required)?** If so, this book is for you.
- **Are you currently enrolled in a calculus course (or soon will be)?** If so, this book is for you.
- **Did you learn calculus long ago and are now seeking a quick refresher on the subject?** If so, this book is for you.

If you answered "no" to all of those questions, this book *might* not be for you. I encourage you to skim it first to see if it may still be an appropriate resource for you. If you answered "yes" to any of those questions, great! Read on.

Reason 1 You Should Use This Book: Its Goldilocks Approach to Calculus

Cognitive scientists have accumulated evidence over the past few decades supporting what is today called the "Goldilocks Effect": we learn best when the content being taught contains just the right amount of challenge and complexity—not too much, and not too little.

Consider now the challenge of learning calculus. The typical calculus student turns to three particular resources for help: a calculus textbook, a calculus professor, and a calculus supplement. Each of these resources, however, has its strengths and weaknesses. I've highlighted three particular dimensions along which to understand those strengths and weaknesses in (a) of Figure 1: level of detail, personalization of content, and depth of insights.

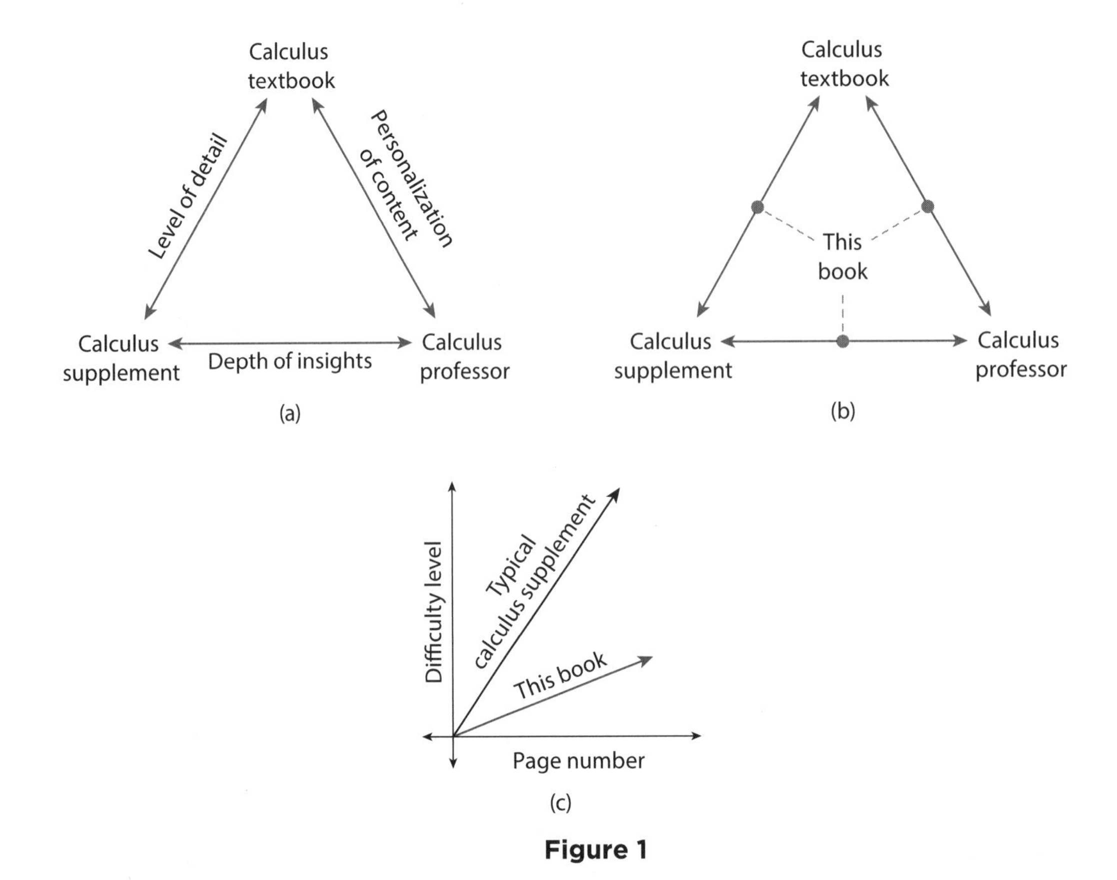

Figure 1

Compare the aforementioned three calculus resources along those three dimensions and here are some of the things you'd notice:

- **Regarding the level of detail.** Most calculus supplements (e.g., *Calculus for Dummies*) are devoid of formal statements of theorems. That means it isn't always clear when one can apply the formulas and techniques discussed (this is clearly articulated by the hypotheses of a theorem). Most calculus textbooks, on the other hand, have the opposite problem—they are replete with formal statements of theorems (and their proofs). The result is a calculus learning experience that feels too formal, where proof and rigor often overcomplicate explanations and obscure the intuition behind the concepts. Conclusion: too little detail may give you a false sense of confidence in your calculus knowledge; too much detail may turn you off from calculus altogether.

- **Regarding the depth of insights.** Most calculus supplements offer only superficial mathematical insights, focusing instead more on teaching computational skills, procedures, and techniques. (Example: "Do *this* when you see *that*.") In fairness, most calculus supplements are *supplements*; the idea is to use them alongside a calculus textbook and/or calculus professor, which furnish those deeper mathematical insights. Often these resources go too far, however, as evidenced by one of the most common nuggets of student feedback in a calculus course: "less theory, more examples, please." Conclusion: too little depth of insight will make learning calculus feel like rote learning; too much depth of insight makes calculus feel too theoretical and impractical.

- **Regarding the personalization of content.** Most calculus textbooks are thousand-page tomes containing way more content than any calculus professor can cover in a Calculus 1 course. The average calculus textbook, therefore, is not at all personalized to your interests. We calculus professors do our best to distill the hundreds of pages of content provided by the calculus textbook into roughly 30 hour-long lessons, ideally taking into consideration the particular interests of the students in the course. This is an improvement over the calculus textbook, for sure. But in a class of *many* students it is still hard to personalize the content to *each* student. Conclusion: too little personalization of content is a wasted opportunity to engage you in learning calculus; however, the amount the average calculus instructor provides, while an improvement over that of a calculus textbook's, is still not personalized enough.

The conclusion: *none of the resources just discussed are "just right" for learning calculus.* That's where this book comes in.

This book takes a "Goldilocks approach" to learning calculus.

As Figure 1(b) is meant to illustrate:

- **This book balances intuition with theory to provide you with just the right level of detail.** Chapter 1 teaches you the core ideas of calculus. What you will learn there will anchor *all* of what you will learn in the rest of the book. That's because the Chapter focuses on developing the intuition behind the main concepts, mindset, and overarching framework of calculus. Subsequent chapters discuss the math of calculus, with just the right balance of formal statements of definitions and theorems, so that you learn the terminology of calculus and understand the *full* story, complete with when and why it works.
- **This book allows you to personalize your calculus adventure.** First of all:

  No prior knowledge of exponential, logarithmic, or trigonometric functions is needed to learn calculus from this book.

  Don't know what $\sin x$ is (or don't yet fully understand it)? Same issue with e^x or $\ln x$? No problem; the calculus of such functions is left until the end of

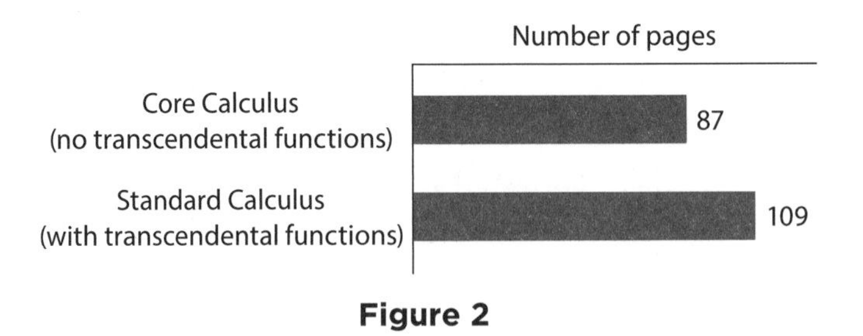

Figure 2

every section (discussed under the heading Transcendental Tales). Include it in your reading if you'd like; skip it if you'd rather not. The choice is yours.

Additionally, sections *do* include discussions of theory, but focus on intuition rather than proof. More technical discussions and proofier content is relegated to the chapter appendixes (which you can download from the book's website). Include it in your reading if you'd like; skip it if you'd rather not. The choice is yours.

Finally, the same approach is used with the more in-depth real-world applications. These are included in Appendix C. Include it in your reading if you'd like; skip it if you'd rather not. The choice is yours.

The net effect of all this: a gentler learning experience, as illustrated in Figure 1(c), and a focus on intuition (hence the *Calculus Simplified* title).

- **This book provides just the right amount of depth to the mathematical insights unearthed.** You will learn both the "how" *and* the "why" of calculus. You will understand why its core concepts are important. You will be exposed to the various other places (e.g., real-world contexts) where calculus concepts show up. And if you want to, you will also understand the historical context that motivated the invention of calculus (some of that content is optional and contained in the chapter appendixes).

Reason 2 You Should Use This Book: The Bonus Features

The Goldilocks approach used in this book is complemented by the following additional features designed to supercharge your learning of calculus.

- **A focus on conciseness.** Excluding the exercises and appendixes:

  This book teaches you calculus in at most 110 pages.

  Exclude transcendental functions (exponentials, logarithms, and trigonometric functions) and you're down to 87 pages (Figure 2). Both of these numbers are considerably less than the roughly 200–300 pages the typical calculus textbook takes to teach the same content (again excluding the exercises and appendixes).

- **Nearly 200 solved examples.** I've included 196 distinct solved examples in this book. I also included more than just the calculations for many of them—I wrote out my thought process too. This will help you learn to think like a mathematician thinks about calculus.

- **Answers to *all* non-proof exercises.** There are 337 exercises in this book. I have included answers to *all* of the exercises that are not proofs or derivations; answers start on page 227.

- **Use of color and boxes to separate content.** In-chapter definitions, theorems, and important takeaways appear in blue boxes to help you easily spot them. End-of-chapter exercises are also color-coded: exercises colored blue are applied exercises. And of course, colors are used to help explain various concepts.

- **Interactive online content linked to book content.** In mathematics, an interactive graph—or a video lesson—is sometimes worth a million words. I've capitalized on the learning power of these digital resources by creating several ones based on the content in this book. You'll find references to them throughout the book; a link to the graphs and other digital resources can be accessed on the book's website:

  https://press.princeton.edu/titles/13351.html

- **Inclusion of references.** Sources for external content used—and for additional useful resources—are provided in the Bibliography. References appear in-text as brackets and use a number (e.g., [3]) to identify the entry in the bibliography being referenced.

- **A direct line to the author.** I wrote this book to help you learn calculus. I stand by this goal. So, *feel free to email me with any questions, comments, or suggestions.* Seriously. Here is my email:

  math@surroundedbymath.com

  I also encourage you to submit feedback on the book via the link below.

  https://goo.gl/forms/yOIFolqTEEdzkVhr2

  Your submission will be anonymous. Your feedback will help me improve the book and will be incorporated into future editions.

Parting Thoughts

Calculus Simplified is a resource for anyone interested in learning (or relearning) Calculus 1. First and foremost, it is an attempt to re-structure the standard presentation of content in a calculus textbook to strike that "just right" balance on the level of detail, depth of insights, and personalization of content illustrated in Figures 1(a) and (b).

Second, *Calculus Simplified* was designed to streamline your learning of calculus. But do not confuse "streamline" with "water down." This book is not a collection of calculus formulas or merely a quick review of calculus concepts (which presumes you already know calculus). It is not an idiot's or dummies guide to calculus (you are neither). *Calculus Simplified* is a college-level Calculus 1 course—based on the notes I created for teaching calculus—streamlined to eliminate excess content that gets in the way of learning calculus, written in more informal prose, including relevant real-world examples, and structured to afford you multiple routes to learning calculus.

Finally, though the vast majority of topics covered in a calculus textbook are also covered in *Calculus Simplified*, this book is not intended to be a comprehensive treatment of Calculus 1. In the present conception of what a calculus textbook consists of, this book is also not intended to be a textbook (though it can certainly be used as such in some settings). At the same time, *Calculus Simplified* is much more than the run-of-the-mill calculus supplement. Explained within the theme of this Preface, I think of *Calculus Simplified* as occupying the "Goldilocks zone" between a calculus textbook and a calculus supplement.

I am excited to begin working with you as you set off on your calculus adventure. Once you are done learning the calculus in this book, I encourage you to read the Epilogue; it contains some useful advice and encouragement for navigating mathematics beyond calculus. See you in Chapter 1!

Oscar E. Fernandez
Brookline, MA

To the Student

Welcome to *Calculus Simplified!* Before you embark on your adventure through this book, I thought I would give you a few practical tips intended to help you conquer calculus.

What You Can Expect to Learn from This Book

This subheading's title might seem silly, given that this is a book about calculus. But research shows that students learn best when they know *beforehand* what they are about to learn and what they should be able to do with that knowledge at the end of the lesson. Rather than insert these learning goals and objectives at the start of each lesson, I have provided them in a supplemental document titled *Calculus Simplified: Learning Goals and Objectives*, available on the book's website. (That document also maps this book's learning goals and objectives to the curriculum framework used in AP Calculus, in case you're enrolled in such a course.) I highly recommend you keep that document handy as you read through each section of the book.

How to Read a Math Book (Including This One)

Though I have done my best to infuse my writing with the elements of a novel—characters, a plot, etc.—this book is not a novel. One thing this implies is that you need to read this book differently than you would a novel. For example, simply *reading* this book won't help you understand calculus. Rather, I recommend you *work through* this book—work out the examples, work out the solutions to the exercises, work through the supplemental content. By *doing* mathematics you will be helping yourself *learn* mathematics. Moreover, jot down questions and comments as you read and work through this book. This will ensure you are learning *actively* rather than passively.

Lastly, let me mention the special role that theorems play in mathematics, and how to ensure you're getting the most out of them. Loosely speaking, a theorem is a statement that has been proven true. A typical theorem has the following structure: preamble, hypotheses, conclusion. Example:

Theorem (Pythagorean Theorem): Consider a right triangle in the plane. Let c denote the length of the hypotenuse of the triangle and a and b the lengths of the other two sides. Then $c^2 = a^2 + b^2$.

In this theorem the first sentence is the preamble; its role is to provide context for what the theorem says. The second sentence in the theorem contains some assumptions (as happened here, sometimes the preamble also contains assumptions). The last sentence contains the conclusion.

Echoing my earlier advice to work through this book, do the same with theorems—whenever you come across a theorem take a moment to understand what it is saying. Try to draw pictures, explain the theorem in words, and imagine removing some of the hypotheses to see how the conclusion might be affected. Doing all this will help you appreciate what the real use of the theorem is, help you remember it, and help you learn when it can (and cannot) be applied.

How to Become a Better Student

I have one last recommendation for you: *employ the latest research from the science of learning while you study*. Study strategies like retrieval practice and interleaving—both backed by cognitive science research—can supercharge your studying. You can read more about these and other research-backed study strategies in the supplemental document titled *Evidence-Based Study Strategies*, available on the book's website.

Alright, that's all I have for you at the moment. Let's get started with your calculus adventure!

To the Instructor

You might be thinking, "Not another calculus book!" But this one is different. My goal is neither to add another calculus textbook to the volumes of such books nor to provide students with an overly simplistic treatment of the subject. Instead, as I wrote in the preface, *Calculus Simplified* is an attempt to re-structure the standard presentation of content in a calculus textbook to strike that "just right" balance on the level of detail, depth of insights, and personalization of content.

Calculus Simplified is also a sign of the times. In the age of Twitter, it has become increasingly clear that shorter and more succinct treatments of calculus are favored by students. The byte-sized explanations utilized in this book, along with the freedom to add or exclude content related to transcendental functions, cut down the time investment necessary for a student to quickly familiarize herself with a calculus concept. As such, I have used this very text successfully not just as a main text in introductory calculus courses, but also as a quick reference text in more rigorous calculus courses (especially due to its excellent breadth of exercises).

One final note that you may find helpful: the explicit connection to the AP Calculus Mathematical Practices curriculum framework. Even if you are not teaching an AP Calculus course, I highly recommend reading the Fall 2016 revision of the AP Calculus curriculum. In addition to discussing a variety of useful teaching techniques for calculus, the document details six broad learning goals for calculus (these are the Mathematical Practices) along with detailed learning objectives for each concept covered in AP Calculus. On the website that accompanies this book, you will find a supplemental document titled *Calculus Simplified: Learning Goals and Objectives* that maps each of the AP Calculus framework's learning goals and objectives to the associated sections and exercises in this book. This makes it especially easy to use this book in just-in-time fashion to provide content and exercises for whatever calculus topic you are teaching on a given day.

Before You Begin . . .

Here is some useful information that will help you navigate this book.

Numbering scheme

- Where applicable, equations in the chapters are numbered in (chapter.equation) format. Example: equation (3.17) refers to the 17th numbered equation in Chapter 3.
- Equation numbers appear flushed right in the book, like this: (3.17)
- Figures and tables in the chapters follow the same numbering scheme as equations, except no parentheses surround the "x.y" reference.
- Equations and figures/tables in appendixes A, B, and C are numbered in (appendix.equation) format. Example: equation (B.7) refers to the 7th numbered equation in Appendix B.
- Equations and figures/tables in the appendix to chapter X are numbered as (AX.y) and AX.y, respectively. Example: equation (A1.5) refers to the 5th numbered equation in the appendix to Chapter 1.

Color-coded alerts

- Definitions and theorems appear next to thin blue vertical rectangles, like this:

  Theorem 3.2 Theorem text . . .

- Appendix C contains applied examples too long to include in the chapters. These are referenced with light blue rectangles, like this:

  Reference Text Short description of applied example.

- Suggested end-of-chapter exercises are referenced like this:

  Related Exercises 3, 4, etc.

- Thin blue-colored rectangles in the margin—like the one to the right—alert you to online interactive versions of the graph(s) or content being discussed (accessible via the book's website, listed in the Preface).

Interactive Figure

- Content involving transcendental functions (exponential, logarithmic, and trigonometric functions) appears under the subheading **Transcendental Tales**.
- In many sections I suggest tips and summarize takeaways; these appear under the subheading **Tips, Tricks, and Takeaways**.

CALCULUS
Simplified

1 The Fast Track Introduction to Calculus

Chapter Preview. *Calculus is a new way of thinking about mathematics. This chapter provides you with a working understanding of the calculus mindset, core concepts of calculus, and the sorts of problems they help solve. The focus throughout is on the* ideas *behind calculus (the big picture of calculus); the subsequent chapters discuss the* math *of calculus. After reading this chapter, you will have an intuitive understanding of calculus that will ground your subsequent studies of the subject. Ready? Let's start the adventure!*

1.1 What Is Calculus?

Here's my two-part answer to that question:

Calculus is a mindset—a dynamics mindset. Contentwise, calculus is the mathematics of infinitesimal change.

Calculus as a Way of Thinking

The mathematics that precedes calculus—often called "pre-calculus," which includes algebra and geometry—largely focuses on *static* problems: problems devoid of change. By contrast, change is central to calculus—calculus is about *dynamics.* Example:

- What's the perimeter of a square of side length 2 feet? ⟵ *Pre-calculus problem.*
- How fast is the square's perimeter *changing* if its side length is *increasing* at the constant rate of 2 feet per second? ⟵ *Calculus problem.*

This statics versus dynamics distinction between pre-calculus and calculus runs even deeper—change is the *mindset* of calculus. The subject trains you to think of a problem in terms of dynamics (versus statics). Example:

- Find the volume of a sphere of radius r. *Pre-calculus mindset:* Use $\frac{4}{3}\pi r^3$ (Figure 1.1(a)).
- Find the volume of a sphere of radius r. *Calculus mindset:* Slice the sphere into a gazillion disks of tiny thickness and then add up their volumes (Figure 1.1(b)). When the disks' thickness is made "infinitesimally small" this approach reproduces the $\frac{4}{3}\pi r^3$ formula. (We will discuss why in Chapter 5.)

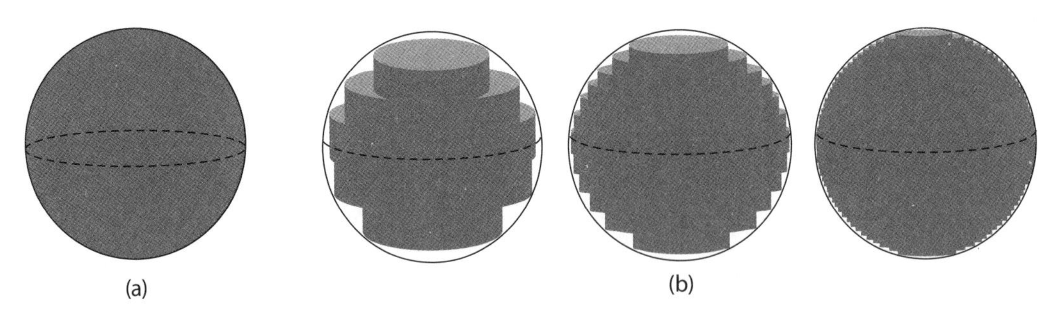

Figure 1.1: Visualizing the volume of a sphere via (a) a pre-calculus mindset and (b) a calculus mindset.

There's that mysterious word again—infinitesimal—and I've just given you a clue of what it might mean. I'll soon explain. Right now, let me pause to address a thought you might have just had: "Why the slice-and-dice approach? Why not just use the $\frac{4}{3}\pi r^3$ formula?" The answer: had I asked for the volume of some random blob in space instead, that static pre-calculus mindset wouldn't have cut it (there is no formula for the volume of a blob). The dynamics mindset of calculus, on the other hand, would have at least led us to a reasonable approximation using the same slice-and-dice approach.

That volume example illustrates the power of the dynamics mindset of calculus. It also illustrates a psychological fact: *shedding the static mindset of pre-calculus will take some time.* That was the dominant mindset in your mathematics courses prior to calculus mathematics courses, so you're accustomed to thinking that way about math. But fear not, young padawan (a *Star Wars* reference), I am here to guide you through the transition into calculus' dynamics mindset. Let's continue the adventure by returning to what I've been promising: insight into infinitesimals.

What Does "Infinitesimal Change" Mean?

The volume example earlier clued you in to what "infinitesimal" might mean. Here's a rough definition:

"Infinitesimal change" means: *as close to zero change as you can imagine, but not zero change.*

Let me illustrate this by way of Zeno of Elea (c. 490–430 BC), a Greek philosopher who devised a set of paradoxes arguing that motion is not possible. (Clearly, Zeno did not have a dynamics mindset.) One such paradox—the Dichotomy Paradox—can be stated as follows:

To travel a certain distance you must first traverse half of it.

Figure 1.2 illustrates this. Here Zeno is trying to walk a distance of 2 feet. But because of Zeno's mindset, with his first step he walks only half the distance: 1 foot

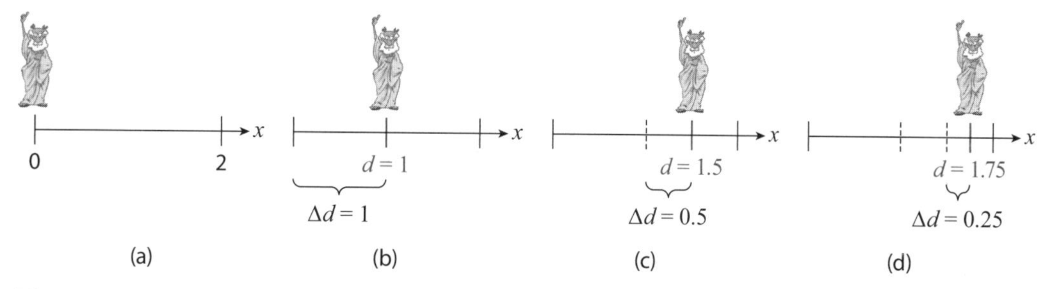

Figure 1.2: Zeno trying to walk a distance of 2 feet by traversing half the remaining distance with each step.

(Figure 1.2(b)). He then walks half of the remaining distance in his second step: 0.5 foot (Figure 1.2(c)). Table 1.1 keeps track of the total distance d, and the change in distance Δd, after each of Zeno's steps.

Table 1.1: The distance d and change in distance Δd after each of Zeno's steps.

Δd	1	0.5	0.25	0.125	0.0625	0.03125	0.015625	0.0078125	...
d	1	1.5	1.75	1.875	1.9375	1.96875	1.984375	1.9921875	...

Each change Δd in Zeno's distance is half the previous one. So as Zeno continues his walk, Δd gets closer to zero but never becomes zero (because each Δd is always half of a positive number). If we checked back in with Zeno after he's taken an infinite amount of steps, the change Δd resulting from his next step would be . . . drum roll please . . . an *infinitesimal change*—as close to zero as you can imagine but not equal to zero.

This example, in addition to illustrating what an infinitesimal change is, also does two more things. First, it illustrates the dynamics mindset of calculus. We discussed Zeno *walking*; we thought about the *change* in the distance he traveled; we visualized the situation with a figure and a table that each conveyed *movement*. (Calculus is full of action verbs!) Second, the example challenges us. Clearly, one *can* walk 2 feet. But as Table 1.1 suggests, that doesn't happen during Zeno's walk—he *approaches* the 2-foot mark with each step yet never *arrives*. How do we describe this fact with an equation? (That's the challenge.) No pre-calculus equation will do. We need a new concept that quantifies our very *dynamic* conclusion. That new concept is the mathematical foundation of calculus: limits.

1.2 Limits: The Foundation of Calculus

Let's return to Table 1.1. One thing you may have already noticed: Δd and d are related. Specifically:

$$\Delta d + d = 2, \qquad \text{or equivalently,} \qquad d = 2 - \Delta d. \tag{1.1}$$

This equation relates each Δd value to its corresponding d value in Table 1.1. Great. But it is not the equation we are looking for, because it doesn't encode the dynamics inherent in the table. The table clearly shows that the distance d traveled by Zeno approaches 2 as Δd approaches zero. We can shorten this to

$$d \to 2 \quad \text{as} \quad \Delta d \to 0.$$

(We are using "$\to$" here as a stand-in for "approaches.") The table also reiterates what we already know: were we to let Zeno continue his walk forever, he would be closer to the 2-foot mark than anyone could measure; in calculus we say: "infinitesimally close to 2." To express this notion, we write

$$\lim_{\Delta d \to 0} d = 2, \tag{1.2}$$

read "the limit of d as $\Delta d \to 0$ (but is never equal to zero) is 2."

Equation (1.2) is the equation we've been looking for. It expresses the intuitive idea that the 2-foot mark is the *limiting* value of the distance d Zeno's traversing. (This explains the "lim" notation in (1.2).) Equation (1.2), therefore, is a statement about the dynamics of Zeno's walk, in contrast to (1.1), which is a statement about the static snapshots of each step Zeno takes. Moreover, the Equation (1.2) reminds us that d is always *approaching* 2 yet never *arrives* at 2. The same idea holds for Δd: it is always *approaching* 0 yet never *arrives* at 0. Said more succinctly:

Limits approach indefinitely (and thus never arrive).

We will learn much more about limits in Chapter 2 (including that (1.2) is actually a "right-hand limit"). But the Zeno example is sufficient to give you a sense of what the calculus concept of limit is and how it arises. It also illustrates this section's title—the limit concept is the foundation on which the entire mansion of calculus is built. Figure 1.3 illustrates the process of building a new calculus concept that we will use over and over again throughout the book: *start with a finite change Δx in a quantity Y that depends on x, then shrink Δx to zero without letting it equal zero (i.e., take the limit as $\Delta x \to 0$) to arrive at a calculus result.* Working through this process—like we just did with the Zeno example, and like you can now recognize in Figure 1.1—is part of what *doing* calculus is all about. This is what I meant earlier when I said that calculus is the mathematics of infinitesimal change—contentwise, calculus is the collection of what results when we apply the workflow in Figure 1.3 to various quantities Y of real-world and mathematical interest.

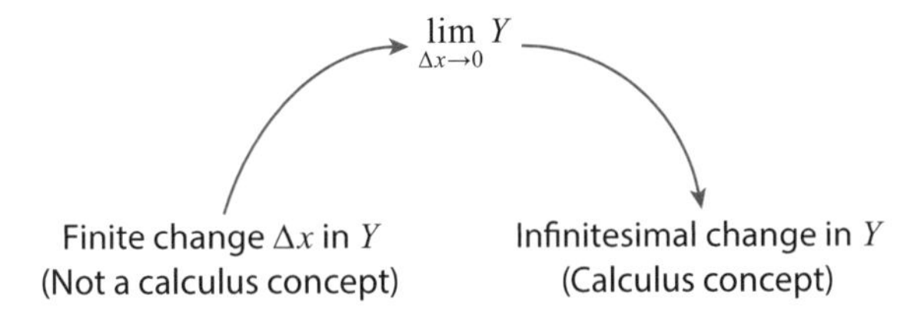

Figure 1.3: The calculus workflow.

Three such quantities drove the historical development of calculus: instantaneous speed, the slope of the tangent line, and the area under a curve. In the next section we'll preview how the calculus workflow in Figure 1.3 solved all of these problems. (We'll fill in the details in Chapters 3–5.)

1.3 The Three Difficult Problems That Led to the Invention of Calculus

Calculus developed out of a need to solve three Big Problems (refer to Figure 1.4):[1]

1. *The instantaneous speed problem:* Calculate the speed of a falling object at a particular instant during its fall. (See Figure 1.4(a).)
2. *The tangent line problem:* Given a curve and a point P on it, calculate the slope of the line "tangent" to the curve at P. (See Figure 1.4(b).)
3. *The area under the curve problem:* Calculate the area under the graph of a function and bounded by two x-values. (See Figure 1.4(c).)

Figure 1.4 already gives you a sense of why these problems were so difficult to solve—the standard approach suggested by the problem itself just doesn't work. For example, you've been taught you need *two* points to calculate the slope of a line. The tangent line problem asks you to calculate the slope of a line using just *one* point (point P in Figure 1.4(b)). Similarly, we think of speed as "change in distance divided by change in time." How, then, can one possibly calculate the speed at an *instant*, for which there is no change in time? These are examples of the sorts of roadblocks that stood in the way of solving the three Big Problems.

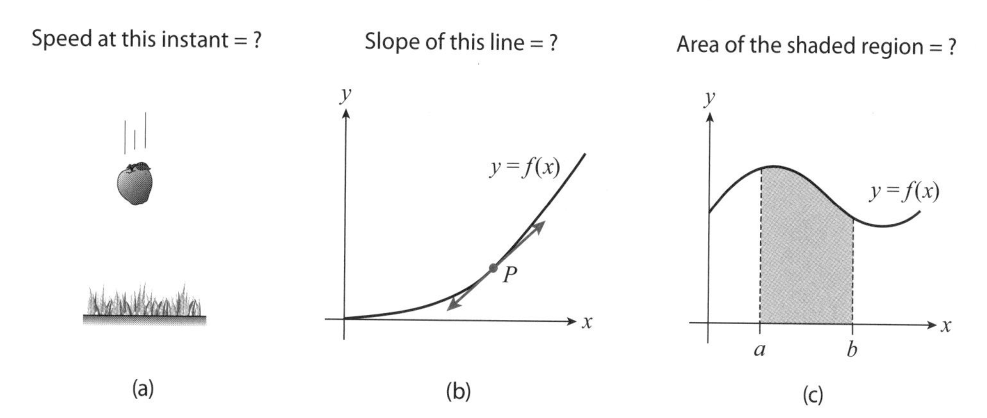

Figure 1.4: The three problems that drove the development of calculus.

[1] These may not seem like important problems. But their resolution revolutionized science, enabling the understanding of phenomena as diverse as gravity, the spread of infectious diseases, and the dynamics of the world economy.

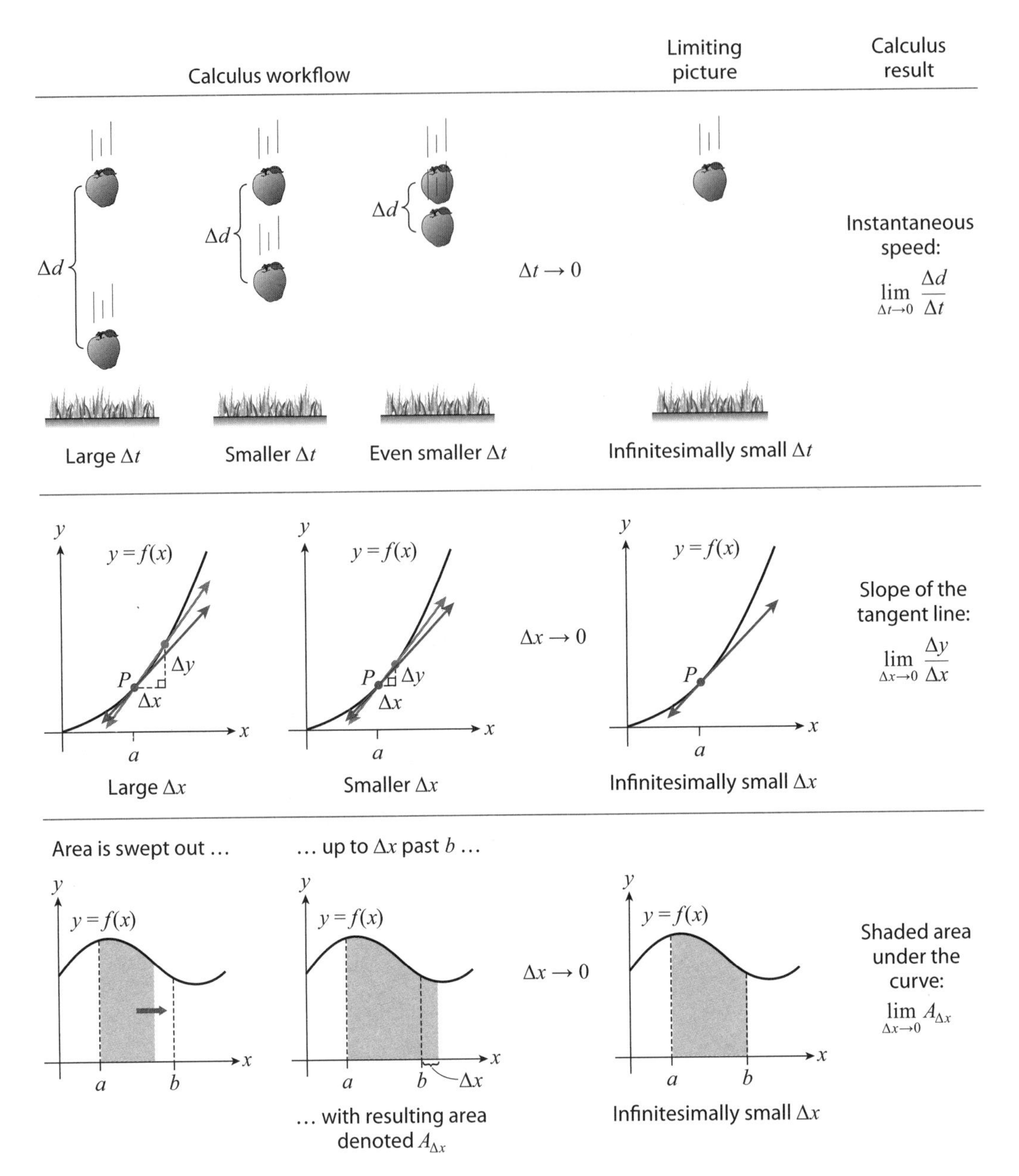

Figure 1.5: The calculus workflow (from Figure 1.3) applied to the three Big Problems.

But recall my first characterization of calculus: calculus is a dynamics mindset. Nothing about Figure 1.4 says "dynamics." Every image is a static snapshot of something (e.g., an area). So let's *calculus* the figure. (Yep, I'm encouraging you to think of calculus as a verb.)

Figure 1.5 illustrates the application of the calculus workflow (from Figure 1.3) to each Big Problem. Each row uses a dynamics mindset to recast the problem as the limit of a sequence of similar quantities (e.g., slopes) involving *finite* changes. Specifically:

- *Row #1:* The instantaneous speed of the falling apple is realized as the limit of its **average speeds** $\frac{\Delta d}{\Delta t}$ (ratios of changes in distance to changes in time) as $\Delta t \to 0$.
- *Row #2:* The slope of the tangent line is realized as the limit of the slopes of the **secant lines** $\frac{\Delta y}{\Delta x}$ (the gray lines in the figure) as $\Delta x \to 0$.
- *Row #3:* The area under the curve is realized as the limit as $\Delta x \to 0$ of the area swept out from $x = a$ up to Δx past b.

The limit obtained in the second row of the figure is called the **derivative** of $f(x)$ at $x = a$, the x-value of point P. The limit obtained in the third row of the figure is called the **definite integral** of $f(x)$ between $x = a$ and $x = b$. Derivatives and integrals round out the three most important concepts in calculus (limits are the third). We will discuss derivatives in Chapters 3 and 4 and integrals in Chapter 5, where we'll also fill in the mathematical details associated with the three limits in Figure 1.5.

This completes my big picture overview of calculus. Looking back now at Figures 1.1, 1.2, and 1.5, I hope I've convinced you of the power of the calculus mindset and the calculus workflow. We will employ both throughout the book. And because the notion of a limit is at the core of the workflow, I'll spend the next chapter teaching you all about limits—their precise definition, the various types of limits, and the myriad techniques to calculate them. See you in the next chapter.

2 Limits: How to Approach Indefinitely (and Thus Never Arrive)

Chapter Preview. *Limits are the foundation of calculus. As stressed in our Calculus Workflow, they are the intermediary between finite and infinitesimal changes, the latter being the type of change calculus is all about. But there are many types of limits—one-sided, two-sided, etc. This chapter takes you on a limit safari to teach you all about this core calculus concept. We'll learn how to visualize, approximate, and calculate limits; we'll learn about the real-world applications of limits; and along the way I'll give you lots of tips and tricks to help you master limits. I'll assume you're comfortable with the content in Appendixes A and B, so skim that first if you haven't already. Ready? Let's start the expedition.*

2.1 One-Sided Limits: A Graphical Approach

Soon after we calculated the limit (1.2) (page 4) in the Zeno example from Chapter 1, I mentioned it was an example of a "right-hand limit." Figure 2.1 illustrates what I mean.

The rightmost graph plots Zeno's distance traveled d versus the change Δd in his distance traveled. (Recall from (1.1) that $d = 2 - \Delta d$.) The graph excludes the point $(0, 2)$ (hence the hole) because $\Delta d \neq 0$ (Zeno never arrives at the 2-foot mark, remember?). But this is the static view of Zeno's walk. Switching to a dynamics mindset produces the other three plots in the figure. As noted, observe how d approaches 2 (the arrows on the d-axis) as Δd approaches 0 (the arrows on the Δd-axis). Since

Interactive Figure

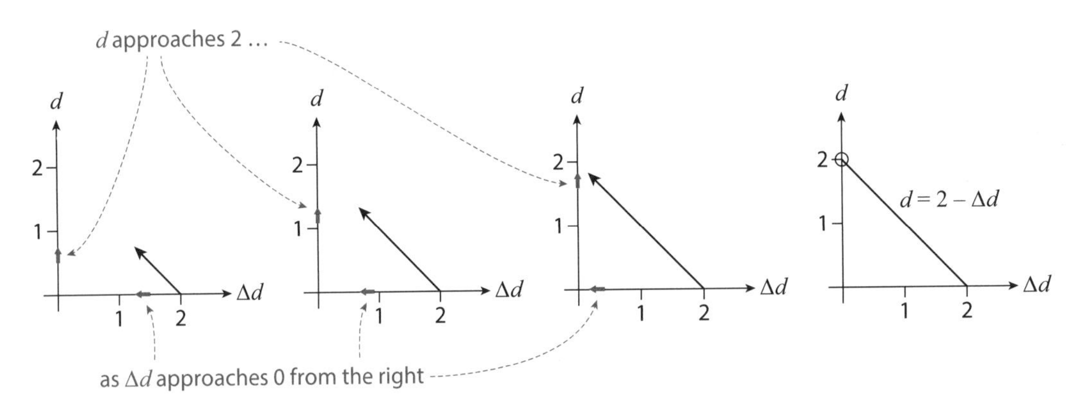

Figure 2.1: Illustrating the right-hand limit $\lim_{\Delta d \to 0^+} d = 2$.

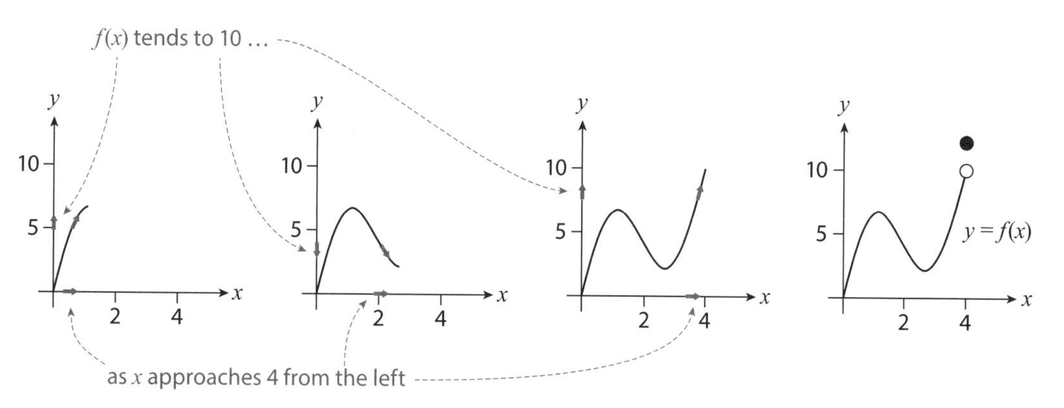

Figure 2.2: Illustrating the left-hand limit $\lim_{x \to 4^-} f(x) = 10$.

Δd is approaching 0 from numbers *to the right* of 0 on the number line, we denote this in limit notation as

$$\lim_{\Delta d \to 0^+} d = 2.$$

We call this limit a **right-hand limit**. As you may have guessed, there are also **left-hand limits**, denoted by

$$\lim_{x \to c^-} f(x).$$

To evaluate these we look at what number the y-values $f(x)$ tend to as x approaches c from numbers *to the left* of c on the x-axis. Figure 2.2 shows an example in which $\lim_{x \to 4^-} f(x) = 10$. Here are important comments on the figure and on left-hand limits in general.

- Note that the answer to the limit in Figure 2.2 is indeed 10 and not the y-value *at* $x = 4$ (the closed circle in the figure); limits approach indefinitely and thus *never arrive*, remember?

- On the topic of "approaching," I say in Figure 2.2 that $f(x)$ *tends* to 10, as opposed to $f(x)$ *approaches* 10. That's because sometimes the y-values are moving *away* from 10 (as in the second plot from the left). That's not a problem, because recall that limits approach indefinitely, so what really matters is where the y-values are headed when the x-values are very close to $x = 4$. Nonetheless, I'll switch to the "tends to" terminology henceforth to remind you that sometimes the y-values may oscillate on their way to their limit.

- Finally, be careful not to confuse 4^- with -4. The notation 4^- indicates the *direction* of approach (i.e., from numbers to the left of 4 on the x-axis). By contrast, -4 denotes the negative number -4.

I encourage you to use the dynamics mindset inherent in Figures 2.1 and 2.2 every time you see a graph (particularly if you're using it to calculate a limit).

Table 2.1: A few values of a fictitious function f

x	$f(x)$
1.9	3.61
1.99	3.9601
1.999	3.99601
$\vdots$	$\vdots$
3.001	6.004001
3.01	6.0401
3.1	6.41

EXAMPLE 2.1 Assuming the patterns in Table 2.1 continue, what answers do you expect for $\lim_{x\to 2^-} f(x)$ and $\lim_{x\to 3^+} f(x)$?

Solution We expect that $\lim_{x\to 2^-} f(x) = 4$ and $\lim_{x\to 3^+} f(x) = 6$. ■

EXAMPLE 2.2 Use Figure 2.3(a) to evaluate the limits:

$$\lim_{x\to -1^-} f(x), \quad \lim_{x\to -1^+} f(x), \quad \lim_{x\to 0^-} f(x), \quad \lim_{x\to 1^+} f(x).$$

Solution From left to right:

$$\lim_{x\to -1^-} f(x) = 1, \quad \lim_{x\to -1^+} f(x) = 0.5, \quad \lim_{x\to 0^-} f(x) = 1.5, \quad \lim_{x\to 1^+} f(x) = 0.5. \quad ■$$

APPLIED EXAMPLE 2.3 Alicia took her daily B complex multivitamin this morning with breakfast. Her body will use up the multivitamin's micronutrients over the next 24 hours. When she takes her next dose of the multivitamin *exactly* 24 hours later, her B vitamin stores will be replenished. Figure 2.3(b) illustrates this process. In it, V denotes the total amount of B vitamins in Alicia's body t hours since she took her vitamin this morning. Assuming the multivitamin was her only source of B vitamins, use the figure to calculate the limits (and interpret

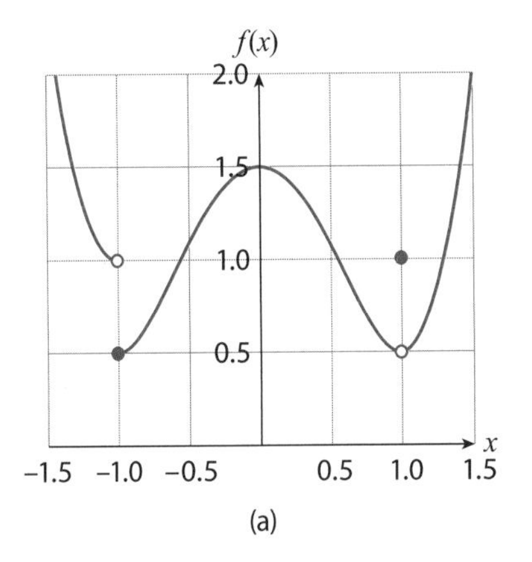

(a)

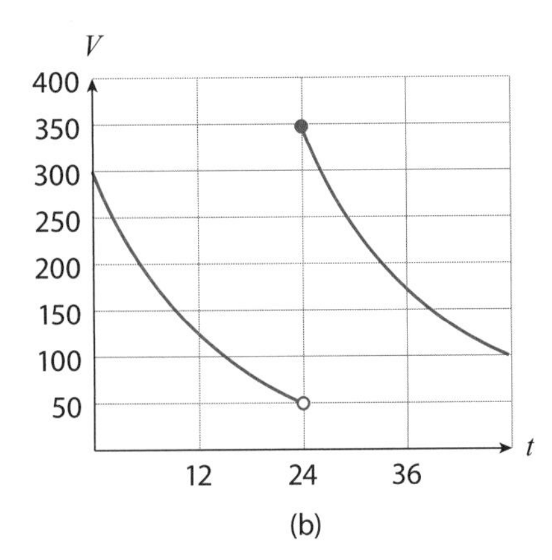

(b)

Figure 2.3

the first one):

$$\lim_{t\to 24^-} V(t), \qquad \lim_{t\to 24^+} V(t).$$

Solution The answers to the limits are 50 and 350, respectively. The first limit's interpretation: before it's been 24 hours since Alicia took her multivitamin, the amount of vitamins in her body is approaching 50 as she approaches the 24-hour mark. ■

Related Exercises 1 and 2(a)(i)–(ii) and (iv)–(v).

Tips, Tricks, and Takeaways

Figure 2.4 condenses the results of this section into a diagram illustrating that the graphs of functions with left-hand or right-hand limits as x approaches a number c fit one of only a handful of templates.

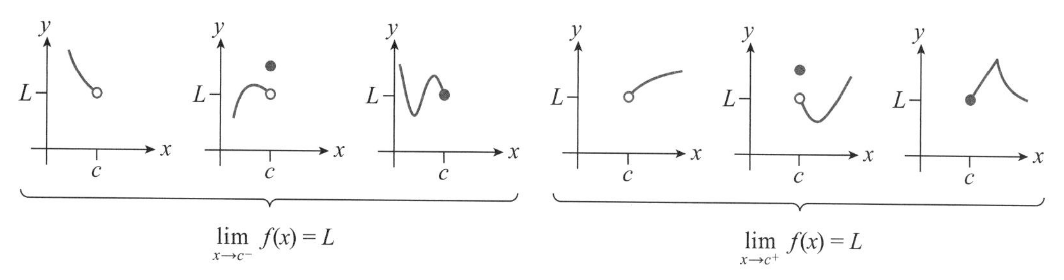

Figure 2.4: Examples of what $\lim_{x\to c^-} f(x) = L$ and $\lim_{x\to c^+} f(x) = L$ look like graphically.

Note again that the y-value *at* $x = c$ does not affect the value of the limit.

2.2 Existence of One-Sided Limits

When evaluating a limit we are looking for the y-value that a function's values $f(x)$ tend to as x approaches a number c from the left, or from the right. (In the next section we'll discuss "two-sided" limits.) But sometimes no such y-value exists; we write "DNE" for the answer—short for Does Not Exist—in these instances.

Two common reasons for a limit not to exist are:

(1) The graph races off to infinity as x approaches c. Infinity is a not a number, so the limit cannot exist.

(2) The graph oscillates wildly as x approaches c, with no discernible y-value as its limiting value.

Figure 2.5(a) illustrates Case (1). We say that

$$\lim_{x\to 0^-} f(x) \text{ DNE } (-\infty), \qquad \lim_{x\to 0^+} f(x) \text{ DNE } (+\infty).$$

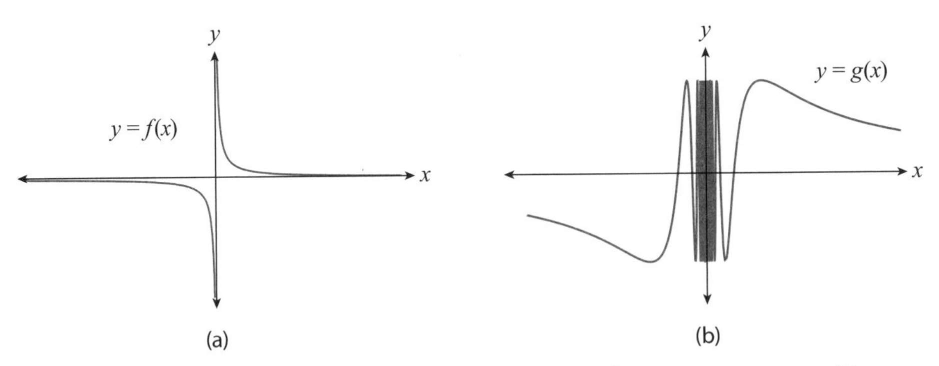

Figure 2.5: Portions of the graphs of: (a) $f(x) = \frac{1}{x}$, and (b) $g(x) = \sin\left(\frac{1}{x}\right)$.

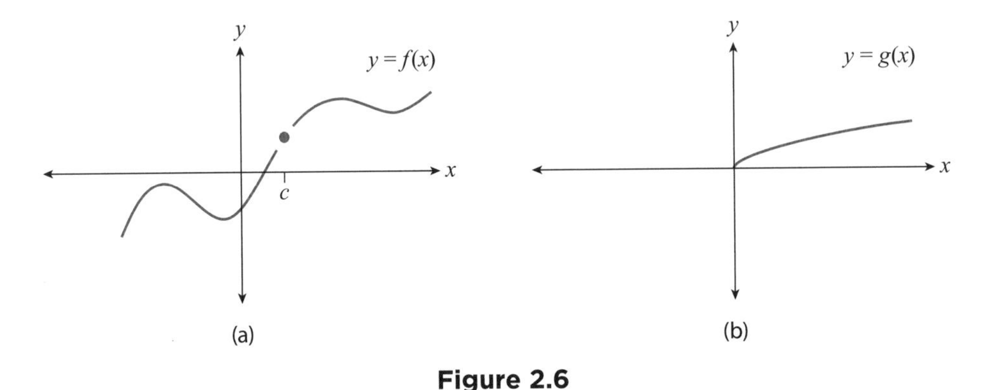

Figure 2.6

Notice that I've kept track of the type of infinity (positive or negative) the graph is racing off to. Some textbooks prefer to use "$= \infty$" and "$= -\infty$." In my experience, this notation confuses people, so I'll stick with the notation above. Figure 2.5(b) illustrates Case (2). We say that

$$\lim_{x \to 0^-} g(x) \text{ DNE}, \qquad \lim_{x \to 0^+} g(x) \text{ DNE}.$$

Our discussion of limits thus far has presumed that we can approach a number $x = c$ on the the graph of a function. But that's not possible in the following two instances:

(3) When the function is defined at $x = c$ but not anywhere else near it. (We cannot then approach $x = c$ on the graph.)

(4) When the function is not defined for $x < c$ (if calculating a left-hand limit) or for $x > c$ (if calculating a right-hand limit).

Figure 2.6(a) illustrates Case (3): we cannot tell what y-value $f(x)$ tends to as x approaches c, since we aren't given information about what happens just to the left (and right) of $x = c$. Figure 2.6(b) illustrates Case (4): there is no portion of the graph for $x < 0$, so we cannot evaluate the limit of $g(x)$ as $x \to 0^-$.

Many calculus textbooks would say that the limits in Figure 2.6 do not exist. I favor referring to these limits as "not even possible to begin to evaluate." But I will follow convention here and say that those limits do not exist also:

$$\lim_{x\to c^-} f(x) \text{ DNE}, \qquad \lim_{x\to c^+} f(x) \text{ DNE}, \qquad \lim_{x\to 0^-} g(x) \text{ DNE},$$

where I'm referring here to Figure 2.6.

EXAMPLE 2.4 Using the graph of the function $y = f(x)$ in Figure 2.7, evaluate the following limits (if possible).

$$\lim_{x\to -2^-} f(x), \quad \lim_{x\to -1^+} f(x), \quad \lim_{x\to 2^-} f(x),$$

$$\lim_{x\to 3^+} f(x), \quad \lim_{x\to 4^+} f(x).$$

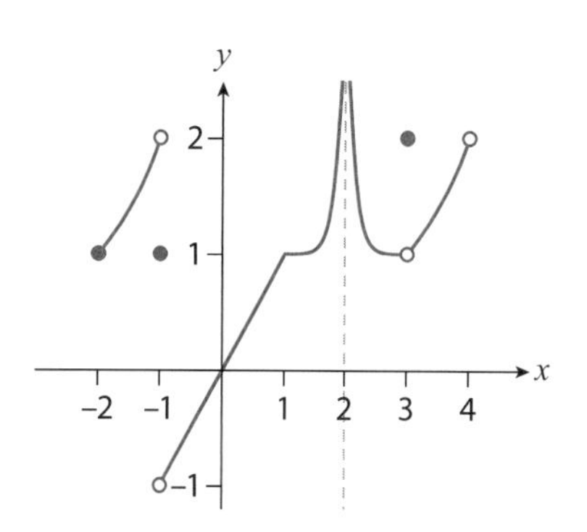

Figure 2.7

Solution The first and last limits DNE (there is no graph to the left of $x = -2$ or to the right of $x = 4$ that we can use to evaluate them). The remaining limits are

$$\lim_{x\to -1^+} f(x) = -1, \qquad \lim_{x\to 2^-} f(x) \text{ DNE } (+\infty), \qquad \lim_{x\to 3^+} f(x) = 1. \quad ■$$

Related Exercises 3(a)(i)–(vi).

Tips, Tricks, and Takeaways

- We cannot evaluate the limit of a function $f(x)$ as $x \to c^-$ (or $x \to c^+$) if the portion of the graph of f directly to the left (respectively, right) is missing (as was the case in Figure 2.6).
- Assuming such portion(s) are supplied, the limit exists only if the values $f(x)$ values tend to a y-value L as $x \to c^-$ or $x \to c^+$ (as opposed to racing off to infinity, for example).
- Finally, direction matters—the left-hand limit need not equal the right-hand limit.

That last takeaway sets us up for the next topic: two-sided limits.

2.3 Two-Sided Limits

A **two-sided limit** is a limit in which we consider *both* directions $x \to c^-$ and $x \to c^+$ in evaluating the limit. If both of these one-sided limits exist and are equal, then the two-sided limit exists and equals their common value. For example, returning to

Figure 2.7, we say that

$$\text{Since} \quad \lim_{x\to 3^-} f(x) = \lim_{x\to 3^+} f(x) = 1, \quad \text{then} \quad \lim_{x\to 3} f(x) = 1.$$

That last limit is the two-sided one; you can tell because the "$x \to 3$" in the limit has no superscript on the 3. We don't need superscripts anymore because $f(x) \to 1$ as x approaches 3 from either side (left or right).

Because two-sided limits require the existence of their one-sided counterparts, a two-sided limit will not exist if one or both of those one-sided limits do not exist. And even if both one-sided limits exist, if they're unequal in value the two-sided limit will again not exist (since in this case the y-value $f(x)$ tends to depends on the direction of approach to c). Here's all that in math speak.

Box 2.1: Criteria for the Existence of a Two-Sided Limit

The limit $\lim_{x\to c} f(x)$ exists only if its left- and right-hand limit counterparts exist and are equal:

$$\lim_{x\to c^-} f(x) = \lim_{x\to c^+} f(x) = L, \qquad \text{where } L \text{ is a number.}$$

If this is the case, then $\lim_{x\to c} f(x) = L$.

We've finally learned enough about limits for me to present you with the definition that follows.

Definition 2.1 Intuitive Definition of the Limit. Let c be a number and f be a function defined on an interval containing c (except possibly at c). Suppose the values $f(x)$ tend to a number L as x approaches c indefinitely (but is never equal to c). We then say "the **limit** of $f(x)$ as x approaches c equals L" and write

$$\lim_{x\to c} f(x) = L.$$

If no number L satisfies this, we say that the limit does not exist (DNE for short).

Note how everything we've talked about thus far in this chapter is contained in this definition:

- The "defined on an interval containing c" avoids the issues raised in Cases (3) and (4) of the previous section.
- The "except possibly at c" and "but is never equal to c" reminds us that limits approach indefinitely (and never arrive).

- Finally, "the values $f(x)$ tend to" reminds us that $f(x)$ may oscillate on its way to L (recall Figure 2.2).

There are even more formal definitions of the limit; these make more precise the idea of "approaches indefinitely (and never arrives)." Section A2.1 in the online appendix to this chapter discusses them, if you're interested. Now on to the examples.

EXAMPLE 2.5 Referring to Figure 2.7, evaluate the following limits (if possible):

$$\lim_{x\to -1} f(x), \quad \lim_{x\to 1} f(x), \quad \lim_{x\to 3} f(x).$$

Solution The answers are: $\lim_{x\to -1} f(x)$ DNE, $\lim_{x\to 1} f(x) = 1$, $\lim_{x\to 3} f(x) = 1$. The first limit is DNE because $f(x) \to 2$ as $x \to -1^-$ but $f(x) \to -1$ as $x \to -1^+$. ■

EXAMPLE 2.6 Determine the x-values at which $\lim_{x\to c} f(x)$ does not exist in Figure 2.7 and explain why.

Solution The only x-values at which the two-sided limit does not exist are $x = -2$, $x = -1$, $x = 2$, and $x = 4$. At $x = -2$ the limit DNE because the left-hand limit as $x \to -2^-$ DNE (there's no graph for $x < -2$). Similarly, at $x = 4$ right-hand limit as $x \to 4^+$ DNE (there's no graph for $x > 4$). We discussed why the limit as $x \to -1$ DNE in the previous example. Finally, at $x = 2$ the limit DNE because the graph races off to infinity. ■

Related Exercises 2(a)(iii) and (vi), 3(a)(vii)–(ix), and 13.

2.4 Continuity at a Point

In Figure 2.4 I presented two sets of graphs that all had the same one-sided limit L. But the last graphs in each set were special: their L-value was the function's y-value at $x = c$. In other words, in those graphs, $L = f(c)$.

Definition 2.2 Continuity at a Point. We say that a function f is

- **Continuous from the left at** $x = c$ if $\lim_{x\to c^-} f(x) = f(c)$
- **Continuous from the right at** $x = c$ if $\lim_{x\to c^+} f(x) = f(c)$
- **Continuous at** $x = c$ when $\lim_{x\to c} f(x) = f(c)$

When none of these equations hold, we say f is **discontinuous at** $x = c$.

In this terminology, the function graphed in the third plot in Figure 2.4 is continuous from the left at $x=c$, and the one graphed in the sixth plot of the figure is continuous from the right at $x=c$. Since continuity at $x=c$ is equivalent to the special case when $L=f(c)$ in a limit, we can use our earlier criteria for the existence of a limit (from page 14) to derive the criteria for continuity.

Box 2.2: Criteria for a Function to Be Continuous at a Point

Let f be a function and c a number in its domain. Then f is continuous at $x=c$ only when

(1) $f(c)$ exists, (2) $\lim_{x\to c} f(x)$ exists, (3) $\lim_{x\to c} f(x)=f(c)$.

When c is the right endpoint of the domain, the limit in criteria (2) and (3) is replaced by "$\lim_{x\to c^-}$." When c is the left endpoint of the domain, the limit in criteria (2) and (3) is replaced by "$\lim_{x\to c^+}$."

EXAMPLE 2.7 Is the function graphed in Figure 2.3(a) continuous at $x=-1$? At $x=0$? At $x=1$?

Solution The function is:

- Discontinuous at $x=-1$. Though the left- and right-hand limits exist, they are are unequal, violating Criterion (2) in Box 2.2.
- Continuous at $x=0$. Both the left- and right-hand limits exist and are equal (to 1.5) and equal $f(0)=1.5$.
- Discontinuous at $x=1$. Though the left- and right-hand limits exist and are equal (to 0.5), they do not equal $f(0)=1$; this violates Criterion (3) in Box 2.2. ■

EXAMPLE 2.8 At what x-values is the function graphed in Figure 2.7 discontinuous, and why?

Solution The function is discontinuous at $x=-1$, $x=2$, $x=3$, and $x=4$. At $x=-1$,

$$\lim_{x\to -1^-} f(x)=2, \qquad \lim_{x\to -1^+} f(x)=-1.$$

These aren't equal, so the two-sided limit as $x\to -1$ DNE. This violates Criterion (2) in Box 2.2. At $x=2$, $f(2)$ does not exist. This violates Criterion (1). At $x=3$,

$$\lim_{x\to 3^-} f(x)=\lim_{x\to 3^+} f(x)=1,$$

but $f(3) = 2$, violating Criterion (3). Finally, $f(4)$ DNE, violating Criterion (1). ■

Related Exercises 2(b), 3(b), and 35(a)–(c).

Tips, Tricks, and Takeaways

For a function to be continuous at $x = c$ it:

- Cannot have a *hole* at $x = c$ (violates Criterion 1)
- Cannot *jump* as we cross $x = c$, like the graph in Figure 2.3(b) does at $t = 24$ (violating Criterion 2)
- Cannot *gap* up or down as we cross $x = c$, like the graph in Figure 2.3(a) does at $x = 1$ (violating Criterion 3)

In sum: the function's graph at $x = c$ must look like we drew it with one *continuous* stroke. (There's that dynamics mindset again.) This leads to the notion of continuity on an *interval*. That's the topic of the next section.

2.5 Continuity on an Interval

Continuity is a pointwise property—it refers to a specific point: $x = c$. But many functions are continuous on an *interval* of points. We use the following terminology when that's the case.

> **Definition 2.3** If a function f is continuous for all x in an interval I, we say that f is **continuous on** I. When f is continuous on $(-\infty, \infty)$, we say that f is **continuous everywhere**, or simply, **continuous**.

Example: the function graphed in Figure 2.7 is continuous on $[-2, -1) \cup (-1, 2) \cup (2, 3) \cup (3, 4)$.[1]

Up to now we've worked exclusively with graphs. But what if I gave you a specific function, say $f(x) = x^3 + 3x$, and asked you for its interval(s) of continuity? It would be an arduous task to check the continuity criteria from Box 2.2 for each real number x. Luckily, we can avoid that via the following results.[2]

> **Theorem 2.1** The following families of functions are continuous at all points in their domain:
>
> polynomials, power functions, rational functions

[1] The $\cup$ symbol means "and," so that $[1, 2] \cup [3, 4]$ represents the set of real numbers between 1 and 2, and between 3 and 4, including all four numbers.

[2] These theorems are proven using a set of rules for limits called the **Limit Laws**; we'll discuss these in the next section.

So, for example, $f(x) = x^3 + 3x$ is continuous everywhere, since f is a polynomial and polynomials have domain all real numbers. Notice that successfully using Theorem 2.1 requires being able to classify a function (e.g., "polynomial") and knowing its domain. Both of these skills are reviewed in Appendix B.

The function $f(x) = x^3 + 3x$ can also be seen as the sum of simpler functions (x^3 and $3x$). The following theorem tells us when combinations of functions are continuous.

Theorem 2.2 Suppose f and g are continuous at c, and that a is a real number. Then the following functions are also continuous at c:

$$f+g, \qquad f-g, \qquad af, \qquad fg, \qquad \frac{f}{g} \quad (\text{provided } g(c) \neq 0).$$

The conclusions of this theorem are easier to remember if you say them in words; for example, the first conclusion is: The sum of continuous functions is continuous.

One combination of functions not discussed in Theorem 2.2 is composite functions. The following theorem addresses that. (The theorem follows from more technical theorems relating to changing variables in limits; see Section A2.2 in the online appendix to this chapter if you're interested.)

Theorem 2.3 Suppose g is continuous at c and f continuous at $g(c)$. Then $f \circ g$ is continuous at c.

This theorem says that the composition of a continuous function with a continuous function is continuous.

One final note: Theorems 2.2 and 2.3 also hold when "continuous" is replaced by "continuous from the right" or "continuous from the left."

EXAMPLE 2.9 On what interval(s) is $f(x) = \sqrt{x} + 1$ continuous?

Solution The domain of the power function $\sqrt{x}$ is $[0, \infty)$ and the domain of the constant function 1 is $(-\infty, \infty)$. The domain of $f(x) = \sqrt{x} + 1$, therefore, is $[0, \infty)$. It follows from Theorem 2.1 that f is continuous on $[0, \infty)$. ■

EXAMPLE 2.10 On what interval(s) is $g(x) = \dfrac{x}{x^2 + 5x + 6}$ continuous?

Solution The domain of the rational function $g(x)$ is all real numbers except those for which

$$x^2 + 5x + 6 = 0 \quad \Longleftrightarrow \quad (x+2)(x+3) = 0 \quad \Longleftrightarrow \quad x = -2, \ x = -3.$$

Thus, Theorem 2.1 tells us that g is continuous on $(-\infty,-3)\cup(-3,-2)\cup(-2,\infty)$. ■

Related Exercises 2(c), 18–20, and 37–38.

EXAMPLE 2.11 Evaluate $\lim_{x\to 1}\sqrt{x^2+1}$ using Theorem 2.3.

Solution Note that $\sqrt{x^2+1}=f(g(x))$, where $f(x)=\sqrt{x}$ and $g(x)=x^2+1$. The function g is continuous everywhere (it's a polynomial), and $g(1)=2$. Since f is continuous at $x=2$, Theorem 2.3 applies:

$$\lim_{x\to 1}\sqrt{x^2+1}=\sqrt{1^2+1}=\sqrt{2}.$$ ■

Einstein's Relativity Applied Example C.1 in Appendix C employs one-sided limits to show that moving objects' length *shrinks to zero* as their speed approaches the speed of light.

Related Exercises 14–17, 33–34, and 35(d).

We've now learned a lot about limits and continuity in the context of algebraic functions. Let's expand our horizon to include exponential, logarithmic, and trigonometric functions (these functions are reviewed in Appendix B).

Transcendental Tales

First up are the results for exponential and logarithmic functions. The "can draw with one stroke of your pencil" view of continuity leads us to the following starter theorem.

Theorem 2.4 Every exponential function is continuous. Every logarithmic function is continuous on $(0,\infty)$.

Section A2.3 in the online appendix to this chapter discusses why this theorem is true. Mathematically, these results tell us that

$$\lim_{x\to c} b^x=b^c,\qquad \lim_{x\to c}\log_b x=\log_b c,$$

where b^x is an exponential function (which requires $b>0$ and $b\neq 1$) and $\log_b x$ is a logarithmic function (which requires the same restrictions on b). Theorem 2.4 can then be combined with the others in this section to help us evaluate limits.

EXAMPLE 2.12 Evaluate $\lim_{x\to 0} e^{-x^2}$.

Solution Note that e^{-x^2} is a composition of continuous functions: $e^{-x^2} = f(g(x))$, where $f(x) = e^x$ and $g(x) = -x^2$. It follows that $f(g(x)) = e^{-x^2}$ is also continuous. Therefore:

$$\lim_{x\to 0} e^{-x^2} = e^{-(0)^2} = e^0 = 1. \quad \blacksquare$$

EXAMPLE 2.13 Evaluate $\lim_{t\to 2} \frac{\log t^2}{t}$.

Solution Here we have a quotient of functions continuous at $t = 2$. And since the value of the denominator at $t = 2$ is not zero, the various theorems in this section apply and

$$\lim_{t\to 2} \frac{\log t^2}{t} = \frac{\log 2^2}{2} = \frac{2\log 2}{2}. \quad \blacksquare$$

EXAMPLE 2.14 Evaluate $\lim_{z\to 3} \ln\sqrt{z^2 - 1}$.

Solution Note that $\ln\sqrt{z^2-1} = f(g(z))$, where $f(x) = \ln x$ and $g(z) = z^2 - 1$. The function g is continuous everywhere and $g(3) = 8$. The function f is continuous at $x = 8$, so

$$\lim_{z\to 3} \ln\sqrt{z^2-1} = \ln\sqrt{8} = \ln(2^{3/2}) = \frac{3}{2}\ln 2. \quad \blacksquare$$

A final note: We could have simplified the functions in Examples 2.13 and 2.14 first before calculating the limits:

$$\frac{\log t^2}{t} = \frac{2\log t}{t}, \qquad \ln\sqrt{z^2-1} = \frac{1}{2}\ln(z^2-1) = \frac{1}{2}\left[\ln(z-1) + \ln(z+1)\right].$$

I leave it to you to verify that we would have arrived at the same answers for the limits.

Let's now discuss the continuity of trigonometric functions. The graphs of $\sin x$ and $\cos x$ (see Figures B.18(a)–(b) in Appendix B) tells us that these are continuous functions:

$$\lim_{x\to c} \sin x = \sin c, \qquad \lim_{x\to c} \cos x = \cos c, \tag{2.1}$$

for any real number c. It follows from Theorem 2.2 that $\tan x = \frac{\sin x}{\cos x}$ is continuous everywhere except where $\cos x = 0$ (i.e, $x = \pm\pi/2, \pm 3\pi/2$, etc.); the graph of $\tan x$ appears in B.18(c) in Appendix B.

Theorem 2.5 The functions $\sin x$ and $\cos x$ are continuous; the function $\tan x$ is discontinuous at $x = k\pi/2$, where k is an odd integer.

These results are sufficient to help us evaluate simple limits involving trigonometric functions.

EXAMPLE 2.15 Evaluate $\lim_{x\to\pi} \dfrac{1}{3x+2\sin x}$.

Solution Here we have a quotient of continuous functions. Since the value of the limit in the denominator is not zero, this section's theorems kick in, and the answer is

$$\lim_{x\to\pi} \frac{1}{3x+2\sin x} = \frac{1}{3\pi+2(0)} = \frac{1}{3\pi}. \quad \blacksquare$$

EXAMPLE 2.16 Evaluate $\lim_{t\to\pi} \dfrac{\tan^2 t}{\sqrt{1+t^2}}$.

Solution Here again we have a quotient of continuous functions, and the value of the limit in the denominator is not zero. Thus, the answer is

$$\lim_{t\to\pi} \frac{\tan^2 t}{\sqrt{1+t^2}} = \frac{\tan^2 \pi}{\sqrt{1+\pi^2}} = 0. \quad \blacksquare$$

Related Exercises 39, 41, 43, 51, and 61

Tips, Tricks, and Takeaways

- *Continuity simplifies limit calculations.* Once you know $f(x)$ is continuous at $x = c$, the answer to $\lim_{x\to c} f(x)$ is just $f(c)$.
- *Continuous functions' graphs are nice and ... continuous.* The graph of any continuous function with domain $[a, b]$ passes through all of the y-values associated with each x-value in the domain. (This is called the **Intermediate Value Theorem**; see Section A2.4 in the online appendix to this chapter if you're interested.) Translation: you can draw the graph of such a function with one stroke of your pencil. The converse is also true: a graph you can draw without lifting your pencil (and without backtracking) is the graph of a continuous function. Figure 2.8 illustrates these ideas visually.

One final note. Our work thus far contains a hidden insight: *when you're given the equation of $f(x)$, the first thing to try when evaluating the limit is to substitute in $x = c$. If you get a real number, that's the answer.* Sometimes, however, this doesn't work. In the next section we'll discuss a set of rules that we'll use to develop more general techniques for evaluating limits.

2.6 The Limit Laws

Many functions we want to calculate limits of are arithmetic combinations of simpler functions. (Example: $f(x) = x + x^2$ is the sum of the simpler functions $g(x) = x$ and $h(x) = x^2$.) In many cases, evaluating the limit of such functions reduces to

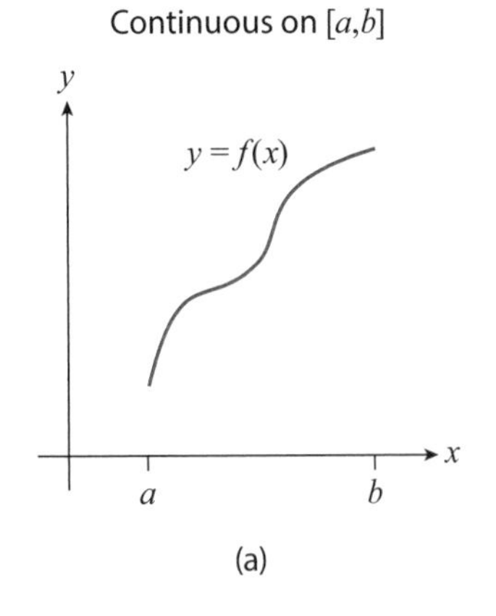

(a)

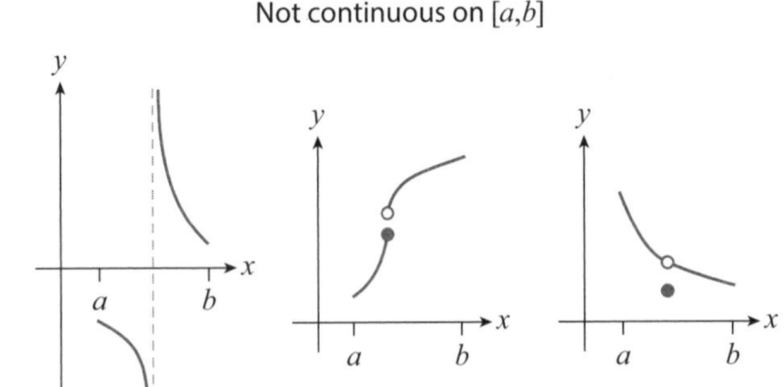

(b) (c) (d)

Figure 2.8

evaluating the limits of the "ingredient functions" (e.g., g and h in the previous example). The **Limit Laws** tell us how to do just that. Here they are.

Theorem 2.6 The Limit Laws. Let "lim" be a stand-in for $\lim\limits_{x\to c}$, $\lim\limits_{x\to c^-}$, or $\lim\limits_{x\to c^+}$. Suppose that f and g are functions, and that $\lim f(x)$ and $\lim g(x)$ both exist. Then

1. $\lim\,[f(x) \pm g(x)] = \lim\, f(x) \pm \lim\, g(x)$
2. $\lim\,[kf(x)] = k[\lim\, f(x)]$, where k is a real number
3. $\lim\,[f(x)g(x)] = [\lim\, f(x)][\lim\, g(x)]$
4. $\lim\, \dfrac{f(x)}{g(x)} = \dfrac{\lim\, f(x)}{\lim\, g(x)}$, provided $[\lim g(x)] \neq 0$
5. For n a positive integer, $\lim\, \sqrt[n]{f(x)} = \sqrt[n]{\lim\, f(x)}$, provided $[\lim f(x)] \geq 0$ if n is even
6. For n a positive integer, $\lim\,[f(x)]^n = [\lim\, f(x)]^n$
7. Supposing f is continuous at $\lim g(x)$, $\lim f(g(x)) = f\left(\lim g(x)\right)$.

These Limit Laws can be stated in English; for example, the first says that the limit of a sum (or difference) of functions is the sum (or difference) of the limits of each function (provided the latter limits exist).

EXAMPLE 2.17 Evaluate $\lim\limits_{x\to 1}\,(2x^2 + x - 1)$.

Solution

$$\lim_{x\to 1}(2x^2+x-1)=\lim_{x\to 1}(2x^2)+\lim_{x\to 1}x+\lim_{x\to 1}(-1) \qquad \text{Limit Law 1}$$

$$=2\left(\lim_{x\to 1}x\right)^2+\lim_{x\to 1}x+\lim_{x\to 1}(-1) \qquad \text{Limit Laws 2 and 6}$$

$$=2(1)^2+1+(-1)=2.$$

■

EXAMPLE 2.18 Evaluate $\lim_{x\to 0^+}\sqrt{x+1}$.

Solution

$$\lim_{x\to 0^+}\sqrt{x+1}=\sqrt{\lim_{x\to 0^+}(x+1)}=\sqrt{\lim_{x\to 0^+}x+\lim_{x\to 0^+}1} \qquad \text{Limit Law 5; Limit Law 1}$$

$$=\sqrt{0+1}=1.$$

■

EXAMPLE 2.19 Evaluate $\lim_{x\to 3}\frac{x^2-9}{x-2}$.

Solution

$$\lim_{x\to 3}\frac{x^2-9}{x-2}=\frac{\lim_{x\to 3}(x^2-9)}{\lim_{x\to 3}(x-2)}=\frac{\left(\lim_{x\to 3}x\right)^2-\lim_{x\to 3}9}{\lim_{x\to 3}x-\lim_{x\to 3}2} \qquad \text{Limit Law 4; Limit Laws 1 and 6}$$

$$=\frac{9-9}{3-2}=0.$$

■

Related Exercises 4, 5–7, and 36.

Transcendental Tales

Let's now apply the Limit Laws to evaluate limits involving exponential, logarithmic, and trigonometric functions.

EXAMPLE 2.20 Evaluate $\lim_{h\to 0}\frac{e^h-1}{h-1}$.

Solution

$$\lim_{h\to 0}\frac{e^h-1}{h-1}=\frac{\lim_{h\to 0}(e^h-1)}{\lim_{h\to 0}(h-1)}=\frac{\lim_{h\to 0}e^h-\lim_{h\to 0}1}{\lim_{h\to 0}h-\lim_{h\to 0}1} \qquad \text{Limit Law 4; Limit Law 1}$$

$$=\frac{1-1}{0-1}=0. \qquad \text{Using continuity of } e^h$$

■

EXAMPLE 2.21 Evaluate $\lim_{x\to 2^+} xe^x$.

Solution

$$\begin{aligned}\lim_{x\to 2^+} xe^x &= \left[\lim_{x\to 2^+} x\right]\cdot\left[\lim_{x\to 2^+} e^x\right] && \text{Limit Law 3}\\ &= 2e^2. && \text{Using continuity of } e^x\end{aligned}$$

■

EXAMPLE 2.22 Evaluate $\lim_{x\to 1} \frac{\ln x}{e^{-x}}$.

Solution

$$\begin{aligned}\lim_{x\to 1} \frac{\ln x}{e^{-x}} &= \frac{\lim_{x\to 1} \ln x}{\lim_{x\to 1} e^{-x}} && \text{Limit Law 4}\\ &= \frac{\ln 1}{e^{-1}} = 0. && \text{Using continuity of } \ln x \text{ and } e^{-x}\end{aligned}$$

■

EXAMPLE 2.23 Evaluate $\lim_{x\to 1} \ln[x(x+1)]$.

Solution

$$\begin{aligned}\lim_{x\to 1} \ln[x(x+1)] &= \lim_{x\to 1} [\ln x + \ln(x+1)] && \text{Rules of Logarithms (Theorem B.1)}\\ &= \lim_{x\to 1} \ln x + \lim_{x\to 1} \ln(x+1) && \text{Limit Law 3}\\ &= 0 + \ln 2 = \ln 2.\end{aligned}$$

■

EXAMPLE 2.24 Evaluate $\lim_{x\to \pi} \frac{\sin x}{\sin x + \cos x}$.

Solution

$$\begin{aligned}\lim_{x\to\pi} \frac{\sin x}{\sin x + \cos x} &= \frac{\lim_{x\to\pi} \sin x}{\lim_{x\to\pi} \sin x + \lim_{x\to\pi} \cos x} && \text{Limit Laws 4 and 1}\\ &= \frac{0}{0+(-1)} = 0. && \text{Using continuity of } \sin x \text{ and } \cos x\end{aligned}$$

■

EXAMPLE 2.25 Evaluate $\lim_{x\to 0} \tan x$.

Solution

$$\lim_{x\to 0} \tan x = \lim_{x\to 0} \frac{\sin x}{\cos x} = \frac{\lim\limits_{x\to 0} \sin x}{\lim\limits_{x\to 0} \cos x} \qquad \text{Since } \tan x = \frac{\sin x}{\cos x}\text{; Limit Law 4}$$

$$= \frac{0}{1} = 0. \qquad \text{Using the continuity of } \sin x \text{ and } \cos x \quad \blacksquare$$

EXAMPLE 2.26 Evaluate $\lim\limits_{x\to \frac{\pi}{4}} (x^2 \sin^2 x)$.

Solution

$$\lim_{x\to \frac{\pi}{4}} (x^2 \sin^2 x) = \left(\lim_{x\to \frac{\pi}{4}} x\right)^2 \cdot \left(\lim_{x\to \frac{\pi}{4}} \sin x\right)^2 \qquad \text{Limit Laws 3 and 6}$$

$$= \left(\frac{\pi}{4}\right)^2 \left(\frac{\sqrt{2}}{2}\right)^2 = \frac{\pi^2}{32}. \qquad \text{Using the continuity of } x \text{ and } \sin x$$

$\blacksquare$

My main goal in this section was to illustrate the Limit Laws. That's why if you go back and look, all of the examples we just worked through could have been done using the "plug in the c-value" approach. (Indeed, all the functions in those examples were continuous at $x = c$.) In the next section I'll add a layer of complexity—illustrating how we can use the Limit Laws to help us evaluate limits of functions *discontinuous* at $x = c$.

2.7 Calculating Limits—Algebraic Techniques

Here's a tricky limit:

$$\lim_{x\to 0} \frac{x}{x}. \tag{2.2}$$

The first thing to try—the "plug in the c-value method"—yields $0/0$. We can't divide by zero, so that method didn't work for us. But you probably already know what to do: simplify $\frac{x}{x}$ to 1. And since

$$\lim_{x\to 0} 1 = 1,$$

we think the answer is 1. Figure 2.9 confirms this.

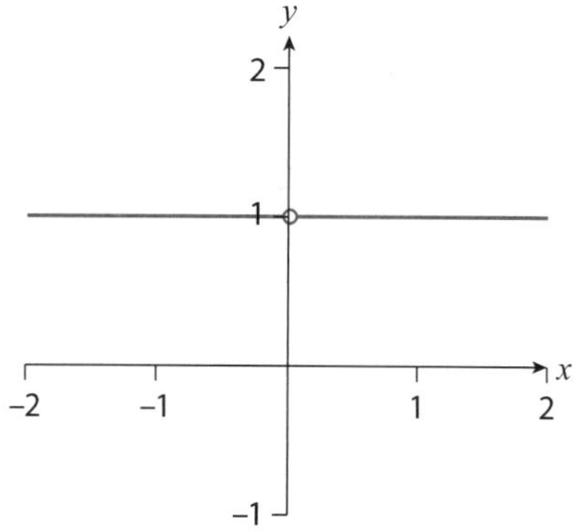

Figure 2.9: $f(x) = \frac{x}{x}$.

There's a lot to learn from this deceptively simple example. But I'll return to that in the Tips subsection. For now, the takeaway is: *Algebra helps us evaluate limits.* Here are other examples of this.

- Consider the limit

$$\lim_{x\to 1} \frac{x^2-x}{x-1}.$$

  Substituting in $x=1$ yields $0/0$, which doesn't help. But because the denominator is contained in the numerator, we can factor and cancel:

$$\lim_{x\to 1} \frac{x^2-x}{x-1} = \lim_{x\to 1} \frac{x(x-1)}{x-1} = \lim_{x\to 1} x = 1.$$

- Consider one more example:

$$\lim_{x\to 1} \frac{1-\sqrt{x}}{1-x}.$$

  Substituting in $x=1$ again yields $0/0$. We *could* factor the denominator, but let me illustrate another approach: "un-rationalization." This refers to multiplying a fraction whose numerator contains a radical by 1 in a way that gets rid of the radical in the numerator. For example,

$$\frac{1-\sqrt{x}}{1-x} = \frac{1-\sqrt{x}}{1-x} \cdot \frac{1+\sqrt{x}}{1+\sqrt{x}} = \frac{1-x}{(1-x)(1+\sqrt{x})} = \frac{1}{1+\sqrt{x}}.$$

  This then helps us to evaluate the limit:

$$\lim_{x\to 1} \frac{1-\sqrt{x}}{1-x} = \lim_{x\to 1} \frac{1}{1+\sqrt{x}} = \frac{1}{1+\sqrt{1}} = \frac{1}{2}.$$

  We refer to the quantity that gets rid of the radical in the numerator as the **conjugate**; often this is the numerator rewritten with the opposite sign. (For instance, $1+\sqrt{x}$ was the conjugate of $1-\sqrt{x}$ in the previous example.)

EXAMPLE 2.27 Evaluate $\lim_{x\to 3} \frac{x^2-9}{x-3}$.

Solution Substituting in $x=3$ yields $0/0$ again. However, we can factor the numerator and cancel:

$$\lim_{x\to 3} \frac{x^2-9}{x-3} = \lim_{x\to 3} \frac{(x+3)(x-3)}{x-3} = \lim_{x\to 3}(x+3) = 6.$$

■

EXAMPLE 2.28 Evaluate $\lim_{x\to 0^+} \frac{x}{\sqrt{x}(x+1)}$.

Solution The $\sqrt{x}$ in the denominator is approaching 0 as $x \to 0^+$, so we again need to work around this division by zero issue. Luckily, all that's needed is to simplify $\sqrt{x}/x$:

$$\frac{x}{\sqrt{x}(x+1)} = \frac{\sqrt{x}}{x+1}, \quad \text{so that} \quad \lim_{x \to 0^+} \frac{x}{\sqrt{x}(x+1)} = \lim_{x \to 0^+} \frac{\sqrt{x}}{x+1} = \frac{\sqrt{0}}{0+1} = 0. \quad ■$$

EXAMPLE 2.29 Evaluate $\lim_{x \to 0} \dfrac{\sqrt{x+1}-1}{x}$.

Solution Substituting in $x = 0$ yields $0/0$. Since there's a radical in the numerator, however, let's try un-rationalization:

$$\begin{aligned} \frac{\sqrt{x+1}-1}{x} &= \frac{\sqrt{x+1}-1}{x} \cdot \frac{\sqrt{x+1}+1}{\sqrt{x+1}+1} \\ &= \frac{(x+1)-1}{x(\sqrt{x+1}-1)} = \frac{x}{x(\sqrt{x+1}-1)} = \frac{1}{\sqrt{x+1}+1}. \end{aligned}$$

It follows that

$$\lim_{x \to 0} \frac{\sqrt{x+1}-1}{x} = \lim_{x \to 0} \frac{1}{\sqrt{x+1}+1} = \frac{1}{\sqrt{1}+1} = \frac{1}{2}. \quad ■$$

Related Exercises 8–12.

Transcendental Tales

Calculating limits of trigonometric functions is tricky. That's partly because there are so many relationships between trigonometric functions (e.g., $\sin^2 x + \cos^2 x = 1$) and partly because we sometimes need some specialized techniques. Often limits involving trigonometric functions make use of two special limits. The first is

$$\lim_{x \to 0} \frac{\sin x}{x} = 1. \tag{2.3}$$

This limit can be deduced from the graph of $\sin x/x$ (see also Section A2.5 in the online appendix to this chapter for a more intuitive argument). The second special limit is derived in Example 2.30.

EXAMPLE 2.30 Show that

$$\lim_{x \to 0} \frac{\cos x - 1}{x} = 0. \tag{2.4}$$

Solution Substituting in $x = 0$ yields $0/0$ again. So let's try something else: pretending $\cos x$ is a radical and using the un-rationalization method:

$$\frac{\cos x - 1}{x} = \frac{\cos x - 1}{x} \cdot \frac{\cos x + 1}{\cos x + 1} = \frac{\cos^2 x - 1}{x(\cos x + 1)} = \frac{-\sin^2 x}{x(\cos x + 1)}, \tag{2.5}$$

where I've used the fact that $\cos^2 x - 1 = -\sin^2 x$, which comes from the identity $\sin^2 x + \cos^2 x = 1$ (this identity is derived in (B.22) in Appendix B). Then,

$$\lim_{x\to 0} \frac{\cos x - 1}{x} = \lim_{x\to 0} \frac{-\sin^2 x}{x(\cos x + 1)} = -\lim_{x\to 0} \left(\frac{\sin x}{x} \cdot \frac{\sin x}{\cos x + 1}\right).$$

It follows from Limit Law 4 (Theorem 2.6) and (2.3) that

$$\begin{aligned}\lim_{x\to 0} \frac{\cos x - 1}{x} &= -\left[\lim_{x\to 0} \frac{\sin x}{x}\right] \cdot \left[\lim_{x\to 0} \frac{\sin x}{\cos x + 1}\right] \\ &= -\left(1 \cdot \frac{\sin 0}{\cos 0 + 1}\right) = -(1 \cdot 0) = 0.\end{aligned}$$

■

EXAMPLE 2.31 Evaluate $\lim_{x\to 0} \frac{\tan x}{x}$.

Solution Substituting in $x = 0$ yields $0/0$ again. But since $\frac{\tan x}{x} = \frac{\sin x}{x \cos x} = \frac{\sin x}{x} \cdot \frac{1}{\cos x}$,

$$\begin{aligned}\lim_{x\to 0} \frac{\tan x}{x} &= \lim_{x\to 0} \left(\frac{\sin x}{x} \cdot \frac{1}{\cos x}\right) \\ &= \left[\lim_{x\to 0} \frac{\sin x}{x}\right] \cdot \left[\lim_{x\to 0} \frac{1}{\cos x}\right] && \text{Limit Law 3} \\ &= 1 \cdot 1 = 1. && \text{Using (2.3) and the continuity of } \cos x\end{aligned}$$

■

EXAMPLE 2.32 Evaluate $\lim_{x\to 0} \frac{\sin(2x)}{x}$.

Solution Substituting in $x = 0$ yields $0/0$ again. So, let's simplify the function first. Note that

$$\frac{\sin(2x)}{x} = 2\frac{\sin(2x)}{2x}.$$

Introducing $u = 2x$, we note that if $x \to 0$ then $u \to 0$ also. So, it should follow that

$$\lim_{x\to 0} \frac{\sin(2x)}{2x} = \lim_{u\to 0} \frac{\sin u}{u} = 1,$$

via (2.3). Thus, we conclude that

$$\lim_{x \to 0} \frac{\sin(2x)}{x} = 2(1) = 2. \qquad \blacksquare$$

The change of variables we used to convert the limit from one in which $x \to 0$ to one in which $u \to 0$ works for other limit calculations too; see Section A2.2 in the online appendix to this chapter for the general result. *Related Exercises* 40 and 52–56.

Tips, Tricks, and Takeaways

1. *If one limit evaluation method fails to yield an answer, try another*. Just because your chosen method didn't work doesn't mean the limit doesn't exist—it may just mean that your original method was inapplicable. For example, substituting in $x = 0$ to evaluate (2.2) doesn't work because $f(x) = \frac{x}{x}$ is not continuous at $x = 0$.

2. *If possible, use multiple approaches to evaluate a limit.* Evaluating a limit using a graph, a table, *and* algebra will help you be confident of the true answer.

3. *A result of* $0/0$ *is never the answer to a limit.* Getting $0/0$ in a limit calculation is a tell-tale sign that you need to try another method.

4. *Most of the time algebra is all you'll need, but not always.* As we found out with trigonometric functions, sometimes a change of variables or combination of that, algebra, and the Limit Laws is required.

Finally, let me return again to (2.2). It's important to note that $\frac{x}{x} \neq 1$. "What?!" you may say. Here's the explanation: the function on the left, $\frac{x}{x}$, is not defined when $x = 0$; the function on the right, 1, *is* (it's equal to 1 when $x = 0$). So $\frac{x}{x}$ cannot equal 1. The proper way to simplify $\frac{x}{x}$ is:

$$\frac{x}{x} = 1, \quad x \neq 0.$$

This says: "x/x is equal to 1 for all nonzero x," which is what's actually true. This explains Figure 2.9 and also explains why

$$\lim_{x \to 0} \frac{x}{x} = 1,$$

since, as I've reiterated, *limits approach indefinitely (and thus never arrive)*. So, it doesn't matter what $\frac{x}{x}$ equals *at* $x = 0$, only what it equals *infinitesimally close to* $x = 0$ (which is 1, by our corrected simplification). Keep this subtle point in mind as you calculate limits.

We've now worked through many ways to calculate limits. The final two sections in this chapter redo what we've done for a new setting: limits involving infinity.

2.8 Limits Approaching Infinity

Limits approaching infinity are limits in which $x \to \infty$ or $x \to -\infty$. In either case, we are looking to see what happens to the values $f(x)$ as x gets very large and positive (in the case of $x \to \infty$) or negative (in the case of $x \to -\infty$). If $f(x) \to L$ in either case we then write

$$\lim_{x\to\infty} f(x) = L \quad \text{or} \quad \lim_{x\to-\infty} f(x) = L.$$

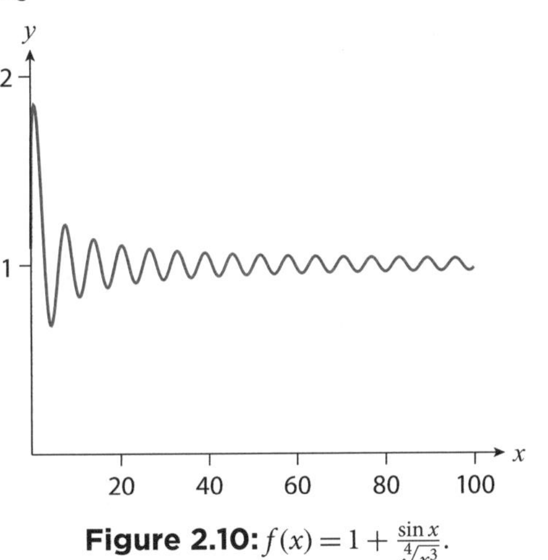

Figure 2.10: $f(x) = 1 + \frac{\sin x}{\sqrt[4]{x^3}}$.

Figure 2.10 shows an example of the first limit. We see a function for which $f(x) \to 0$ as $x \to \infty$. You may already know the name for $y = 0$ in the figure: a "horizontal asymptote." Indeed, we can define horizontal asymptotes in terms of limits.

> **Definition 2.4 Horizontal Asymptote.** Let "lim" be a stand-in for $\lim_{x\to\infty}$ or $\lim_{x\to-\infty}$, and suppose f is a function. If
>
> $$\lim f(x) = L,$$
>
> then we call the line $y = L$ a **horizontal asymptote** of the curve $y = f(x)$.

Takeaway: You can calculate a function's horizontal asymptotes using limits!

Finally, let me mention that all of the theorems discussed thus far in this chapter (including those in the online appendix to this chapter) apply to limits approaching infinity. That is, their results remain valid if we substitute "lim" for "$\lim_{x\to\infty}$" or "$\lim_{x\to-\infty}$" in those theorems. This is particularly useful because unlike in the previous sections, we cannot "plug in ∞" to calculate the limits in this section (though I will give you some tips and tricks later).

EXAMPLE 2.33 Evaluate the limit: (a) $\lim_{x\to\infty} \frac{1}{x}$ (b) $\lim_{x\to\infty} \frac{1}{x^2}$.

Solution

(a) As x becomes a larger and larger positive number (say, 10^{100}), the reciprocal $\frac{1}{x}$ becomes a very tiny positive number (continuing, 10^{-100}). Therefore, we suspect that

$$\lim_{x\to\infty} \frac{1}{x} = 0.$$

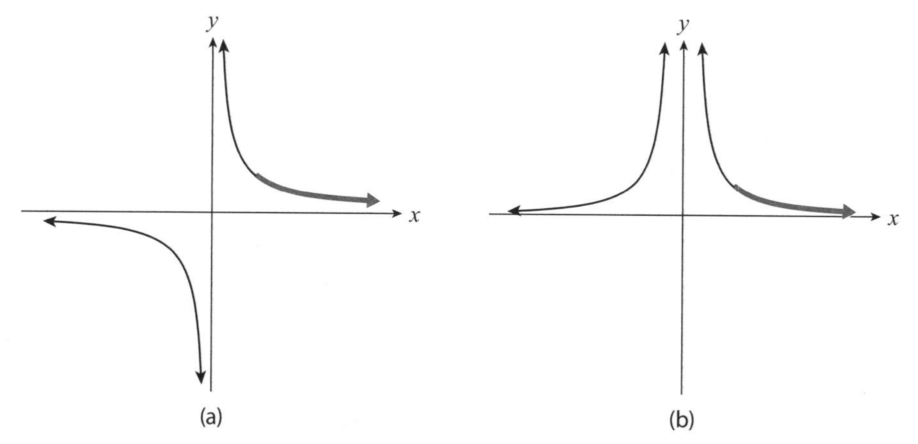

Figure 2.11: Portions of the graphs of (a) $f(x) = \frac{1}{x}$ and (b) $f(x) = \frac{1}{x^2}$. The arrows indicate that $f(x) \to 0$ as $x \to \infty$.

(This tells us that $y = 0$ is a horizontal asymptote of the function.) This is illustrated by the blue-colored arrow in Figure 2.11(a).

(b) Using a similar argument as before, as x becomes large x^2 becomes even larger and $\frac{1}{x^2}$ becomes a very tiny number. We therefore suspect that

$$\lim_{x\to\infty} \frac{1}{x^2} = 0.$$

(This tells us that $y = 0$ is a horizontal asymptote of the function.) This is illustrated by the blue-colored arrow in Figure 2.11(b). ■

Exercise 30 generalizes this example's results to show that

$$\lim_{x\to\infty} \frac{1}{x^r} = 0 \quad \text{for any rational number } r > 0. \tag{2.6}$$

Let's next discuss how to use this result to calculate the limit as $x \to \infty$ of more complicated rational functions. In the process, I'll introduce you to a useful technique when dealing with limits approaching infinity of rational functions: *divide the numerator and denominator by the highest power of x in the denominator.*

EXAMPLE 2.34 Evaluate $\lim_{x\to\infty} \frac{x}{x-1}$.

Solution The approach we used in the previous example won't work here, because as x becomes large so does $x - 1$. Here's one technique that will help: divide the numerator and denominator by the highest power of x in the denominator (this assumes $x \neq 0$, but since $x \to \infty$ we're fine), which in this case is x:

$$\frac{x}{x-1} = \frac{\frac{x}{x}}{\frac{x-1}{x}} = \frac{1}{1-\frac{1}{x}}, \quad x \neq 0.$$

Therefore, using this, Limit Laws 1 and 4 (from Theorem 2.6), and the $r=1$ case of (2.6), we have

$$\lim_{x\to\infty} \frac{x}{x-1} = \lim_{x\to\infty} \frac{1}{1-\frac{1}{x}} = \frac{\lim\limits_{x\to\infty} 1}{\lim\limits_{x\to\infty} 1 - \lim\limits_{x\to\infty} \frac{1}{x}} = \frac{1}{1-0} = 1.$$

(This tells us that $y=1$ is a horizontal asymptote of the function.) ■

Related Exercises 23–27, and 28–29 (only horizontal asymptotes).

Transcendental Tales

The first thing to know here is that e, Euler's number and the base of the natural exponential function e^x, is itself defined in terms of a limit approaching infinity (see Section A2.6 in the online appendix to this chapter if you're interested in the back story):

$$e = \lim_{x\to\infty} \left(1 + \frac{1}{x}\right)^x.$$

Exercise 45 relates this definition of e to the one given earlier in equation (A2.2) of Section A2.3 of the online appendix to this chapter by employing a change of variables. Now on to examples of limits approaching infinity of transcendental functions.

EXAMPLE 2.35 Evaluate $\lim\limits_{x\to\infty} \ln\left(1 + \frac{1}{x}\right)$.

Solution Employing Limit Laws 7 and 1 (from Theorem 2.6):

$$\lim_{x\to\infty} \ln\left(1 + \frac{1}{x}\right) = \ln\left[\lim_{x\to\infty}\left(1 + \frac{1}{x}\right)\right] = \ln\,[1+0] = \ln 1 = 0.$$

(This tells us that $y=0$ is a horizontal asymptote of the function.) ■

EXAMPLE 2.36 Evaluate $\lim\limits_{x\to\infty} e^{-x}$.

Solution Since $e^{-x} = \frac{1}{e^x}$, Limit Law 4 (Theorem 2.6) implies that

$$\lim_{x\to\infty} e^{-x} = \frac{\lim\limits_{x\to\infty} 1}{\lim\limits_{x\to\infty} e^x} = 0,$$

since $e^x \to \infty$ as $x \to \infty$ (since $e > 1$, e^x is an exponentially growing function). (This tells us that $y = 0$ is a horizontal asymptote of the function.) ■

Related Exercises 45–48 and 50.

Tips, Tricks, and Takeaways

- Though we cannot plug in ∞ (or $-\infty$) for x while calculating a limit, here are some useful rules:
$$\text{“}\frac{1}{\infty} = 0\text{”} \qquad \text{“}\frac{1}{-\infty} = 0\text{”}.$$
I put these in quotes because I want you to understand them as results about limits, not as actual equations. I'd interpret these as dividing 1 by a bigger and bigger number (positive or negative) yields numbers closer and closer to zero.

- Limits as $x \to \pm\infty$ can be converted into the one-sided limits not involving infinity:
$$\lim_{x\to\infty} f(x) = \lim_{t\to 0^+} f\left(\frac{1}{t}\right), \tag{2.7}$$
$$\lim_{x\to-\infty} f(x) = \lim_{t\to 0^-} f\left(\frac{1}{t}\right), \tag{2.8}$$
provided all limits involved exist. (Exercise 31 guides you through the proof.) These results are especially useful for limits approaching infinity of exponential and trigonometric functions; see Exercises 59 and 60.

2.9 Limits Yielding Infinity

We set the stage for this section back in Figure 2.5(a). I've reproduced that figure as Figure 2.12. The graph of the rational function $f(x) = \frac{1}{x}$ plotted therein shows that as $x \to 0^+$ (the blue portion) the values $f(x)$ get larger and larger, without bound. We concluded in Section 2.2 that

$$\lim_{x\to 0^+} \frac{1}{x} \quad \text{DNE } (+\infty).$$

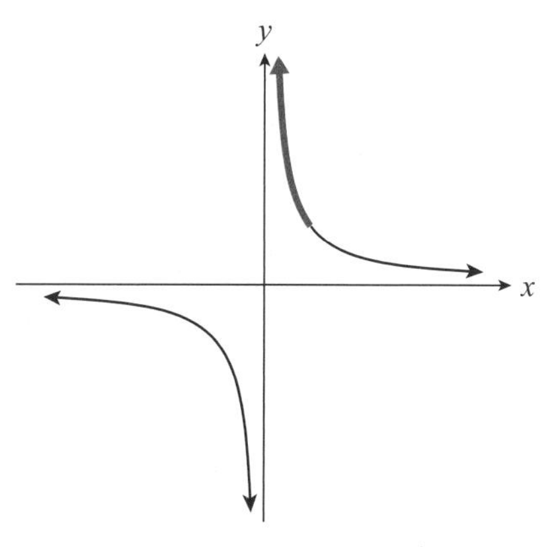

Figure 2.12: $f(x) = \frac{1}{x}$.

As in the previous section, you likely already know the name for the vertical line $x = 0$ in Figure 2.12: a vertical asymptote. That's right, we can define vertical asymptotes using limits.

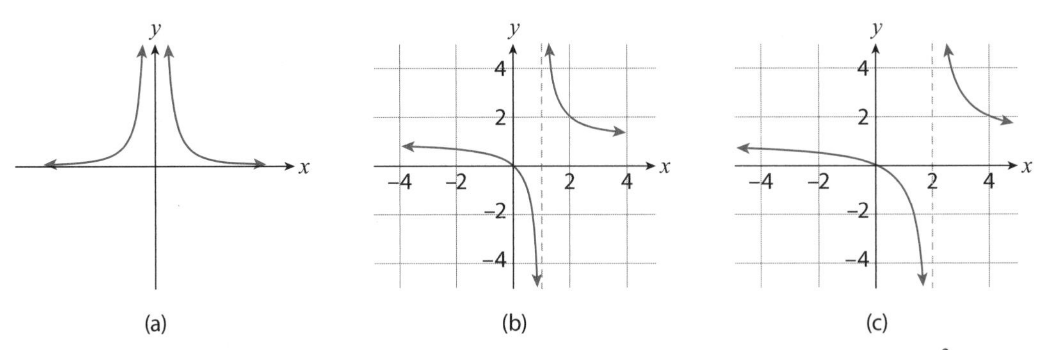

Figure 2.13: Portions of the graphs of (a) $f(x) = \dfrac{1}{x^2}$, (b) $f(x) = \dfrac{x}{x-1}$, and (c) $f(x) = \dfrac{x^2+2x}{x^2-4}$.

Definition 2.5 Vertical Asymptote. Let "lim" be a stand-in for $\lim\limits_{x\to c}$, $\lim\limits_{x\to c^-}$, or $\lim\limits_{x\to c^+}$, and suppose f is a function. If

$$\lim f(x) \text{ DNE } (+\infty) \quad \text{or} \quad \lim f(x) \text{ DNE } (-\infty),$$

then we call the line $x = c$ a **vertical asymptote** of f.

One final note: A function's graph *can* cross a horizontal asymptote (as Figure 2.10, illustrates), but it *cannot* cross a vertical asymptote (since the graph races off to infinity as x approaches the asymptote).

EXAMPLE 2.37 Evaluate the infinite limits: (a) $\lim\limits_{x\to 0} \dfrac{1}{x^2}$ (b) $\lim\limits_{x\to 1} \dfrac{x}{x-1}$ (c) $\lim\limits_{x\to 2^+} \dfrac{x^2+2x}{x^2-4}$.

Solution

(a) As $x \to 0$, $f(x) = 1/x^2$ becomes unbounded and positive; see Figure 2.13(a). Thus,

$$\lim_{x\to 0} \frac{1}{x^2} \text{ DNE } (+\infty).$$

(b) As $x \to 1^-$, $f(x) = x/(x-1)$ approaches 1 divided by a tiny negative number, which yields a large negative number. As $x \to 1^+$, $f(x) = x/(x-1)$ approaches 1 divided by a tiny positive number, which yields a large positive number. (See Figure 2.13(b) for an illustration of both limits.) Thus,

$$\lim_{x\to 1^-} \frac{x}{x-1} \text{ DNE } (-\infty), \qquad \lim_{x\to 1^+} \frac{x}{x-1} \text{ DNE } (+\infty),$$

so that, $\lim\limits_{x\to 1} \dfrac{x}{x-1}$ DNE.

(c) As $x \to 2^+$, the numerator approaches $2^2 + 2(2) = 8$ but the denominator approaches a tiny positive number. The quotient of these numbers is large (and positive); see Figure 2.13(c). We conclude that

$$\lim_{x \to 2^+} \frac{x^2 + 2x}{x^2 - 4} \quad \text{DNE } (+\infty).$$ ■

Note that in the solution to Example 2.37 (b) there is no parenthetical information next to "DNE." That's because the associated two-sided limit races off to infinities with different signs depending on which direction we approach 1 from.

Related Exercises 21–22, 28–29 (only vertical asymptotes), and 32.

Einstein's Relativity Applied Example C.2 in Appendix C explores the phenomenon that moving objects become more massive the closer to the speed of light they travel.

Transcendental Tales

EXAMPLE 2.38 Find the limit: (a) $\lim_{x \to 0^+} \ln x$ (b) $\lim_{x \to 0^+} e^{-1/x}$.

Solution

(a) The graph of $f(x) = \ln x$ is similar to that of Figure B.12 in Appendix B. We conclude from that graph that

$$\lim_{x \to 0^+} \ln x \quad \text{DNE } (-\infty).$$

Thus, $x = 0$ is a vertical asymptote of $\ln x$.

(b) First note that

$$e^{-1/x} = \frac{1}{e^{1/x}}. \tag{2.9}$$

Then, recalling from Figure 2.12 that $\frac{1}{x} \to \infty$ as $x \to 0^+$, we see that $e^{1/x} \to \infty$ as $x \to 0^+$. Thus, we are dividing by a larger and larger number in (2.9), producing a number closer and closer to zero. We conclude that

$$\lim_{x \to 0^+} e^{-1/x} = 0.$$

(This makes $y = 0$ a horizontal asymptote of $e^{-1/x}$.) ■

Related Exercises 42 and 44.

Vertical asymptotes are also present in several of the trigonometric functions you likely studied in a pre-calculus course. Here are some examples of that.

EXAMPLE 2.39 Calculate the limits: (a) $\lim_{x\to\frac{\pi}{2}^-} \tan x$ (b) $\lim_{x\to\frac{\pi}{2}^+} \tan x$ (c) $\lim_{x\to 0^+} \dfrac{1}{\sin x}$

Solution

(a) Recall that $\tan x = \frac{\sin x}{\cos x}$. Now, $\cos x \to 0^+$ as $x \to \frac{\pi}{2}^-$ (recall the graph of $y = \cos x$ in Figure B.18(b) in Appendix B). Meanwhile, $\sin x \to 1$ as $x \to \frac{\pi}{2}^-$ (recall the graph of $y = \sin x$ in Figure B.18(a) in Appendix B). Thus, the ratio of these numbers, which is $\tan x$, is approaching 1 divided by a tiny positive number. We conclude that

$$\lim_{x\to\frac{\pi}{2}^-} \tan x \ \text{ DNE } (+\infty),$$

and that $x = \frac{\pi}{2}$ is a vertical asymptote for $\tan x$. (This is reflected in the graph of $y = \tan x$ in Figure B.18(c) in Appendix B.)

(b) The situation here is similar, except that $\cos x \to 0^-$ as $x \to \frac{\pi}{2}^+$. Thus, $\tan x$ approaches 1 divided by a tiny negative number. We conclude that

$$\lim_{x\to\frac{\pi}{2}^+} \tan x \ \text{ DNE } (-\infty),$$

and again that $x = \frac{\pi}{2}$ is a vertical asymptote for $\tan x$. (This is also reflected in the graph of $y = \tan x$ in Figure B.18(c) in Appendix B.)

(c) Since $\sin x \to 0^+$ as $x \to 0^+$, we conclude that

$$\lim_{x\to 0^+} \frac{1}{\sin x} \ \text{ DNE } (+\infty),$$

meaning that $x = 0$ is a vertical asymptote of the graph of $1/\sin x$. ■

The reciprocal function in part (c) is one of the trio of reciprocal trigonometric functions:

$$\sec x = \frac{1}{\cos x}, \qquad \csc x = \frac{1}{\sin x}, \qquad \cot x = \frac{1}{\tan x}.$$

As part (c) of the preceding example suggests, these functions have plenty of vertical asymptotes.

Tips, Tricks, and Takeaways

- *We can now use limits to definitively identify vertical asymptotes.* Some students learn that we find vertical asymptotes by setting the denominator of a quotient to zero and solving for the resulting x-values. *But this doesn't always work.* For instance, this method of finding vertical asymptotes fails for the function $f(x) = \frac{x}{x}$. As Figure 2.9 shows, that function does not have a vertical asymptote at $x = 0$. This is confirmed by calculating the limits (since $f(x) \to 1$, not infinity, as $x \to 0$).

- *Dividing 1 by a tiny number yields a huge number.* Like I did in the previous section, here are two "equations" that summarize the action in this section:

$$\text{“}\frac{1}{0^+} = +\infty\text{,”} \quad \text{“}\frac{1}{0^-} = -\infty\text{.”}$$

2.10 Parting Thoughts

We've now learned A LOT about limits. We introduced the concept in Chapter 1 to help us express the notion of "infinitesimal change." Now we have a more thorough understanding of what a limit is, how to calculate limits, and how they show up in real-world contexts.

In Chapter 1 we also previewed how limits form the foundation for the solution to the three Big Problems that drove the development of calculus. In the next chapter we'll use limits to tackle two of those three problems—instantaneous speed and the slope of the tangent line problem.

CHAPTER 2 EXERCISES

1. You're told that $\lim_{x \to c^-} f(x)$ and $\lim_{x \to c^+} f(x)$ are equal to some number. Describe the graph of $y = f(x)$. What changes if the limits equal different numbers?

2. Below is the graph of a function $y = f(x)$.

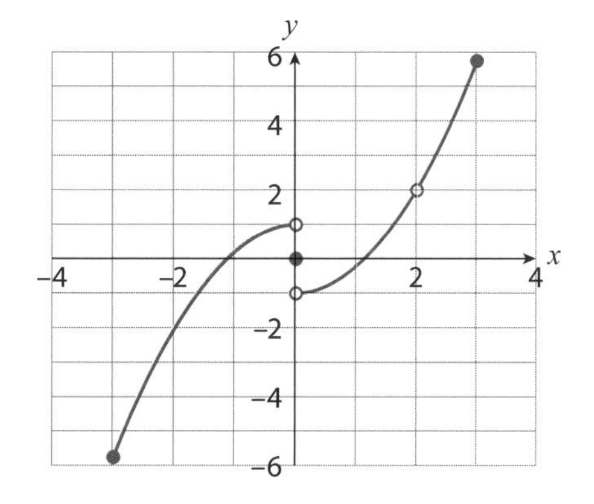

(a) Evaluate the limit, or explain why it does not exist.

(i) $\lim_{x \to 0^-} f(x)$ (ii) $\lim_{x \to 0^+} f(x)$

(iii) $\lim_{x \to 0} f(x)$ (iv) $\lim_{x \to 2^-} f(x)$

(v) $\lim_{x \to 2^+} f(x)$ (vi) $\lim_{x \to 2} f(x)$

(b) True or False: The function is continuous at $x = 2$.

(c) For what x-values in the interval $(-1, 3)$ is the function continuous?

3. Below is the graph of a function $y = f(x)$.

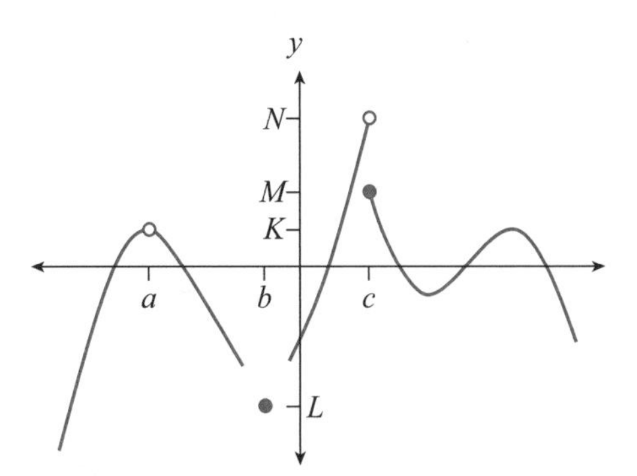

(a) Evaluate the limit, or explain why it does not exist.

(i) $\lim_{x\to a^-} f(x)$ (ii) $\lim_{x\to a^+} f(x)$

(iii) $\lim_{x\to b^-} f(x)$ (iv) $\lim_{x\to b^+} f(x)$

(v) $\lim_{x\to c^-} f(x)$ (vi) $\lim_{x\to c^+} f(x)$

(vii) $\lim_{x\to a} f(x)$ (viii) $\lim_{x\to b} f(x)$

(ix) $\lim_{x\to c} f(x)$

(b) True or False: The function is continuous at $x = c$.

4. Suppose $\lim_{x\to c} f(x) = 1$ and $\lim_{x\to c} g(x) = 2$. Evaluate the following limits.

(a) $\lim_{x\to c} [f(x) + g(x)]$ (b) $\lim_{x\to c} [2f(x)]$

(c) $\lim_{x\to c} [f(x)g(x)]$ (d) $\lim_{x\to c} \dfrac{f(x)}{g(x)}$

(e) $\lim_{x\to c} \sqrt{f(x)}$

(f) $\lim_{x\to c} [(x-c)f(x) - g(x)]$

5–10: Evaluate the limit with the help of algebra, if the limit exists.

5. $\lim_{x\to 0} (x^2 - 2x + 1)$ **6.** $\lim_{x\to 9} (\sqrt{x} - 3)$

7. $\lim_{x\to 1^-} \sqrt{x^2+1}$ **8.** $\lim_{x\to 0} \dfrac{x^3 - x}{x}$

9. $\lim_{x\to 1^+} \dfrac{1}{x-1}$ **10.** $\lim_{x\to 4} \dfrac{\sqrt{x}-2}{x^2-4}$

11. $\lim_{h\to 0} \dfrac{\sqrt{2h+1}-1}{h}$

12. $\lim_{x\to 0} \left(\dfrac{1}{x} - \dfrac{1}{x^2+x}\right)$

13. For what value of a is $\lim_{x\to 1} f(x) = 1$ for the function below?

$$f(x) = \begin{cases} ax + 2, & \text{if } x \le 1, \\ x^2, & \text{if } x > 1. \end{cases}$$

14–17: (a) Find the domain of the function, and then (b) use Theorems 2.1–2.3 to determine the interval(s) where the function is continuous.

14. $f(x) = \dfrac{x}{x^2 + 2x + 1}$

15. $g(x) = \sqrt{x}(1 - \sqrt[3]{x})$

16. $h(x) = x^2 + \sqrt{x}$

17. $h(t) = (\sqrt{t} + \sqrt{1-t})^3$

18–20: Determine the interval(s) on which the piecewise function is continuous. It may help to graph the function.

18. $f(x) = \begin{cases} x, & \text{if } x \le 0, \\ x^2, & \text{if } x > 0. \end{cases}$

19. $g(x) = \begin{cases} x^3, & \text{if } x \le 1, \\ \sqrt{x+1} & \text{if } x > 1. \end{cases}$

20. $h(x) = \begin{cases} \dfrac{1}{x}, & \text{if } x \le 1, \\ x+1 & \text{if } x > 1. \end{cases}$

21–26: These limits involve infinity. Evaluate those that exist, and determine $\pm\infty$ for those that do not.

21. $\lim_{x\to 3^-} \dfrac{x+7}{x-3}$ **22.** $\lim_{x\to 2} \dfrac{1-x}{(x-2)^2}$

23. $\lim_{x\to\infty} \dfrac{2}{3x+2}$ **24.** $\lim_{x\to -\infty} \dfrac{-3x^2+4x}{x^2+1}$

25. $\lim_{x\to\infty} \frac{x+1}{\sqrt{3x^2+7}}$ **26.** $\lim_{x\to-\infty} \frac{\sqrt[3]{x}}{x+1}$

27. Evaluate $\lim_{x\to\infty} (\sqrt{x+1}-\sqrt{x})$.

Hint: Multiply the function by $\frac{\sqrt{x+1}+\sqrt{x}}{\sqrt{x+1}+\sqrt{x}}$.

28–29: Find the function's vertical and horizontal asymptotes.

28. $f(x) = \frac{3x+4}{x-3}$ **29.** $f(x) = \frac{x^2+1}{2x^2-2}$

30. Use the result of Example 2.33(a) and Theorem 2.6 to show that

$$\lim_{x\to\infty} \frac{1}{x^r} = 0,$$

where $r > 0$ is any rational number.

31. This problem uses Theorem A2.1 (in the online appendix to this chapter) to derive (2.7). Let f be a function and define $g(x) = \frac{1}{x}$. We showed that $g(x) \to \infty$ as $x \to 0^+$, so let's verify the two remaining assumptions of the Theorem to derive (2.7).

(a) Explain why $g(x) \neq 0$ for any $x \neq 0$ in an interval containing 0.

(b) Supposing $\lim_{t\to\infty} f(t)$ exists, show that (A2.1) in the online appendix to this chapter (with $g(x) = 1/x$) yields (2.7) (with the variables x and t swapped).

32. Relativity In 1905 Albert Einstein discovered that measurements of some physical quantities—such as time and length—depend on the frame of reference used. For example, suppose you are riding on a train moving at velocity v. Einstein's *Special Theory of Relativity* says that the passage of t seconds *relative to you* is equivalent to the passage of T *relative to a stationary observer* outside the train, where

$$T(v) = \frac{t}{\sqrt{1-v^2/c^2}},$$

where c is the speed of light. (This is called **time dilation**.)

(a) Evaluate and interpret $T(0)$ and $T(0.5c)$.

(b) Show that $T(v) \to \infty$ as $v \to c^-$, and interpret your result.

(c) Why was a left-hand limit necessary?

33. Everyday Continuity Which functions below are continuous?

(a) The height of a person as a function of his or her age

(b) A student's high school GPA (supposing it's not constant) as a function of time

(c) The balance on one of your credit cards as a function of time

34. Cab Fare Suppose a taxi in New York City charges \$2.50 to pick you up and \$2.50 for each mile traveled.

(a) Let C denote the total cost of traveling x miles in the taxi. Sketch a graph of $C(x)$ for $0 \le x \le 4.5$.

(b) Is $C(x)$ a continuous function? Briefly explain.

(c) Are the discontinuities of $C(x)$ gap discontinuities or jump discontinuities? Briefly explain.

35. Newton's Law of Gravity Let F be the gravitational force exerted by the Earth on an object of mass m placed a distance $r \ge 0$ from the center of the Earth. Assuming that the Earth is a perfect sphere of radius R and that it has mass M, Newton's Universal Law of Gravity then tells us that

$$F(r) = \begin{cases} \dfrac{GMmr}{R^3}, & \text{if } r < R, \\ \dfrac{GMm}{r^2}, & \text{if } r \ge R, \end{cases}$$

where $G > 0$ is a constant (the *gravitational constant*).

(a) Briefly explain how the gravitational force felt by the mass m changes as r changes.
(b) Calculate $\lim_{r \to R^-} F(r)$ and $\lim_{r \to R^+} F(r)$.
(c) Is $F(r)$ continuous at $r = R$? Briefly explain.
(d) On what interval(s) is F continuous?

36. Use the limit laws to prove that if $f(x)$ is a polynomial, then $\lim_{x \to c} f(x) = f(c)$.

37. What a-value(s) make the function continuous?

$$f(x) = \begin{cases} x+1, & \text{if } x \leq 1, \\ ax^2, & \text{if } x > 1. \end{cases}$$

38. What a-value(s) make the function continuous?

$$g(x) = \begin{cases} x^2+3, & \text{if } x \leq a, \\ 4x, & \text{if } x > a. \end{cases}$$

Exercises Involving Exponential and Logarithmic Functions

39–44: Evaluate the limit, if it exists.

39. $\lim_{x \to 0} e^{-x}$

40. $\lim_{h \to 0} \dfrac{\sqrt{1+3^h} - 1}{3^h}$

41. $\lim_{t \to 2} t^2 2^t$

42. $\lim_{z \to 1^+} e^{-(z-1)^{-1}}$

43. $\lim_{x \to 1} e^x \ln x$

44. $\lim_{x \to 1^-} \left[x^2 e^{-x} + \ln(1-x)\right]$

45. This exercise derives the definition of e given in (A2.2) from the original definition given in A2.7.

(a) First, show that the substitution $x = 1/n$ converts

$$\left(1 + \frac{1}{n}\right)^n \quad \text{to} \quad (1+x)^{1/x}.$$

(b) Next, show that the substitution $x = 1/n$ converts the limit as $n \to \infty$ into the limit as $x \to 0^+$.
(c) Plot the graph of $f(x) = (1+x)^{1/x}$ to see that $f(x)$ approaches the same y-value as $x \to 0^-$ and $x \to 0^+$. Equation (A2.2) then follows.

46. Compounding Interest in a Savings Account This exercise will derive equation (A2.5), and an associated formula for the continuous compounding case. To begin, suppose $M(t)$ is the balance (in dollars) of a savings account t years after opening it with an initial deposit of M_0 dollars (and that no subsequent deposits are made). Let r denote the yearly interest rate earned—where r is expressed as a decimal—and suppose that the savings account compounds interest n times per year.

(a) Show that the balance of the account right after the first compounding of interest is

$$M_0 \left(1 + \frac{r}{n}\right).$$

(b) Show that the balance of the account at the end of the first year is

$$M_0 \left(1 + \frac{r}{n}\right)^n.$$

(c) Generalize the result of part (b) to derive the formula for $M(t)$.
(d) Letting $x = r/n$, show that

$$M(t) = M_0 \left[(1+x)^{1/x}\right]^{rt}.$$

(e) Show that

$$\lim_{x \to 0^+} M(t) = M_0 e^{rt},$$

and explain this result in practical terms.

47. Doubling Time and the Rule of 70 Returning to the previous exercise, suppose the savings account's interest is compounded yearly, so that

$$M(t) = M_0(1+r)^t.$$

(a) Let T denote the time required for the balance to equal double the initial deposit (i.e., $M(T) = 2M_0$). T is called the **doubling time**. Show that

$$T = \frac{\ln 2}{\ln(1+r)}.$$

(b) Plot the graph of $f(r) = \ln(1+r) - r$ for $-0.5 \leq r \leq 0.5$, and deduce from it that

$$\lim_{r \to 0} f(r) = 0.$$

(c) Real-world interest rates for savings accounts are typically in the range $0 \leq r \leq 0.1$. Thus, since part (b) tells us that $\ln(1+r) \approx r$ for r-values close to zero, return to part (a) and show that

$$T \approx \frac{0.7}{r}, \quad \text{for } r \text{ close to zero.}$$

(d) Let $R = 100r$. (R is then the interest rate expressed as a number between 0 and 100.) Use part (c) to show that

$$T \approx \frac{70}{R}.$$

This is known as the **Rule of 70**.

48. Terminal Velocity of a Falling Raindrop In Exercise 43 of Appendix B we discussed the following velocity function for a falling raindrop:

$$v(t) = 13.92(1 - e^{-2.3t}),$$

Calculate $\lim_{t \to \infty} v(t)$ to find the raindrop's terminal velocity.

49. Prove Theorem 2.5. *Hint:* Rewrite an arbitrary exponential function ab^x in the form $ae^{rx} = a(e^x)^r$ and then use a couple of theorems in this chapter.

50. Suppose $f(x) = ab^x$ is an exponential function. As we've discussed, we can rewrite f as $f(x) = ae^{rx}$. Show that either $\lim_{x \to \infty} f(x)$ or $\lim_{x \to -\infty} f(x)$ produces zero, regardless of the signs of a and r.

Exercises Involving Trigonometric Functions

51–56: Evaluate the limit.

51. $\lim_{x \to 0} \sqrt{1+\sin x}$

52. $\lim_{x \to \frac{\pi}{4}} \dfrac{\sin^2 x - \cos^2 x}{\sin x - \cos x}$

53. $\lim_{x \to 0} \dfrac{\sin(3x)}{x}$

54. $\lim_{x \to 0} \dfrac{\cos x - 1}{\sin x}$

55. $\lim_{h \to 0} \dfrac{\sin(2h)}{\sin(6h)}$

56. $\lim_{t \to 0} \dfrac{\sin t}{t + \tan t}$

57. Let $f(x) = \sin\left(\frac{1}{x}\right)$. Explain why $-1 \leq f(x) \leq 1$. Then explain how this helps deduce that for any $d > 0$, $|xf(x)| \leq d$ whenever $|x| \leq d$.

58. What a-value(s) make the following function continuous?

$$f(x) = \begin{cases} \dfrac{\sin(2x)}{x}, & \text{if } x \neq 0, \\ a^2, & \text{if } x = 0. \end{cases}$$

59. Consider the function $f(x) = x\sin\left(\frac{1}{x}\right)$, graphed in Figure (A2.1) (in the online appendix to this chapter) for $-0.1 \leq x \leq 0.1$. Use the substitution $t = 1/x$ and (2.3) to show that

$$\lim_{x \to \infty} f(x) = 1.$$

60. Exercise 60 of Appendix B illustrates how to approximate the area of a circle of radius r using inscribed triangles. Show that

$$\lim_{n \to \infty} A(n) = \pi r^2$$

by suitably modifying the result of the preceding exercise.

61. The Acceleration due to Gravity as a Function of Latitude In the chapter we used the approximation 32 $\mathrm{ft/s^2}$ for the acceleration due to gravity, denoted by g. (That approximation is roughly 9.8 $\mathrm{m/s^2}$ in SI units.) A more accurate formula for g is the Geodetic Reference Formula of 1967:

$$g(x) = a(1 + b\sin^2 x - c\sin^4 x) \quad \mathrm{m/s^2},$$

where x is the latitude (in degrees) north or south of the equator, and

$$a = 9.7803185$$

$$b = 0.005278895$$

$$c = 0.000023462$$

(a) Plot $g(x)$ for $-\frac{\pi}{2} \leq x \leq \frac{\pi}{2}$. At what latitude(s) is g largest? Smallest?

(b) Calculate $\lim_{x\to 0} g(x)$ and interpret your result.

3 Derivatives: Change, Quantified

Chapter Preview. *Cambridge University, August 1665. An outbreak of plague across England has forced the university to close down. One of its students—Isaac Newton—returns to his countryside home, Woolsthorpe Manor (Figure 3.1). As he later described to his friends, one day Newton witnesses an apple falling from a nearby tree and asks: Does the same force that pulls the apple to the ground (gravity) also pull on other objects, like the Moon? This question kick-starts Newton's work on what would become his Law of Universal Gravitation. But Newton soon encounters a conceptual hurdle: instantaneous speed. You see, gravity continuously accelerates objects (such as the apple) and therefore changes their speed from instant to instant. To understand gravity, then, you need a mathematical theory of instantaneous speed. There was none at the time. So Newton invented one. We'll follow in Newton's footsteps in this chapter to solve the Instantaneous Speed Problem, and later discover—as he did—that the results can be generalized. In an epic 2-for-1 deal, we'll learn that the result—the derivative—also solves a second Big Problem from Chapter 1: the Tangent Line Problem.*

3.1 Solving the Instantaneous Speed Problem

Let's go back to Figure 1.4(a), the snapshot of an apple falling from Newton's apple tree. Let's assume that the snapshot is 1 second into the apple's fall (let's denote that by $t=1$) and that the tree is 30 feet tall. Question: Can we calculate the apple's speed at that instant?

A reasonable starting point is to remember that speed is distance divided by time. More precisely,

$$\text{average speed} = \frac{\text{change in distance}}{\text{change in time}} = \frac{\Delta d}{\Delta t}, \tag{3.1}$$

Figure 3.1: Woolsthorpe Manor (background) and the famous apple tree (foreground).

where Δd denotes the change in distance and Δt the change in time. But there is no passage of time in the instant depicted in Figure 1.4(a). Therefore, $\Delta t = 0$. That's a problem, because Δt is in the denominator of (3.1), and one commandment of mathematics is: *Thou shalt never divide by zero.* We've reached an impasse.

But recall my first characterization of calculus: calculus is a dynamics mindset. Nothing about Figure 1.4(a) says "dynamics." That's why we switched to the more dynamic picture in the top row of Figure 1.5. Let's quantify those snapshots of the apple's fall and see how the limit as $\Delta t \to 0$ solves the instantaneous speed problem.

Several decades before Newton started thinking about gravity, Galileo Galilei (1564–1642) and his contemporaries had figured out how to describe mathematically the distance traveled by a sufficiently heavy object (e.g., an apple) dropped from rest:

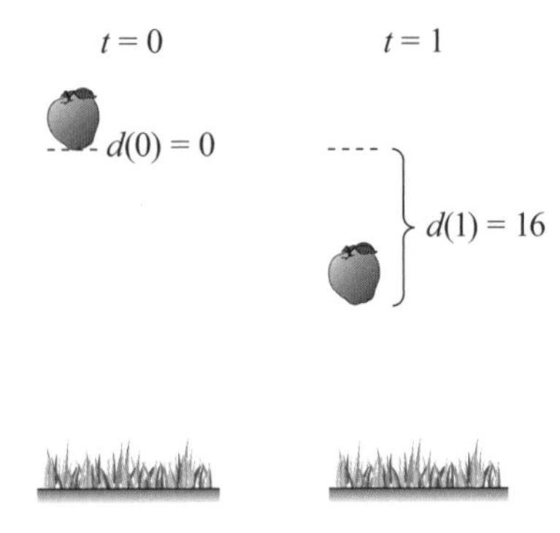

Figure 3.2: Two snapshots of a falling apple

$$d(t) = 16t^2, \tag{3.2}$$

where d is the distance measured in feet, t the seconds elapsed since the object was dropped, and air resistance is ignored (see Figure 3.2). (Galileo established (3.2) through experiments with inclined planes.) Now, since

- $d(1)$ yields the distance the apple has traveled 1 second into its fall, and
- $d(1 + \Delta t)$ yields the distance the apple has traveled $1 + \Delta t$ seconds into its fall,

then the apple's change in distance Δd is: $\Delta d = d(1 + \Delta t) - d(1)$. Inserting this into (3.1) yields

$$\text{apple's average speed} = \frac{d(1 + \Delta t) - d(1)}{\Delta t}. \tag{3.3}$$

Using (3.2) now to evaluate the numerator:

$$\begin{aligned}
&\text{apple's average speed} \\
&= \frac{16[1 + \Delta t]^2 - 16(1)^2}{\Delta t} && \text{Using } d(1 + \Delta t) = 16[1 + \Delta t^2] \\
&= \frac{16[1 + 2(\Delta t) + (\Delta t)^2] - 16}{\Delta t} \\
&= \frac{32(\Delta t) + 16(\Delta t)^2}{\Delta t} && \text{Distributing the 16 and simplifying} \\
&= 32 + 16(\Delta t), \quad \Delta t \neq 0. && \text{Canceling the } \Delta t
\end{aligned}$$

This suggests that the apple's average speeds approach the limiting value 32 as $\Delta t \to 0$:

$$\lim_{\Delta t\to 0}\frac{\Delta d}{\Delta t}=32.$$

This is the most reasonable answer to the question, "What's the apple's instantaneous speed at $t=1$?" No *finite* change in time Δt works to answer that question—you'd be calculating the *average* speed then. By examining an *infinitesimal* change in time, however, we've uncovered the apple's true *instantaneous* speed at $t=1$.

We can easily generalize what we've just done. Starting from the notion that

$$\text{apple's instantaneous speed} = \lim_{\Delta t\to 0}(\text{apple's average speed}), \tag{3.4}$$

if we substitute (3.3) into (3.4) we get

$$\text{apple's instantaneous speed} = \lim_{\Delta t\to 0}\frac{d(1+\Delta t)-d(1)}{\Delta t}.$$

Denoting the apple's instantaneous speed 1 second into it's fall by $\mathcal{s}(1)$, the equation above becomes

$$\mathcal{s}(1)=\lim_{\Delta t\to 0}\frac{d(1+\Delta t)-d(1)}{\Delta t}. \tag{3.5}$$

This is the equation for the apple's instantaneous speed at $t=1$.[1]

But there was nothing special about $t=1$ for the purposes of this analysis; we could just as easily have calculated the apple's instantaneous speed at $t=0.5$, for example. Therefore, if we replace "1" with a, some other t-value for which the apple is still in motion, (3.5) yields the following definition.

Definition 3.1 The **instantaneous speed at time** $t=a$ of an object with distance function $d(t)$, denoted by $\mathcal{s}(a)$, is defined by

$$\mathcal{s}(a)=\lim_{\Delta t\to 0}\frac{d(a+\Delta t)-d(a)}{\Delta t}. \tag{3.6}$$

EXAMPLE 3.1 Find $\mathcal{s}(1)$ (ignoring units) for an object with distance function $d(t)=3t+5$.

Solution

$$\mathcal{s}(1)=\lim_{\Delta t\to 0}\frac{d(1+\Delta t)-d(1)}{\Delta t} \qquad \text{Equation (3.6) with } a=1$$

$$=\lim_{\Delta t\to 0}\frac{3(1+\Delta t)+5-(3(1)+5)}{\Delta t} \qquad \text{Using } d(1+\Delta t)=3(1+\Delta t)+5$$

[1] A note about notation here. I've used the cursive $\mathcal{s}$ to denote the instantaneous speed. That's because by convention s refers to an object's **position function** $s(t)$. We will discuss $s(t)$ and its instantaneous change—**velocity**—in Chapter 5.

$$
\begin{aligned}
&= \lim_{\Delta t \to 0} \frac{3\Delta t}{\Delta t} && \text{Distributing and simplifying} \\
&= \lim_{\Delta t \to 0} 3 = 3. && \text{Canceling } \Delta t \text{ and evaluating the limit} \quad \blacksquare
\end{aligned}
$$

EXAMPLE 3.2 Find $\mathscr{s}(2)$ (ignoring units) for an object with distance function $d(t) = t^2$.

Solution

$$
\begin{aligned}
\mathscr{s}(2) &= \lim_{\Delta t \to 0} \frac{d(2+\Delta t) - d(2)}{\Delta t} && \text{Equation (3.6) with } a = 2 \\
&= \lim_{\Delta t \to 0} \frac{(2+\Delta t)^2 - 2^2}{\Delta t} && \text{Using } d(2+\Delta t) = (2+\Delta t)^2 \\
&= \lim_{\Delta t \to 0} \frac{4(\Delta t) + (\Delta t)^2}{\Delta t} && \text{Squaring out } (2+\Delta t)^2 \text{ and simplifying} \\
&= \lim_{\Delta t \to 0} (4 + \Delta t) = 4. && \text{Canceling } \Delta t \text{ and evaluating the limit} \quad \blacksquare
\end{aligned}
$$

Finally, let's apply what we've learned to completely solve the instantaneous speed problem for Newton's falling apple.

APPLIED EXAMPLE 3.3 Calculate $\mathscr{s}(a)$ for the falling apple's distance function $d(t) = 16t^2$.

Solution

$$
\begin{aligned}
\mathscr{s}(a) &= \lim_{\Delta t \to 0} \frac{d(a+\Delta t) - d(a)}{\Delta t} && \text{Equation (3.6)} \\
&= \lim_{\Delta t \to 0} \frac{16(a+\Delta t)^2 - 16a^2}{\Delta t} && \text{Using } d(a+\Delta t) = 16(a+\Delta t)^2 \\
&= \lim_{\Delta t \to 0} \frac{[16a^2 + 32a(\Delta t) + 16(\Delta t)^2] - 16a^2}{\Delta t} && \text{Squaring out } 16(a+\Delta t)^2 \\
&= \lim_{\Delta t \to 0} [32a + 16(\Delta t)] = 32a. && \text{Simplifying, canceling } \Delta t\text{, and evaluating the limit} \quad \blacksquare
\end{aligned}
$$

Related Exercises 10–11.

Note that because we left the a-value undetermined, we've actually calculated the instantaneous speed *at any time* $t = a$. This is a great, concrete example of the power of calculus. We solved a problem (instantaneous speed) that stumped scientists for *millennia* in just a few lines. What's more, the solution is really simple: the apple's instantaneous speed a seconds into its fall is $32a$ ft/s.

3.2 Solving the Tangent Line Problem—The Derivative at a Point

Recall that the Tangent Line Problem is the problem of finding the slope of the line tangent to the graph of a function $y=f(x)$ at a point P on the graph (recall Figure 1.4(b)). By "tangent line" we mean that the line shares point P with the curve, and also shares the curve's "inclination" at P. (Trace the curve in Figure 1.4(b) with your finger and when you get to point P your finger will be moving in the direction of the tangent line.) What makes the tangent line problem so difficult is this: we need *two* points to calculate the slope of a line, but we only have *one* (point P). Once again, we're stuck. Yet once again—sounding like a broken record by now—the problem is the static mindset inherent in Figure 1.4(b). So let's *calculus* the figure. (Yep, as we said in Chapter 1, calculus is also a verb!)

Figure 3.3 shows the dyanmics mindset take. The gray lines passing through points P and Q are *secant* lines, so named because of the word's Latin root: *secare*, which means to cut. The slope of each secant line is

$$\text{slope of the secant line } \overleftrightarrow{PQ} = \frac{\Delta y}{\Delta x}. \tag{3.7}$$

If we assume point P has x-coordinate a, then the change Δy in y-values between P and Q is

$$\Delta y = \text{ } y\text{-value at } Q \text{ } - \text{ } y\text{-value at } P = f(a+\Delta x) - f(a).$$

Inserting this into (3.7) yields

$$\text{slope of the secant line } \overleftrightarrow{PQ} = \frac{f(a+\Delta x) - f(a)}{\Delta x}. \tag{3.8}$$

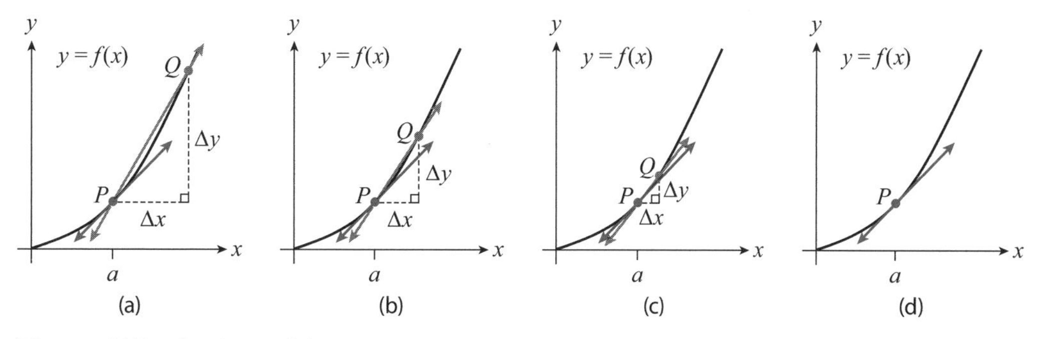

Figure 3.3: The slope of the gray (secant) lines approaches the slope of the blue (tangent) line as $\Delta x \to 0$.

Interactive Figure

Finally, as Figure 3.3 suggests: the slope of the *tangent* line is the limiting value of the slopes of the secant lines as $\Delta x \to 0$:

$$\text{slope of the tangent line at } P = \lim_{\Delta x \to 0} \frac{f(a+\Delta x) - f(a)}{\Delta x}. \tag{3.9}$$

Today we shorten "slope of the tangent line at P" to "**derivative** at a" and use the notation $f'(a)$ (read "f prime at a"), so that (3.9) becomes

$$f'(a) = \lim_{\Delta x \to 0} \frac{f(a+\Delta x) - f(a)}{\Delta x}. \tag{3.10}$$

Et voilá! The tangent line problem is solved. *And* we've discovered something important:

> The derivative *of a function* f *at* $x = a$ *is just the slope of the line tangent to the graph of* f *at* $x = a$.

The derivative is an important concept, so let me give you the formal definition and then mention a few comments before working through examples.

Definition 3.2 The Derivative at a Point. Let f be a function. The **derivative of** f **at** $x = a$, denoted by $f'(a)$, is defined by

$$f'(a) = \lim_{\Delta x \to 0} \frac{f(a+\Delta x) - f(a)}{\Delta x}, \tag{3.11}$$

provided the limit exists.

EXAMPLE 3.4 Calculate $f'(1)$ for $f(x) = x^2$.

Solution

$$\begin{aligned}
f'(1) &= \lim_{\Delta x \to 0} \frac{f(1+\Delta x) - f(1)}{\Delta x} && \text{Equation (3.11) with } a = 1 \\
&= \lim_{\Delta x \to 0} \frac{(1+\Delta x)^2 - 1^2}{\Delta x} && \text{Using } f(1+\Delta x) = (1+\Delta x)^2 \\
&= \lim_{\Delta x \to 0} \frac{2(\Delta x) + (\Delta x)^2}{\Delta x} && \text{Squaring out } (1+\Delta x)^2 \text{ and simplifying} \\
&= \lim_{\Delta x \to 0} [2 + (\Delta x)] = 2. && \text{Simplifying, canceling } \Delta x \text{, and evaluating the limit}
\end{aligned}$$

■

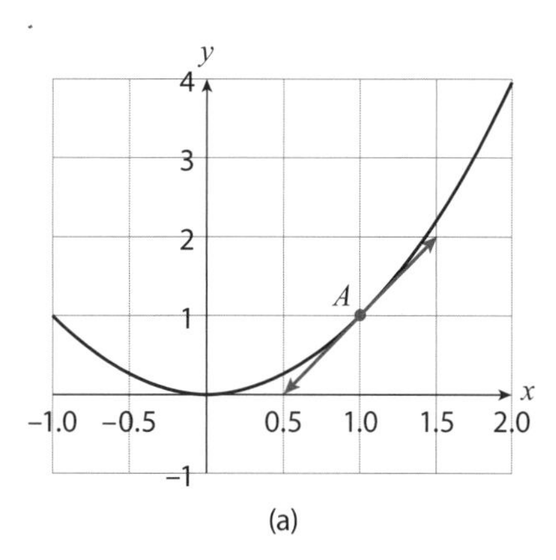

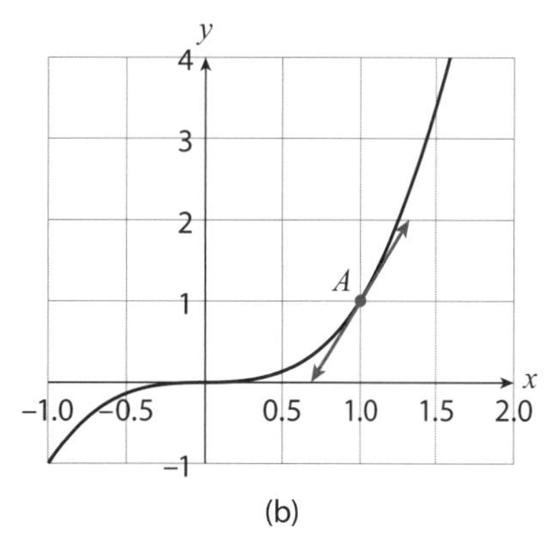

Figure 3.4: Portions of the graphs of (a) $f(x)=x^2$ (a) and (b) $f(x)=x^3$ and their tangent lines at the point $(1,1)$.

EXAMPLE 3.5 Find the equation of the tangent line at point $(1,1)$ on the graph of $f(x)=x^2$.

Solution We just calculated the slope of that tangent line: $f'(1)=2$. It follows from the point–slope equation (reviewed in (B.5) of Appendix B) that the equation of the tangent line is:

$$y-1=2(x-1).$$

Simplifying yields $y=2x-1$. I've plotted that line and the graph of $f(x)$ in Figure 3.4(a). ■

EXAMPLE 3.6 Calculate $f'(1)$ for $f(x)=x^3$.

Solution

$$\begin{aligned}
f'(1) &= \lim_{\Delta x\to 0}\frac{f(1+\Delta x)-f(1)}{\Delta x} && \text{Equation (3.11) with } a=1\\
&= \lim_{\Delta x\to 0}\frac{(1+\Delta x)^3-1^3}{\Delta x} && \text{Using } f(1+\Delta x)=(1+\Delta x)^3\\
&= \lim_{\Delta x\to 0}\frac{1+3(\Delta x)+3(\Delta x)^2+(\Delta x)^3-1}{\Delta x} && \text{Expanding } (1+\Delta x)^3 \text{ via formula (A.10) (from Appendix A)}\\
&= \lim_{\Delta x\to 0}[3+3(\Delta x)+(\Delta x)^2]=3. && \text{Simplifying, canceling } \Delta x, \text{ and evaluating the limit}
\end{aligned}$$

■

EXAMPLE 3.7 Find the equation of the tangent line at point $(1,1)$ on the graph of $f(x)=x^3$.

Solution We just calculated the slope of that tangent line: $f'(1) = 3$. It follows from the point–slope equation that the equation of the tangent line is:

$$y - 1 = 3(x - 1).$$

Simplifying yields $y = 3x - 2$, plotted in Figure 3.4(b). ■

Related Exercises 1–9.

Tips, Tricks, and Takeaways

The definition of $f'(a)$ (Definition 3.2) looks like the definition of $\mathscr{s}(a)$ (Definition 3.1) but with different notation and terminology. Indeed, in the next section we'll exploit these parallels to develop new, useful interpretations of $f'(a)$.

3.3 The Instantaneous Rate of Change Interpretation of the Derivative

The workflow in Section 3.2 is very similar to that of Section 3.1. Figure 3.5 provides a visual comparison. In both cases, we first calculate the **average rate of change** of the function we're interested in—distance in Section 3.1, a general function f in Section 3.2—over a certain interval Δx (or Δt). Then, we find the limit of that *average* rate of change as $\Delta x \to 0$ and get an *instantaneous* rate of change. This interplay between instantaneous speed $\mathscr{s}(a)$ and $f'(a)$ yields the following insights:

	Speed	Rate of change
Average	$\frac{\Delta d}{\Delta t}$	$\frac{\Delta y}{\Delta x}$
Instantaneous	$\mathscr{s}(a) = \lim_{\Delta t \to 0} \frac{\Delta d}{\Delta t}$	$f'(a) = \lim_{\Delta x \to 0} \frac{\Delta y}{\Delta x}$

Figure 3.5: The definition of the derivative is a generalization of the definition of instantaneous speed.

1. *The derivative at $x = a$, $f'(a)$, measures the instantaneous rate of change of f at $x = a$.*

2. *The instantaneous speed of an object at time $t = a$, $\mathscr{s}(a)$, is the derivative of the object's distance function d at $t = a$: $\mathscr{s}(a) = d'(a)$.*

3. *The units of the derivative at $x = a$, $f'(a)$, are the ratio of the units of $f(x)$ to the units of x.*

Insight 3 follows from the right-hand side of (3.8) being a ratio of changes in $f(x)$ with changes in x. I will call Insight 1 the **rate of change interpretation of the derivative** and its "slope of the tangent line" interpretation the **geometric interpretation of the derivative**.

EXAMPLE 3.8 Use the results of Examples 3.4 and 3.6 to answer the following question: Which function's y-values are increasing faster at the point $(1, 1)$, $f(x) = x^2$ or $f(x) = x^3$?

Solution The answer is $f(x) = x^3$, since we calculated in Example 3.6 that $f'(1) = 3$, whereas we got $f'(1) = 2$ for $f(x) = x^2$ in Example 3.4. ■

EXAMPLE 3.9 Returning to Figure 3.4(a), at what x-value(s) in the graph shown is the instantaneous rate of change of the function zero?

Solution Only at $x = 0$, because $f'(0) = 0$ (the tangent line is horizontal there, and so has zero slope) and $f'(a) \neq 0$ for every other a-value (all such tangent lines have nonzero slope). ■

Note how these examples had us switching between the different interpretations of $f'(a)$. The suggested exercises at the end of the chapter help you further practice that skill.

Related Exercises 12 and 44.

We've profited a lot from comparing Definitions 3.1 and 3.2. Yet there is one important *difference* between the two—the qualifier "provided the limit exists" in the latter definition. That suggests that $f'(a)$ doesn't always exist. The next section explores this additional piece of the puzzle that is $f'(a)$.

3.4 Differentiability: When Derivatives Do (and Don't) Exist

The geometric interpretation of $f'(a)$ is that it's the slope of the tangent line to the graph of f at $x = a$. But this presumes such a tangent line exists, which is not always the case. The three graphs in Figure 3.6 illustrate what may happen. The graph in Figure 3.6(a) has a "corner" or "kink" at point A. The tangent line is supposed to

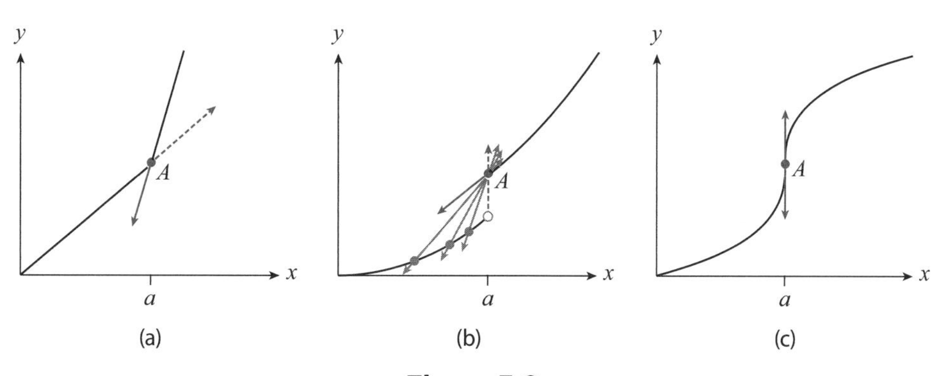

Figure 3.6

share the "inclination" of the graph at the point of tangency. But the inclination of the graph in Figure 3.6(a) just before point A, represented by the dotted blue line, is not the same as just after point A (represented by the solid blue line). Because the tangent line at point A does not exist, the slope of that line, $f'(a)$, also does not exist.

The graph in Figure 3.6(b) illustrates another problematic situation. There, if we draw secant lines through point A and another point on the graph and *left* of point A, as $x \to a^-$ those secant lines become more and more vertical (illustrated by the gray lines in the figure) and approach the dashed blue line, the "inclination" of the graph as we approach point A from the left. That "inclination" is not the same as that obtained by considering only the portion of the graph *right* of A, as illustrated by the solid blue line. Because the tangent line at point A does not exist, $f'(a)$, the supposed slope of that line, also does not exist.

Figure 3.6(c) illustrates the last possibility. That graph *does* have a tangent line at point A, but it's a vertical line and so has infinite slope. We again conclude that $f'(a)$ does not exist (infinity is not a number). The following definition introduces the terminology we use when $f'(a)$ does (or does not) exist.

Definition 3.3 Let f be a function defined on an open interval I including a. We call f **differentiable at** $x = a$ if $f'(a)$ exists. If f is differentiable for all x inside I, we say that f is **differentiable on** I. When f is differentiable on $(-\infty, \infty)$ we say that f is **differentiable everywhere**, or simply, **differentiable**.

EXAMPLE 3.10 Consider the function graphed in Figure 3.7.

(a) At what points in the interval $(-2, 4)$ is f not differentiable?

(b) At what points in the interval $(-2, 4)$ is f discontinuous?

Solution

(a) At $x = -1$ (same problem as in Figure 3.6(b)), $x = 1$ (kink, as in Figure 3.6(a)), $x = 2$ (function isn't even defined there), and $x = 3$ (same problem as in Figure 3.6(b)).

(b) From Example 2.8: The function is discontinuous at $x = -1$, $x = 2$, and $x = 3$. ■

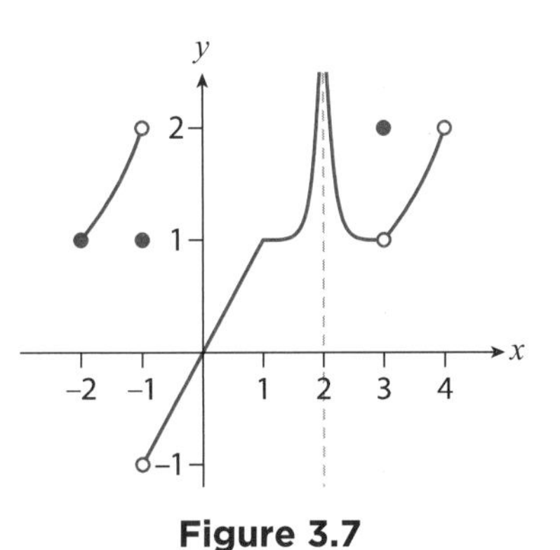

Figure 3.7

Related Exercises 14–15 and 48–49.

Tips, Tricks, and Takeaways

The 5-second summary of differentiability is this: $f'(a)$ exists only when the tangent line to f at $x=a$ exists and has finite slope. As we saw, this means in particular that the graph cannot have kinks (i.e., is "smooth") or gaps. If that reminds you of continuity, it should, because the two notions are related.

Theorem 3.1 If f is differentiable at a, then f is continuous at a.

- The "contrapositive" of the theorem—the statement that if f is not continuous at a then it's not differentiable at a—*is true*. Figure 3.6(b) illustrates this; that graph has a jump discontinuity at $x=a$ and $f'(a)$ does not exist.
- The "converse" of the theorem—the statement that if f is continuous at a then it's differentiable at a—*is not true*. Figure 3.6(a) illustrates this; that graph is continuous at $x=a$ yet $f'(a)$ does not exist.

The takeaway: *Continuity is a* necessary *condition for differentiability, but not a* sufficient *one.* The results also help quickly rule out differentiability for a function—if it's not continuous at $x=a$, it isn't differentiable at $x=a$.

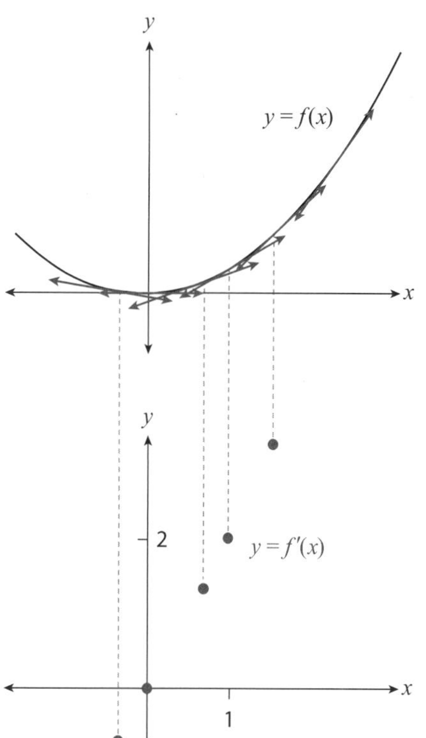

Figure 3.8: Plotting the slopes of the tangent lines to $f(x)$ as the points $(x, f'(x))$.

3.5 The Derivative: A Graphical Approach

We now have a solid understanding of what $f'(a)$ measures, when it does and does not exist, and how to visualize it—as a tangent line superimposed on a graph (as in Figure 3.4). But that visualization method gets clunky if we want to visualize $f'(a)$ for various a-values. Figure 3.8 (top) illustrates this for $f(x)=x^2$. The tangent line soup in that figure obscures the information those lines are providing us with. Part of the problem is that we're plotting everything on the same graph. Let's remedy this.

Back in Example 3.4 we calculated that $f'(1)=2$ for $f(x)=x^2$. Let's plot that on a new graph as the point $(1,2)$; I've put that point in Figure 3.8 (bottom plot). I've

Interactive Figure

Interactive Figure

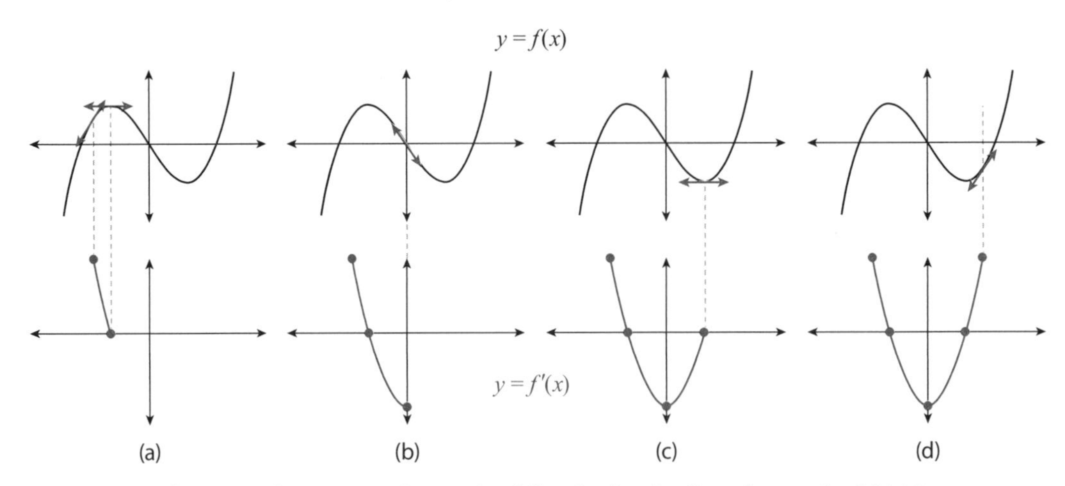

Figure 3.9: Drawing the graph of $f'(x)$ by "surfing" on the graph of $f(x)$."

also added a few other points. Like the point $(1, 2)$, each of those points' y-value is $f'(a)$ for the point's x-value a. Were I to have added even more points, you could likely predict the type of function that would emerge: a *linear* function. (We'll verify this in the next section.) We've discovered a new way to visualize the derivative: *as a function $y = f'(x)$ whose y-values are the slopes of the tangent lines to f at each x-value in the domain of f.*

EXAMPLE 3.11 Sketch the derivative of $f(x) = x^3 - 3x$.

Solution Figure 3.9 shows a play-by-play of the sketching process. For each x-value a we draw a tangent line to the graph of f (top) and calculate its slope $f'(a)$. Then on a new graph (bottom) we plot the point $(a, f'(a))$. When we do this from left to right—Figure 3.9(a) to (d)—the graph of $y = f'(x)$ emerges. ■

Tips, Tricks, and Takeaways

A technical note before giving you tips on sketching f' given f: since $f'(a)$ is defined as a limit and limits are unique (if they exist; this can be proven using the Limit Laws), it follows that $y = f'(x)$ is indeed a function (in the sense of the definition of a function; see Definition B.1). Okay, on to the tips for sketching f':

- Look for horizontal tangent lines on the graph of f. These plot as points on the x-axes on the graph of f'.
- If the slopes of the tangent lines to f are positive then the y-values of the graph of f' must be positive. Conclusion: if the graph of f is "sloping up" then the graph of f' is above the x-axis. Replacing "positive" with "negative" in the first sentence replaces "up" and "above" with "down" and "below" in the second sentence.

You may have noticed that in Figure 3.8 f is a degree 2 polynomial and f' seems to be a degree 1 polynomial (a linear function). Similarly, in Figure 3.9 f is a degree 3 polynomial and f' seems to be a degree 2 polynomial. These are manifestations of a general rule we'll discuss in Section 3.7. In the next section we will confirm our observation by calculating $f'(x)$.

Related Exercises 16–18.

3.6 The Derivative: An Algebraic Approach

Algebraically, the definition of the derivative function $f'(x)$ is just (3.11) with a replaced by x. I'll also replace Δx with h, yielding the following more common formula for $f'(x)$ you will see in calculus textbooks.

Definition 3.4 The Derivative Function. Let f be a function. The **derivative function $f'(x)$ of f**, denoted by $f'(x)$, is defined by

$$f'(x) = \lim_{h\to 0} \frac{f(x+h) - f(x)}{h}, \tag{3.12}$$

for all x-values for which the limit exists.

EXAMPLE 3.12 Calculate $f'(x)$ for $f(x) = x^2$.

Solution

$$\begin{aligned}
f'(x) &= \lim_{h\to 0} \frac{f(x+h) - f(x)}{h} && \text{Equation (3.12)} \\
&= \lim_{h\to 0} \frac{(x+h)^2 - x^2}{h} && \text{Using } f(x+h) = (x+h)^2 \\
&= \lim_{h\to 0} \frac{2xh + h^2}{h} && \text{Squaring out } (x+h)^2 \text{ and simplifying} \\
&= \lim_{h\to 0} [2x + h] = 2x. && \text{Simplifying, canceling } h\text{, and evaluating the limit}
\end{aligned}$$

■

EXAMPLE 3.13 Calculate $f'(x)$ for $f(x) = x^3 - 3x$.

Solution

$$f'(x) = \lim_{h\to 0} \frac{f(x+h) - f(x)}{h} \qquad \text{Equation (3.12)}$$

$$= \lim_{h\to 0} \frac{[(x+h)^3 - 3(x+h)] - [x^3 - 3x]}{h} \quad \text{Using } f(x+h) = (x+h)^3 - 3(x+h)$$

$$= \lim_{h\to 0} \frac{3x^2h + 3xh^2 + h^3 - 3h}{h} \quad \text{Using (A.10) to expand } (x+h)^3$$

$$= \lim_{h\to 0} [3x^2 - 3 + 3xh + h^2] = 3x^2 - 3. \quad \text{Simplifying, canceling } h, \text{ and evaluating the limit} \quad \blacksquare$$

These calculations confirm our earlier observations: the derivative of the quadratic function $f(x) = x^2$ graphed in Figure 3.8 is the linear function $f'(x) = 2x$, and the derivative of the cubic function $f(x) = x^3 - 3x$ graphed in Figure 3.9 is the quadratic function $f'(x) = 3x^2 - 3$. Two (simpler) calculations you can work out on your own are

$$f(x) = x \implies f'(x) = 1, \qquad f(x) = b \implies f'(x) = 0, \tag{3.13}$$

where b is any real number.

Related Exercises 19, 21, and 25.

APPLIED EXAMPLE 3.14 Loosely speaking, an individual's **maximum heart rate** (MHR) is the highest heart rate that can be sustained during prolonged exercise. An accurate formula for MHR was developed in [2]:

$$M(t) = 192 - 0.007t^2.$$

(a) Calculate $M'(t)$.

(b) Calculate $M'(20)$ and interpret your results using the derivative's rate of change interpretation.

Solution

(a) Employing (3.12):

$$M'(t) = \lim_{h\to 0} \frac{M(t+h) - M(t)}{h} \quad \text{Equation (3.12)}$$

$$= \lim_{h\to 0} \frac{[192 - 0.007(t+h)^2] - [192 - 0.007t^2]}{h} \quad \text{Using } M(t+h) = 192 - 0.007(t+h)^2$$

$$= \lim_{h\to 0} \frac{0.007t^2 - 0.007(t+h)^2}{h}$$

$$= \lim_{h\to 0} \frac{0.007t^2 - 0.007t^2 - 0.014th - 0.007h^2}{h}$$ Expanding $(t+h)^2$ and simplifying

$$= \lim_{h\to 0}[-0.014t - 0.007h] = -0.014t.$$ Simplifying, canceling h, and evaluating the limit

(Recall from Section 3.3 that the units of the derivative are the units of the output, in this case bpm, divided by the units of the input, in this case years.)

(b) $M'(20) = -0.014(20) = -0.28$ bpm/year. The rate of change interpretation is: at the instant an individual is 20 years old, his or her MHR is decreasing by 0.28 bpm per year. ("Decreasing" because the rate is negative.) ■

Related Exercises 13.

Transcendental Tales

Let's begin by discussing how to differentiate $f(x) = e^x$ using (3.12):

$$f'(x) = \lim_{h\to 0} \frac{e^{x+h} - e^x}{h} = \lim_{h\to 0} \frac{e^x(e^h - 1)}{h}$$
$$= (e^x)\left(\lim_{h\to 0} \frac{e^h - 1}{h}\right), \quad (3.14)$$

where the last equality follows from the fact that e^x remains e^x as $h \to 0$ (because e^x doesn't depend on h). Table 3.1 suggests the value of the limit in the parentheses in (3.14) is 1. (See Section A2.1 in the online appendix to this chapter for an alternative derivation.) Using this in (3.14) yields the following.

Table 3.1: The values of $\frac{e^h - 1}{h}$ approach one as h approaches zero from either side.

h	$\frac{e^h - 1}{h}$
−0.01	0.99502
−0.001	0.99950
−0.0001	0.99995
⋮	⋮
0.0001	1.00005
0.001	1.00050
0.01	1.00502

Theorem 3.2 $(e^x)' = e^x$.

In other words: *e^x is its own derivative*; this reinforces how special the base e is for an exponential function. I will defer discussion of the derivative of other exponential functions, as well as logarithmic functions, to the next section; those calculations are best done with the derivative shortcuts we'll develop in that section.

Turning now to trigonometric functions, let's begin by returning to the graphs of $\sin x$ and $\cos x$ (Figure B.20). The smoothness of those graphs suggests that they're

Interactive Figure

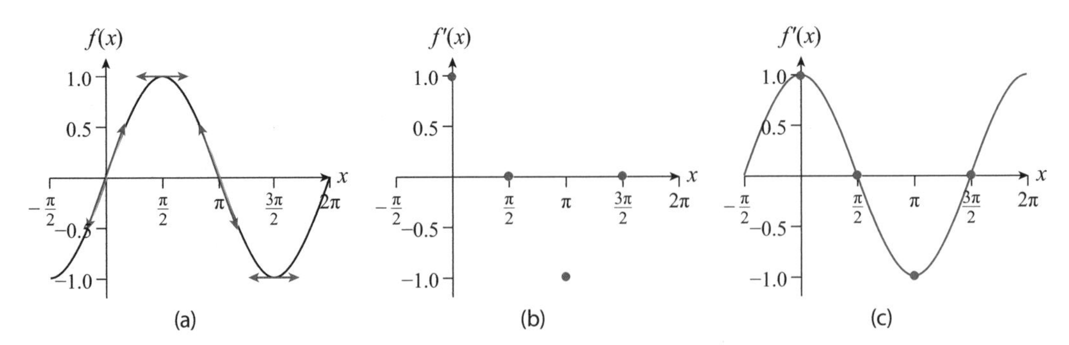

Figure 3.10: (a) $f(x) = \sin x$ and four of its tangent lines, (b)–(c): plotting the slopes of all tangent lines to f.

differentiable everywhere. In Figure 3.10 I've applied our graphical approach to constructing $f'(x)$ for $f(x) = \sin x$ (as we did with Figure 3.9). The $f'(x)$ graph looks like another trigonometric function you might already recognize. A similar graph would have emerged had we started with $f(x) = \cos x$ instead—another familiar trigonometric function. Let's verify these hunches and calculate the actual derivatives.

EXAMPLE 3.15 Show that

$$(\sin x)' = \cos x, \qquad (\cos x)' = -\sin x. \tag{3.15}$$

Solution We have

$$(\sin x)' = \lim_{h\to 0} \frac{\sin(x+h) - \sin(x)}{h}$$

Equation (3.12) for $f(x) = \sin x$

$$= \lim_{h\to 0} \frac{\sin(x)\cos(h) + \sin(h)\cos(x) - \sin(x)}{h}$$

Using (B.23)

$$= \lim_{h\to 0} \frac{\sin(x)(\cos(h) - 1)}{h} + \lim_{h\to 0} \frac{\sin(h)\cos(x)}{h}$$

Rearranging and using Limit Law 1

$$= \sin(x)\left[\lim_{h\to 0} \frac{\cos(h) - 1}{h}\right] + \cos(x)\left[\lim_{h\to 0} \frac{\sin(h)}{h}\right]$$

$\sin x$ and $\cos x$ don't depend on h

$$= \sin(x)\cdot 0 + \cos(x)\cdot 1 = \cos x$$

Using (2.3) and (2.4)

The similar calculation of $(\cos x)' = -\sin x$ is left as an exercise (see Exercise 75).

■

I will defer the calculation of the derivative of $\tan x$ to the next section; it is easiest to carry out that calculation via one of the derivative shortcuts we'll discuss therein.

Let me finish this tour of the derivative function by briefly discussing an alternative notation for it: Leibniz's notation. (Gottfried Leibniz is the co-inventor of calculus.)

Leibniz Notation

Leibniz's notation arises from expressing $f'(x)$ as

$$f'(x) = \lim_{\Delta x \to 0} \frac{\Delta y}{\Delta x}, \tag{3.16}$$

where $\Delta y = f(x+h) - f(x)$. Recall our interpretation of $\Delta x \to 0$ from Chapter 1—an infinitesimal change in x. Leibniz introduced the symbol dx to represent that notion.[2] Accordingly, dy represented the resulting infinitesimal change in the y-values $f(x)$. Leibniz then wrote (3.16) as

$$\frac{dy}{dx} = \lim_{\Delta x \to 0} \frac{\Delta y}{\Delta x}, \qquad \text{so that} \qquad f'(x) = \frac{dy}{dx}. \tag{3.17}$$

The new dy/dx notation is meant to remind you of the derivative's slope-of-the-tangent-line origin.

Today we also use just the "d/dx" portion of Leibniz notation. We understand that to be a stand-in for "the derivative with respect to x of . . ." For example, we've shown in this chapter that

$$\frac{d}{dx}\left(x^2\right) = 2x, \qquad \frac{d}{dx}\left(e^x\right) = e^x, \qquad \frac{d}{dt}\left(192 - 0.007t^2\right) = -0.014t.$$

One drawback to Leibniz's notation is the fraction bar. Leibniz indeed thought of the derivative as the ratio of two infinitesimally small quantities (the differentials dy and dx), and dy/dx certainly *encapsulates* that. But to the untrained eye, dy/dx is literally "dy divided by dx." The problem with this is that dx and dy are not themselves numbers—they are stand-ins for the idea of an "infinitesimally small" quantity—so we can't calculate dy and dx, and therefore the ratio dy/dx should not be thought of as the quotient of two numbers.[3] The takeaway: View dy/dx as merely notation for the derivative that better embodies its slope-of-the-tangent-line origin.

Next up on our agenda is the development of shortcuts for calculating $f'(x)$ that avoid the lengthy limit calculations we've executed thus far in this chapter. All of the differentiation shortcuts I'm about to teach you can be proven using the Limit Laws

[2]See pages 134–144 of [13] to read Leibniz in his own words writing about dx.

[3]We will make some progress on this notation issue in Chapter 5.

(Theorem 2.6) and the formula defining $f'(x)$ (equation (3.12)). First up: the basic rules of derivatives.

3.7 Differentiation Shortcuts: The Basic Rules

Theorem 3.3 The Sum, Difference, and Constant Multiple Rules. Suppose f and g are differentiable functions and c is a real number. Then,

1. **The Sum Rule:** $(f+g)' = f' + g'$
2. **The Difference Rule:** $(f-g)' = f' - g'$
3. **The Constant Multiple Rule:** $(cf)' = cf'$

The first two rules say that the derivative of the sum (or difference) of two functions is the sum (or difference) of their derivatives; the third says that the derivative of a number multiplied by a function is the product of that number with the derivative of the function. Let's now use Theorem 3.3 to help us differentiate the simplest family of functions: linear functions.

EXAMPLE 3.16 Differentiate $f(x) = 3x + 5$.

Solution

$$\begin{aligned} f'(x) &= \frac{d}{dx}(3x+5) \\ &= 3\frac{d}{dx}(x) + 5\frac{d}{dx}(1) && \text{Sum and Constant Multiple Rules, Theorem 3.3} \\ &= 3(1) + 5(0) = 3. && \text{Using (3.13)} \end{aligned}$$

■

EXAMPLE 3.17 Differentiate $g(x) = mx + b$.

Solution

$$\begin{aligned} g'(x) &= \frac{d}{dx}(mx+b) \\ &= m\frac{d}{dx}(x) + b\frac{d}{dx}(1) && \text{Sum and Constant Multiple Rules, Theorem 3.3} \\ &= m(1) + b(0) = m. && \text{Using (3.13)} \end{aligned}$$

■

These results make perfect sense in the context of the slope-of-the-tangent-line interpretation of the derivative: the graph of $g(x) = mx + b$ is a line with slope m, so every line tangent to that graph has slope m, implying that $g'(x) = m$.

APPLIED EXAMPLE 3.18 A person's Resting Metabolic Rate (RMR) is defined as the amount of calories his or her body burns while awake but at rest. RMR is usually calculated for 24-hour periods, thereby providing a daily minimum energy expenditure estimate.[4] Mathematical models of RMR often estimate the quantity using the person's weight, height, and age. The Mifflin–St. Jeor equations, the most accurate RMR formulas currently available (see [9] for a study comparing different RMR equations), are an example. The Mifflin–St. Jeor equation for women is

$$\text{RMR}_{\text{women}} = 4.5x + 15.9h - 5t - 161, \tag{3.18}$$

Here x is weight in pounds, h height in inches, and t age in years.[5]

(a) Suppose $h = 66$ and $t = 20$. Write down the function of x that results.

(b) Find the derivative of that function at $x = 150$ and include the units.

(c) Use the rate of change interpretation of the derivative to interpret the derivative value you found.

Solution

(a) Substituting the values into (3.18) yields $W(x) = 4.5x + 788.4$. (I replaced $\text{RMR}_{\text{women}}$ with W.)

(b) Since $W(x)$ is a linear function, it follows from Example 3.17(b) that $W'(x) = 4.5$, so that $W'(150) = 4.5$ calories/pound ($W(x)$ has units of calories and x units of pounds).

(c) $W'(150) = 4.5$ cal/lb tells us that the RMR of a 20-year-old 5'6" tall woman who weighs 150 pounds is increasing at the instantaneous rate of 4.5 cal/lb. ■

Related Exercises 19, 35.

3.8 Differentiation Shortcuts: The Power Rule

Recall from Definition B.4 (Appendix B) that power functions have the form ax^b. Let's set $a = 1$ and consider b to be a positive integer so that we're studying the power functions x, x^2, x^3, etc. Our prior work in this chapter has established the derivatives of these power functions; I have put those results in the second column of Table 3.2. The third column reexpresses those derivatives

Table 3.2

$f(x)$	$f'(x)$	$f'(x)$
x^1	1	$1x^{1-1}$
x^2	$2x$	$2x^{2-1}$
x^3	$3x^2$	$3x^{3-1}$

[4]"Minimum" because any activity not included in RMR (e.g., walking) adds to the day's total energy expenditure.

[5]These equations are examples of **multilinear functions**. See [7] for more information, as well as a discussion of the limitations of the Mifflin–St. Jeor model.

in a manner that suggests a general pattern. See the pattern? (I'll bet you can now predict what the derivative of x^4 will be.)

The insights from Table 3.2 suggest that the derivative of x^n, where n is a positive integer, follows the rule: "bring down the power and subtract one from the exponent." Mathematically, this leads us to conjecture that if $f(x) = x^n$ then $f'(x) = nx^{n-1}$. This turns out to be true. And that's not all. As you can verify using (3.12):

$$f(x) = x^{-2} \implies f'(x) = -2x^{-3}, \qquad g(x) = x^{1/2} \implies g'(x) = \frac{1}{2}x^{-1/2}.$$

This suggests that the "bring down the power and subtract one from the exponent" rule may also hold for fractional and negative powers of x. This also turns out to be true. In fact, this rule for differentiating x^n is true when n is any real number; it's called the **Power Rule**. (We will later prove the Power Rule with the help of another derivative rule.)

Theorem 3.4 The Power Rule. Let n be a real number. Then,

$$\frac{d}{dx}\left(x^n\right) = nx^{n-1}.$$

EXAMPLE 3.19 Differentiate $f(x) = x^3 - 3x$.

Solution

$$\begin{aligned} f'(x) &= \frac{d}{dx}\left(x^3 - 3x\right) \\ &= \frac{d}{dx}\left(x^3\right) - 3\frac{d}{dx}(x) && \text{Difference and Constant Multiple Rules, Theorem 3.3} \\ &= 3x^2 - 3. && \text{Power Rule and (3.13)} \end{aligned}$$

■

EXAMPLE 3.20 Differentiate $g(x) = 10x^9 - 3\sqrt{x}$.

Solution

$$\begin{aligned} g'(x) &= \frac{d}{dx}\left(10x^9 - 3x^{1/2}\right) && \text{Rewriting } \sqrt{x} = x^{1/2} \\ &= 10\frac{d}{dx}\left(x^9\right) - 3\frac{d}{dx}\left(x^{1/2}\right) && \text{Difference and Constant Multiple Rules, Theorem 3.3} \\ &= 10\left(9x^8\right) - 3\left(\frac{1}{2}x^{-1/2}\right) = 90x^8 - \frac{3}{2\sqrt{x}}. && \text{Power Rule; simplifying} \end{aligned}$$

■

EXAMPLE 3.21 Differentiate $h(x) = \dfrac{2}{x^3} + 5x^{1.2}$.

Solution

$$h'(x) = \frac{d}{dx}\left(2x^{-3} + 5x^{1.2}\right) \qquad \text{Rewriting } 2/x^3 = 2x^{-3}$$

$$= 2\frac{d}{dx}\left(x^{-3}\right) + 5\frac{d}{dx}\left(x^{1.2}\right) \qquad \text{Sum and Constant Multiple Rules, Theorem 3.3}$$

$$= 2\left(-3x^{-4}\right) + 5\left(1.2x^{0.2}\right) = 6\left(x^{0.2} - \frac{1}{x^4}\right). \qquad \text{Power Rule; simplifying} \quad \blacksquare$$

EXAMPLE 3.22 Verify the result of $M'(t)$ for Applied Example 3.14.

Solution Using (3.13) and the Difference, Constant Multiple, and Power Rules:

$$M'(t) = 192\frac{d}{dt}(1) - 0.007\frac{d}{dt}\left(t^2\right) = -0.007(2t) = -0.014t. \quad \blacksquare$$

APPLIED EXAMPLE 3.23 Weather stations often report outdoor temperature together with the "wind chill temperature," which takes into account how much colder it feels due to the wind. The National Weather Service (NWS) has come up with the following model (see [10]) for wind chill temperature:

$$C = 35.74 + 0.6215T + (0.42475T - 35.75)v^{0.16},$$

where C is the wind chill temperature and T the air temperature (both measured in Fahrenheit), v the wind velocity (in miles per hour), and $T \leq 50°$ F and $v \geq 3$ mph.

(a) Calculate the function $C(v)$ that results when $T = 30°$ F.

(b) Using the function from (a), calculate $C(10)$ and interpret your answer.

(c) Using the function from (a), calculate $C'(10)$ and interpret your answer.

Solution

(a) $C(v) = 54.385 - 23.0075v^{0.16}$.

(b) $C(10) \approx 21.13°$ F. This means that, according to the NWS model, when it's 30° F outside a 10 mph wind gust will make it feel about 21° F.

(c) Using the Difference, Constant Multiple, and Power Rules:

$$C'(v) = -23.0075\left(0.16v^{-0.84}\right) = -3.6812v^{-0.84} = -\frac{3.6812}{v^{0.84}} \quad \frac{°\text{F}}{\text{mph}}.$$

Therefore, $C'(10) = -3.6812(10)^{-0.84} \approx -0.53$ °F/mph. Using the rate of change interpretation of the derivative, we can say that, according to the NWS model, at the instant the wind gusts to 10 mph on a 30° F day the wind chill temperature is decreasing at the rate of about half a degree Fahrenheit per mph. ■

Related Exercises 19–24, 30, 34, and 51.

As promised, Theorem 3.3 and the Power Rule make it easier to differentiate polynomials (as well as other functions consisting of power functions). But these results don't address differentiating products, quotients, or compositions of functions. Let's talk next about those differentiation shortcuts.

3.9 Differentiation Shortcuts: The Product Rule

As an illustration of the calculus we've learned thus far, let's derive the rule for differentiating a product of two functions using a geometric argument. Here's the question: What's the instantaneous rate of change of the area of the blue rectangle in Figure 3.11 if the side lengths l and w are changing with time?

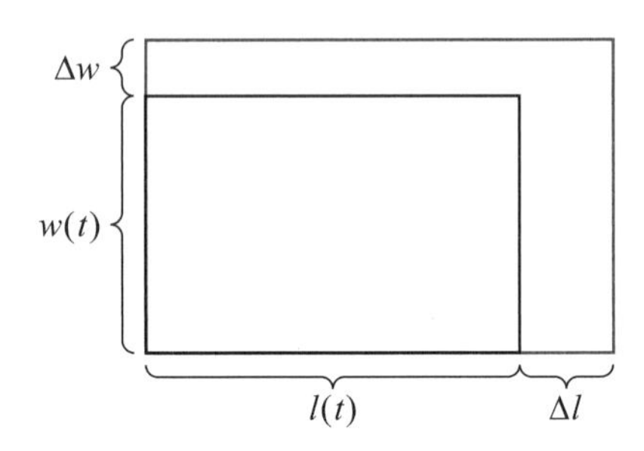

Figure 3.11

We start with the fact that $A(t) = l(t)w(t)$ is the area of the black rectangle. We're looking to calculate $A'(t)$. Employing a dynamics mindset, we envision the rectangle's sides increasing a bit, leading to the larger rectangle in Figure 3.11. The change in area ΔA (the area between the two rectangles) is

$$\begin{aligned}\Delta A &= [l(t) + \Delta l][w(t) + \Delta w] - l(t)w(t)\\ &= l(t)\Delta w + w(t)\Delta l + \Delta l \Delta w.\end{aligned}$$

Recalling now (3.16):

$$\begin{aligned}A'(t) = \lim_{\Delta t \to 0} \frac{\Delta A}{\Delta t} &= \lim_{\Delta t \to 0} \left[\frac{l(t)\Delta w + w(t)\Delta l + \Delta l \Delta w}{\Delta t}\right]\\ &= \lim_{\Delta t \to 0} \left[l(t)\frac{\Delta w}{\Delta t} + w(t)\frac{\Delta l}{\Delta t} + \Delta l \frac{\Delta w}{\Delta t}\right]\\ &= l(t)\left[\lim_{\Delta t \to 0} \frac{\Delta w}{\Delta t}\right] + w(t)\left[\lim_{\Delta t \to 0} \frac{\Delta l}{\Delta t}\right] + \lim_{\Delta t \to 0}\left[\Delta l \frac{\Delta w}{\Delta t}\right]\\ &= l(t)w'(t) + w(t)l'(t),\end{aligned}$$

since $\Delta l \to 0$ as $\Delta t \to 0$. Conclusion: $A'(t) = l'(t)w(t) + l(t)w'(t)$. This is known as the **Product Rule**.

Theorem 3.5 The Product Rule. Suppose f and g are both differentiable. Then,

$$[f(x)g(x)]' = f'(x)g(x) + f(x)g'(x).$$

EXAMPLE 3.24 Differentiate $h(x) = (2x - 3)(4x^3 - 1)$.

Solution

$$\begin{aligned} h'(x) &= (2x-3)'(4x^3-1) + (2x-3)(4x^3-1)' && \text{Product Rule} \\ &= (2)(4x^3-1) + (2x-3)(12x^2) && \text{Difference, Constant Multiple, and Power Rules} \\ &= 32x^3 - 36x^2 - 2. && \text{Simplifying} \end{aligned}$$

■

EXAMPLE 3.25 Differentiate $h(x) = (3x - 1)^2$.

Solution First we rewrite $h(x)$ as a product of two functions: $h(x) = (3x - 1)(3x - 1)$. Then,

$$\begin{aligned} h'(x) &= (3x-1)'(3x-1) + (3x-1)(3x-1)' && \text{Product Rule} \\ &= (3)(3x-1) + (3x-1)(3) && \text{Difference and Constant Multiple Rules, and (3.13)} \\ &= 6(3x-1). && \text{Simplifying} \end{aligned}$$

■

Related Exercises 24 and 27.

3.10 Differentiation Shortcuts: The Chain Rule

Let me follow the previous section's approach and give a geometric illustration of the rule for differentiating a composition of two functions. The question is the same as before, except that now we'll consider a square: What's the instantaneous rate of change of the area of the blue square in Figure 3.12 if the side length x is changing with time?

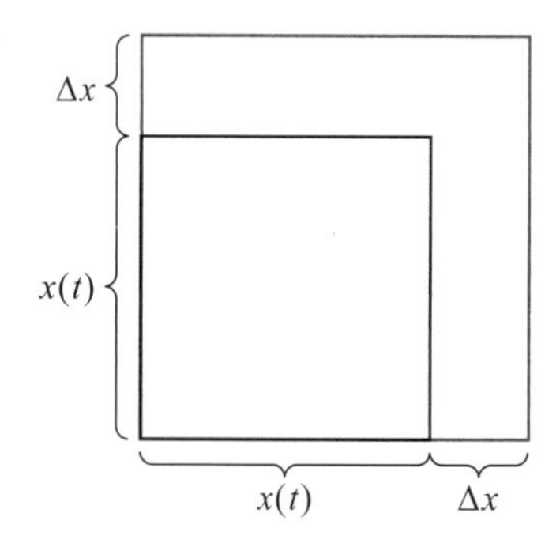

Figure 3.12

We start with the fact that $A(x) = x^2$ is the area of the black rectangle. But since x depends on time, so does A, so we're interested in differentiating $A(x(t)) = [x(t)]^2$ with respect to time t. Employing a dynamics mindset again, we envision the square's sides increasing a bit, leading to the larger square in Figure 3.12. The change in area ΔA (the area between the two squares) is:

$$\begin{aligned} \Delta A &= [x(t) + \Delta x]^2 - [x(t)]^2 \\ &= 2x(t)\Delta x + (\Delta x)^2. \end{aligned}$$

Recalling again (3.16):

$$\frac{d}{dt}[A(x(t))] = \lim_{\Delta t\to 0}\frac{\Delta A}{\Delta t} = \lim_{\Delta t\to 0}\left[\frac{2x(t)\Delta x + (\Delta x)^2}{\Delta t}\right]$$

$$= \lim_{\Delta t\to 0}\left[2x(t)\frac{\Delta x}{\Delta t} + \Delta x\frac{\Delta x}{\Delta t}\right] = 2x(t)\left[\lim_{\Delta t\to 0}\frac{\Delta x}{\Delta t}\right] + \lim_{\Delta t\to 0}\left[\Delta x\frac{\Delta x}{\Delta t}\right]$$

$$= 2x(t)x'(t), \tag{3.19}$$

since $\Delta x \to 0$ as $\Delta t \to 0$. Now, the Power Rule applied to $A(x) = x^2$ gives $A'(x) = 2x$. So, $A'(x(t)) = 2x(t)$. This allows us to rewrite (3.19) as

$$\frac{d}{dt}[A(x(t))] = A'(x(t))x'(t).$$

This is a particular example of the more general rule known as the **Chain Rule**.

Theorem 3.6 The Chain Rule. Suppose f and g are both differentiable. Then,

$$\frac{d}{dx}[f(g(x))] = f'(g(x))g'(x).$$

In the terminology of compositions of functions, $f(x)$ is the "outer function" and $g(x)$ the "inner function." The Chain Rule then says: the derivative of the composite function $f(g(x))$ is the derivative of the outer function evaluated at the inner function (that's $f'(g(x))$) multiplied by the derivative of the inner function (that's $g'(x)$).

EXAMPLE 3.26 Differentiate $h(x) = (3x-1)^2$.

Solution Let's express the function as $h(x) = f(g(x))$, where $f(x) = x^2$ is the outer function and $g(x) = 3x - 1$ the inner function. Then,

$h'(x) = f'(g(x))g'(x)$	Chain Rule
$= f'(3x-1)g'(x)$	Using $g(x) = 3x - 1$
$= f'(3x-1)(3)$	Using $g'(x) = 3$
$= 2(3x-1)(3)$	Since $f'(x) = 2x, f'(3x-1) = 2(3x-1)$
$= 6(3x-1).$	Simplifying

■

EXAMPLE 3.27 Differentiate $h(x) = \sqrt{x^2+1}$.

Solution Let's express the function as $h(x)=f(g(x))$, where $f(x)=\sqrt{x}=x^{1/2}$ is the outer function and $g(x)=x^2+1$ the inner function. Then,

$$\begin{aligned} h'(x) &= f'(g(x))g'(x) && \text{Chain Rule} \\ &= f'(x^2+1)g'(x) && \text{Using } g(x)=x^2+1 \\ &= f'(x^2+1)(2x) && \text{Using } g'(x)=2x \\ &= \left[\frac{1}{2}(x^2+1)^{-1/2}\right](2x) && \text{Using } f'(x)=\frac{1}{2}x^{-1/2} \\ &= \frac{x}{\sqrt{x^2+1}}. && \text{Simplifying} \end{aligned}$$ ■

Related Exercises 25–26, 29, 32–33, 45–47, and 52.

The Chain Rule is easier to remember when expressed in Leibniz's notation. If we let $y=f(g(x))$ and $u=g(x)$ (so that $y=f(u)$), then the Chain Rule is equivalent to

$$\frac{dy}{dx}=\frac{dy}{du}\cdot\frac{du}{dx}, \tag{3.20}$$

where we substitute $u=g(x)$ in as the last step. Let me illustrate this with an example.

EXAMPLE 3.28 Differentiate $h(x)=\sqrt{x^2+1}$ using (3.20).

Solution Using the same inner function, $u=g(x)=x^2+1$:

$$\begin{aligned} \frac{d}{dx}\left(\sqrt{x^2+1}\right) &= \frac{d}{du}\left(\sqrt{u}\right)\cdot\frac{d}{dx}\left(x^2+1\right) && \text{Chain Rule} \\ &= \left[\frac{1}{2}u^{-1/2}\right]\cdot\frac{d}{dx}\left(x^2+1\right) && \text{Power Rule applied to } \sqrt{u}=u^{1/2} \\ &= \left[\frac{1}{2}u^{-1/2}\right](2x) && \text{Power Rule applied to } x^2+1 \\ &= \left[\frac{1}{2}(x^2+1)^{-1/2}\right](2x)=\frac{x}{\sqrt{x^2+1}}. && \text{Substituting in } u=x^2+1 \text{ and simplifying} \end{aligned}$$ ■

Tips, Tricks, and Takeaways

The Chain Rule tends to cause students the most trouble. My recommendation: practice, practice, and more practice.

Let me also add one more remark regarding (3.20). Notice that in Leibniz notation the Chain Rule's proof seems simple: just cancel out the *du*'s. I've cautioned you not to take the fraction bar in Leibniz's dy/dx literally. But recalling (3.17), when the *d*'s in (3.20) are replaced by Δ's the fraction bar *is* a real fraction bar (since then we're talking about *finite*, not infinitesimal, changes) and we would then cancel the Δu's. Takeaway: (3.20) is a useful and intuitive way to remember the Chain Rule but not a proof of the Chain Rule.

Finally, a note about u's. Since function compositions are so ubiquitous, derivative rules are often stated in what many call "u form." For example, you will often find the Power Rule stated as

$$\frac{d}{dx}\left(u^n\right) = nu^{n-1}u' \tag{3.21}$$

in calculus textbooks. This "u form" makes the Power Rule more widely applicable than the $u = x$ version in Theorem 3.4. For instance, it allows us to solve Example 3.26 in one line:

$$\frac{d}{dx}\left[(3x-1)^2\right] = 2(3x-1)^1(3x-1)' = 2(3x-1)(3) = 6(3x-1).$$

3.11 Differentiation Shortcuts: The Quotient Rule

The rule for differentiating a quotient of two functions (i.e., $f(x)/g(x)$) can be derived using the Product and Chain Rules (see Exercise 50). Here is the formula that results.

Theorem 3.7 The Quotient Rule. Suppose f and g are both differentiable, and $g(x) \neq 0$. Then,

$$\frac{d}{dx}\left[\frac{f(x)}{g(x)}\right] = \frac{f'(x)g(x) - f(x)g'(x)}{[g(x)]^2}.$$

EXAMPLE 3.29 Differentiate $h(x) = \dfrac{x^2-1}{x^3+1}$.

Solution

$$h'(x) = \frac{(x^2-1)'(x^3+1) - (x^2-1)(x^3+1)'}{(x^3+1)^2}$$ Quotient Rule with $f(x) = x^2 - 1$ and $g(x) = x^3 + 1$

$$= \frac{(2x)(x^3+1) - (x^2-1)(3x^2)}{(x^3+1)^2}$$ Sum/Difference and Power Rules

$$= -\frac{x(x^3-3x-2)}{(x^3+1)^2}.$$ Simplifying ■

EXAMPLE 3.30 Differentiate $h(x) = \dfrac{x+1}{x}$.

Solution

$$h'(x) = \frac{(x+1)'(x) - (x+1)(x)'}{x^2} \qquad \text{Quotient Rule with } f(x) = x+1 \text{ and } g(x) = x$$

$$= \frac{(1)(x) - (x+1)(1)}{x^2} = -\frac{1}{x^2}. \qquad \text{Sum/Difference and Power Rules; simplifying}$$

■

Related Exercises 25, 28, 31, and 33.

Tips, Tricks, and Takeaways

This last example could have been solved *without* the Quotient Rule by simplifying the function first:

$$\frac{x+1}{x} = \frac{x}{x} + \frac{1}{x} = 1 + \frac{1}{x} = 1 + x^{-1}, \qquad x \neq 0.$$

Applying the Sum and Power Rules then yields the same derivative ($-x^{-2}$). Takeaway: *It may help to simplify the problem before setting off to solve it.*

The derivative shortcuts we've now covered enable us to calculate derivatives quickly, provided you can determine which derivative rule(s) to apply when. This is a skill you'll acquire as you practice using the differentiation rules discussed thus far. To that end, I encourage you to revisit Exercises 19–34 and differentiate those functions using the simplest approach. As a general rule, your goal should be to *use the derivative rule that yields the simplest calculation.*

Now that we've discussed all the derivative rules, let's apply them to transcendental functions.

3.12 (Optional) Derivatives of Transcendental Functions

First, let's calculate the derivative of non-e base exponential functions: $f(x) = b^x$. Since $f(x) = b^x = e^{rx}$, where $r = \ln b$, we can view this as the composition $f(x) = g(h(x))$, where $g(x) = e^x$ and $h(x) = rx$. Then,

$$f'(x) = g'(h(x))h'(x) \qquad \text{Chain Rule}$$

$$= g'(rx)(r) \qquad \text{Using } h(x) = rx \text{ and } h'(x) = r$$

$$= e^{rx}(r) \qquad \text{Using } g'(x) = e^x \text{ (Theorem 3.2) and } g'(rx) = e^{rx}$$

$$= b^x \ln b. \qquad \text{Using } e^{rx} = b^x \text{ and } r = \ln b$$

■

(Note that when $b = e$ we recover the result in Theorem 3.2.) We've derived a new derivative rule.

Theorem 3.8 Derivative of Exponential Function. If b^x is an exponential function, then

$$\frac{d}{dx}(b^x) = b^x \ln b. \tag{3.22}$$

EXAMPLE 3.31 Differentiate $f(x) = 2^x$.

Solution Using (3.22) with $b = 2$ yields $f'(x) = 2^x \ln 2$. ■

EXAMPLE 3.32 Differentiate $g(x) = xe^x$.

Solution

$$\begin{aligned} g'(x) &= (x)'e^x + x(e^x)' && \text{Product Rule} \\ &= e^x + xe^x && \text{Power Rule for } (x)' \text{ and Theorem 3.2} \\ &= (x+1)e^x. && \text{Simplifying} \end{aligned}$$ ■

EXAMPLE 3.33 Differentiate $h(x) = \dfrac{3^x}{2x}$.

Solution

$$\begin{aligned} h'(x) &= \frac{(3^x)'(2x) - (3^x)(2x)'}{(2x)^2} && \text{Quotient Rule} \\ &= \frac{(3^x \ln 3)(2x) - (3^x)(2)}{(2x)^2} && \text{Equation (3.22) with } b = 3 \text{ and the Power Rule} \\ &= \frac{3^x(x \ln 3 - 1)}{2x^2}. && \text{Simplifying} \end{aligned}$$ ■

EXAMPLE 3.34 Differentiate $h(t) = e^{-t^2}$.

Solution Writing $h(t) = f(g(t))$, with $f(t) = e^t$ and $g(t) = -t^2$, we have:

$$\begin{aligned} h'(t) &= f'(g(t))g'(t) && \text{Chain Rule} \\ &= f'(-t^2)(-2t) && \text{Using } g(t) = -t^2 \text{ and } g'(t) = -2t \\ &= e^{-t^2}(-2t) = -2te^{-t^2}. && \text{Using } f'(t) = e^t \text{ and } f'(-t^2) = e^{-t^2}\text{; simplifying} \end{aligned}$$ ■

Related Exercises 53–56 and 61.

APPLIED EXAMPLE 3.35 Suppose that the average rate of occurrence of an event is λ times per minute.[6] In some cases the probability P of waiting at most

[6] For example, the event might be a bus arriving at a bus stop, and λ might be $1/4$, so that on average one bus arrives every 4 minutes.

t minutes for the event to occur can be accurately modeled by

$$P(t) = 1 - e^{-t/\lambda}, \quad \lambda > 0.$$

(a) Suppose that the "event" is a human customer service agent answering your call and that $\lambda = 1/3$. Calculate $P(t)$.

(b) Calculate $P(\lambda)$ (using the information in part (a)) and interpret your result.

(c) Calculate $P'(t)$ for the function from part (a) and interpret $P'(1)$ using the rate of change interpretation of the derivative.

(d) Calculate $\lim_{t\to\infty} P(t)$ and interpret your result.

Solution

(a) $P(t) = 1 - e^{-3t}$.

(b) $P(\lambda) = 1 - e^{-3\lambda} = 1 - e^{-1}$, since $\lambda = 1/3$. The result of $P(\lambda) = 1 - e^{-1} \approx 0.63$ tells us that the probability of waiting at most the average wait time is about 63%. (Thus, it is likely that your call will be answered before the average wait time.)

(c) $P'(t) = -e^{-3t}(-3) = 3e^{-3t}$. From here we see that $P'(1) = 3e^{-3} \approx 0.15$. Interpretation: When you've already been waiting for 1 minute, the probability that your call will be answered is increasing at the rate of about 15% per minute.

(d) Since $e^{-3t} \to 0$ as $t \to \infty$, we conclude that $P(t) \to 1$ as $t \to \infty$. Interpretation: The probability that your call will be answered approaches 100% as the time you are willing to wait gets very large. ■

Related Exercises 63–65.

Let's now discuss how to differentiate logarithmic functions. First up, let's try to differentiate $\ln x$. We'll start with the following fact (discussed in Appendix B.8):

$$e^{\ln x} = x \quad \text{for all } x \text{ in the interval } (0, \infty).$$

Let's now differentiate this equation. The right-hand side's derivative is 1. For the left-hand side:

$$\frac{d}{dx}\left(e^{\ln x}\right) = 1$$

$e^{\ln x}(\ln x)' = 1$ Writing $e^{\ln x} = f(g(x)), f(x) = e^x, g(x) = \ln x$ and applying the Chain Rule

$x(\ln x)' = 1$ Using $e^{\ln x} = x$

Solving this for $(\ln x)'$ yields the following result. (Exercise 62 guides you through an alternative derivation using the limit definition of the derivative, (3.12).)

Theorem 3.9

$$\frac{d}{dx}(\ln x) = \frac{1}{x}.$$

Moreover, since

$$\log_a x = \frac{\log_e x}{\log_e a} = \frac{\ln x}{\ln a}$$

(see (B.15) from Appendix B), the Constant Multiple Rule (Theorem 3.3) yields the following more general version of Theorem 3.9.

Theorem 3.10

$$\frac{d}{dx}(\log_a x) = \frac{1}{x(\ln a)}.$$

EXAMPLE 3.36 Differentiate $f(x) = x\ln x$.

Solution

$$f'(x) = (x)'\ln x + x(\ln x)' \qquad \text{Product Rule}$$

$$= \ln x + x\left(\frac{1}{x}\right) = \ln x + 1. \qquad \text{Power Rule and Theorem 3.9; simplifying}$$

■

EXAMPLE 3.37 Differentiate $h(x) = \ln(x^2 + 2)$.

Solution Writing $h(x) = f(g(x))$, where $f(x) = \ln x$ and $g(x) = x^2 + 2$, we have:

$$h'(x) = f'(g(x))g'(x) \qquad \text{Chain Rule}$$

$$= f'(x^2 + 2)(2x) \qquad \text{Using } g(x) = x^2 + 2 \text{ and } g'(x) = 2x$$

$$= \left(\frac{1}{x^2 + 2}\right)(2x) = \frac{2x}{x^2 + 2}. \qquad \text{Since } f'(x) = \frac{1}{x} \text{ (Theorem 3.9); simplifying}$$

■

EXAMPLE 3.38 Differentiate $h(t) = \ln\sqrt{t^2 + 2}$.

Solution We can use the rules of logarithms (Theorem B.1) to simplify h to $h(t) = \frac{1}{2}\ln(t^2 + 2)$. Then, using the Constant Multiple Rule and the result of the previous example:

$$h'(t) = \frac{1}{2}\left(\frac{2t}{t^2 + 2}\right) = \frac{t}{t^2 + 2}.$$

■

Related Exercises 57–60 and 66.

Turning now to trigonometric functions, let's start by going back to $\tan x$ and calculating its derivative. Since $\tan x = \frac{\sin x}{\cos x}$, we can use the Quotient Rule (Theorem 3.7). The result is (see Exercise 76)

$$\frac{d}{dx}(\tan x) = \frac{1}{\cos^2 x}. \tag{3.23}$$

It's customary to rewrite this result in terms of the reciprocal trigonometric functions defined by

$$\csc x = \frac{1}{\sin x}, \qquad \sec x = \frac{1}{\cos x}, \qquad \cot x = \frac{1}{\tan x}. \tag{3.24}$$

From left to right, these are the **cosecant**, **secant**, and **cotangent** functions. Now, since $\cos^2 x = (\cos x)^2$, the result (3.23) becomes

$$\frac{d}{dx}(\tan x) = \sec^2 x. \tag{3.25}$$

Now that we've calculated the derivatives of the basic trio of trigonometric functions, let's get some more practice using them by differentiating functions involving trigonometric functions.

EXAMPLE 3.39 Differentiate $f(x) = x^2 - \tan x$.

Solution Using (3.25) and the Power and Product Rules: $f'(x) = 2x - \sec^2 x$. ■

EXAMPLE 3.40 Differentiate $h(x) = \sin^2 x$.

Solution Writing $h(x) = f(g(x))$, where $f(x) = x^2$ and $g(x) = \sin x$:

$$\begin{aligned} h'(x) &= f'(g(x))g'(x) && \text{Chain Rule} \\ &= f'(\sin x)(\cos x) && \text{Using } g(x) = \sin x \text{ and } g'(x) = \cos x \\ &= 2\sin x \cos x. && \text{Using } f(x) = x^2 \text{ and } f'(x) = 2x \end{aligned}$$ ■

EXAMPLE 3.41 Differentiate $h(x) = \sec x$.

Solution Writing $h(x) = (\cos x)^{-1} = f(g(x))$, where $f(x) = x^{-1}$ and $g(x) = \cos x$:

$$\begin{aligned} h'(x) &= f'(g(x))g'(x) && \text{Chain Rule} \\ &= f'(\cos x)(-\sin x) && \text{Using } g(x) = \cos x \text{ and } g'(x) = -\sin x \\ &= \left(-\frac{1}{\cos^2 x}\right)(-\sin x) && \text{Using } f(x) = x^{-1} \text{ and } f'(x) = -x^{-2} \\ &= \frac{\sin x}{\cos^2 x} = \sec x \tan x. && \text{Simplifying} \end{aligned}$$ ■

We just calculated that $(\sec x)' = \sec x \tan x$. The derivatives of $\csc x$ and $\cot x$ are calculated in Exercise 77.

Related Exercises 67–74 and 77–80.

Tips, Tricks, and Takeaways

The examples in this section reinforce my two running takeaways from using the derivative rules:

- *Know which rule to use when.* This requires identifying the type of function you're asked to differentiate (e.g., a product of two functions), because then there's likely a rule for differentiating it (e.g., the Product Rule).
- *Simplifying and/or rewriting the function first often helps.* This was the case in Examples 3.38 and 3.41.

We've now studied the derivative rules and can quickly calculate $f'(x)$. In fact: *We could use the rules over and over again to calculate derivatives of derivatives*! Such derivatives are called **higher-order derivatives**, and they're the topic of this chapter's last section.

3.13 Higher-Order Derivatives

Thus far we've focused on differentiating a function f, producing another function f'. If we now consider f' as "the function," we can differentiate *it*; the result: $(f')' = f''$. We've differentiated the original function f twice, so we call f'' the **second derivative** of f (and by extension f' the **first derivative** of f). We can keep doing this to produce higher-order derivatives: f''' (third derivative), f'''' (fourth derivative), and in general, $f^{(n)}$ (the n-th derivative, where n is a natural number).

Let's now discuss the Leibniz notation for higher-order derivatives. Suppose $y = f(x)$, then $f'(x) = \frac{dy}{dx}$ in Leibniz notation. The derivative of *that* in Leibniz notation is

$$\frac{d}{dx}\left(\frac{dy}{dx}\right) = \frac{d^2y}{dx^2} \quad \Longrightarrow \quad f''(x) = \frac{d^2y}{dx^2}.$$

You can then imagine the general pattern for the n-th derivative function:

$$f^{(n)}(x) = \frac{d^ny}{dx^n}.$$

Now that we've defined higher-order derivatives, let me note a fact: *Every differentiation rule we developed for f' works for $f^{(n)}$*. We need only replace all instances of f with f' in said rules. Let's work through a few examples.

EXAMPLE 3.42 Calculate $f^{(n)}(x)$ for $f(x) = x^3$.

Solution By repeated application of the Power Rule, $f'(x) = 3x^2$, $f''(x) = 6x$, $f'''(x) = 6$, and $f^{(n)}(x) = 0$ for every natural number $n \geq 4$. ■

EXAMPLE 3.43 Calculate $g''(x)$ for $g(x) = \sqrt{x+1}$.

Solution Rewriting g as $g(x) = (x+1)^{1/2}$, the Chain Rule yields $g'(x) = \frac{1}{2}(x+1)^{-1/2}$. Differentiating this using the Chain Rule again yields

$$g''(x) = -\frac{1}{4}(x+1)^{-3/2} = -\frac{1}{4\sqrt{(x+1)^3}}.$$ ■

Related Exercises 36–41.

In addition to the derivative rules extending to higher-order derivatives, our interpretation of the derivative does too. That means $f''(a)$ is the instantaneous rate of change of $f'(x)$ at $x = a$; similarly, $f''(a)$ is the slope of the line tangent to the graph of $f'(x)$ at $x = a$. The next example explores these results in the context of speed.

APPLIED EXAMPLE 3.44 In Example 3.3 we used a falling apple's distance function $d(t) = 16t^2$ to derive its instantaneous speed function, $\mathscr{s}(t) = 32t$. Find $d''(t)$ and interpret it physically.

Solution By the Power Rule, $d'(t) = 32t$ and so $d''(t) = 32$ ft/s^2. Since $d'(t) = \mathscr{s}(t)$, $d''(t) = \mathscr{s}'(t)$. That makes $d''(t)$ the instantaneous rate of change of the object's speed. ■

You may have thought of a name for $d''(t)$ already: *acceleration*. But acceleration is the instantaneous rate of change of *velocity*, which is itself the instantaneous rate of change of the object's *position* (not its distance covered). We'll return to the subtle distinction between velocity and speed, and their instantaneous rates of change, in Chapter 5. For these reasons, I'll leave the interpretation of $d''(t)$ as the instantaneous rate of change of the object's speed. Viewed this way, if $d''(t) > 0$ we expect the object's speed to increase, whereas if $d''(t) < 0$ we expect its speed to decrease. (Exercise 43 explores this train of thought further.)

Related Exercises 42–43.

3.14 Parting Thoughts

We can now calculate $f'(x)$ (and its higher-order cousins) and interpret and visualize it in various ways. At the heart of all these accomplishments was Equation (3.11), our original definition of $f'(a)$. Viewed from Leibniz's point of view, $f'(a)$ is the ratio of infinitesimal changes arising from the limit of the slopes of secant lines. Read that sentence again and you'll once more appreciate my description in Chapter 1 of calculus as the mathematics of infinitesimal change.

The derivative, being one way to quantify infinitesimal change, was an important advance in mathematics. In the next chapter we'll apply what we've learned to real-world contexts to discover that derivatives are just as important outside of mathematics.

CHAPTER 3 EXERCISES

1–6: Calculate $f'(1)$ using equation (3.11).

1. $f(x)=(x-1)^2$

2. $f(x)=\frac{x^2}{2}+5$

3. $f(x)=x^2+2x+1$

4. $f(x)=\frac{1}{x^2}$

5. $f(x)=\frac{x+2}{x-2}$

6. $f(x)=\sqrt{x}$. *Hint:* It will help to multiply the numerator and denominator by $\sqrt{x+\Delta x}+\sqrt{x}$.

7. The limit $\lim_{\Delta x\to 0}\frac{\sqrt{16+\Delta x}-4}{\Delta x}$ represents $f'(a)$ for some $f(x)$ and some a-value; what are some possibilities for $f(x)$ and a?

8. Find the equations of the tangent lines at the point $(1,f(1))$ for the functions in Exercises 1 and 2.

9. Suppose $y=2x+4$ is the line tangent to the graph of some function $f(x)$ at $x=2$. Find $f'(2)$ and $f(2)$.

10. Average Speed Let $d(t)=16t$ be a distance function. Find the average speed over the time intervals (a) $1\leq t\leq 2$ and (b) $2\leq t\leq 3$.

11. Instantaneous Speed Use (3.6) to calculate the instantaneous speed $\mathscr{s}(a)$ for the distance functions that follow.

(a) $d(t)=10$. (Explain why your answer is reasonable.)

(b) $d(t)=t^2+1$.

(c) $d(t)=t^3$. (Formula (A.10) will help.)

12. Instantaneous Speed Let $d(t)=4-2t$ be a distance function.

(a) Identify $\mathscr{s}(a)$ without doing any calculations.

(b) Confirm your answer using (3.6).

13. Maximum Heart Rate Let's return to Applied Example 3.14. A simpler model for MHR is $H(t)=220-t$.

(a) Calculate the equation of the tangent line at $t=20$ for $H(t)$.

(b) Calculate the equation of the tangent line at $t=20$ for $M(t)$ (given in Applied Example 3.14).

(c) Discuss briefly why, given your results, $M(t)$ is a more realistic model of MHR than $H(t)$.

14–15: Determine where f is not differentiable.

14. f is the function graphed in Figure 2.10, considering only the subset $(0,100)$ of its domain.

15. f is the function from Exercise 2 of Chapter 2.

16. Sketch the graph of f' from the graph of f given below.

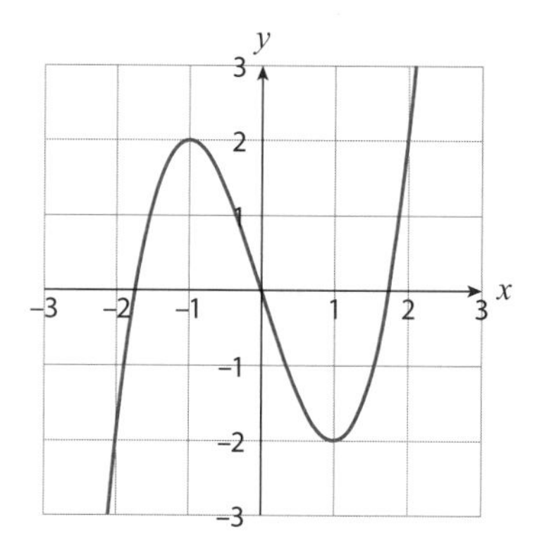

17. Sketch the graph of f' from the graph of f given below.

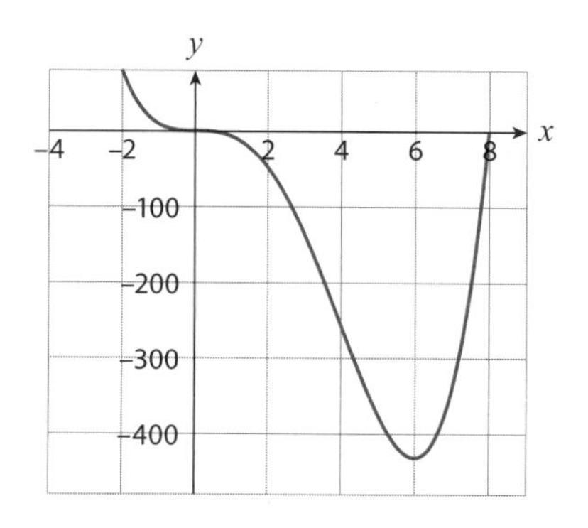

18. Sketch the graph of f' from the graph of f given below.

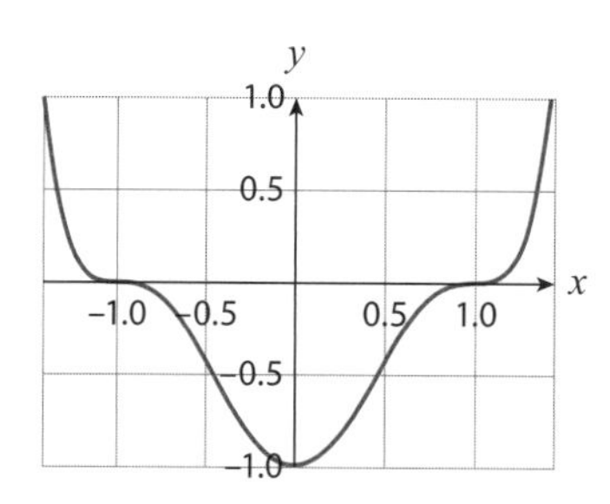

19–34: Find the derivative.

19. $f(x) = \pi$

20. $g(x) = x^{50}$

21. $f(t) = 16t^{1/2}$

22. $h(s) = s^7 - 2s^3$

23. $f(x) = 4\sqrt{x} - 10\sqrt[3]{x}$

24. $h(s) = s^{3/2}(1+s)$

25. $g(x) = \dfrac{1}{x+1}$

26. $h(t) = \sqrt{1-t}$

27. $g(x) = (x^2+7)(\sqrt{x} - 14x)$

28. $f(x) = \dfrac{x-1}{x}$

29. $h(x) = \sqrt{(1+x^2)^2+1}$

30. $g(t) = t^{\pi}$

31. $h(x) = \dfrac{\sqrt{x}}{x+1}$

32. $f(x) = \left(x^3 + \dfrac{2}{x}\right)^3$

33. $f(s) = \dfrac{1}{(3s-7)^2}$

34. $g(t) = 15t^{4/5} - t(t^2+1)$

35. Let $f(x) = \sqrt{x} + x$.

(a) Find the instantaneous rate of change at $x = 1$.
(b) Are the y-values of f increasing at a faster or slower rate at $x = 2$ (compared to $x = 1$)? Explain.
(c) Calculate $f(2) - f(1)$ and compare it to your answer in part (b).
(d) Interpret your answer to part (a) using the geometric interpretation of the derivative.
(e) Find the equation of the line tangent to the graph of f at $(1, 2)$.

36–39: Calculate $f''(x)$.

36. $f(x) = 2x^3 - 3x^2 - 12x$

37. $f(x) = 2 + 3x - x^3$

38. $f(x) = \sqrt{x+3}$

39. $f(x) = x\sqrt{x+3}$

40. Let $f(x) = x^{4/3}$. Is f differentiable at $x = 0$? Is f *twice* differentiable at $x = 0$? Briefly explain. (Conclusion: not all higher-order derivatives exist at every point.)

41. Suppose $f'(x) = 0$ for all x. What can you conclude about $f(x)$? Suppose $f''(x) = 0$. What can you conclude about $f(x)$? Briefly explain.

42. The derivative of acceleration—the "jerk" In physics, the derivative of the acceleration function $a(t)$ of an object with position function $s(t)$ is called the *jerk*: $j(t) = a'(t)$. The terminology is appropriate, because $j(t)$ represents an instantaneous change in acceleration, which the object would feel as a "jerk" in the motion. Suppose an amusement park ride's position function is $s(t) = t^3 + t$, where s is measured in miles and t in hours. Find the jerk at $t = 1$. What are its units?

43. Unemployment Let $U(t)$ denote the unemployment rate at time t in a country. Suppose a politician makes the following claim: "The rate of decrease in the unemployment rate is slowing." Translate this into a statement about U or its derivatives.

44. Student Loans Suppose a student takes out a student loan with a yearly interest rate of r%. Let $C = f(r)$ be the total cost (in $) of repaying that loan.

(a) What does $f(0.05) = \$10,000$ mean?

(b) What are the units of $f'(0.05)$? What would $f'(0.05) = \$1,000$ mean?

(c) Do you expect $f'(r)$ to be positive for all $r > 0$ or negative? Briefly explain your answer.

45. Acceleration due to Gravity Assume that the Earth is a perfect sphere, and suppose an individual stands at a height h above the ground. Gravity pulls the individual toward the center of the Earth with a force equal to mg (this is the individual's **weight**), where m is the individual's mass and g the acceleration due to gravity. In the notation of Newton's Universal Law of Gravity (Exercise 35 in Chapter 2), it follows that $mg = F(R + h)$, which yields a formula for the acceleration due to gravity as a function of the height h (in meters) above the ground (and independent of the object's mass m):

$$g(h) = \frac{GM}{(R+h)^2} \quad \text{m/s}^2.$$

(a) Using $GM \approx 3.98 \times 10^{14}$ and $R \approx 6.37 \times 10^6$, calculate $g(0)$. (We use 9.806 m/s^2 as the "standard acceleration due to gravity.")

(b) Calculate $g'(h)$ and $g'(0)$.

46. Measuring the Acceleration due to Gravity Using a Pendulum Consider a pendulum of length l meters. The **period** of the pendulum is the time T (in seconds) it takes for the pendulum to undergo a complete oscillation. For small oscillations, T is well approximated by the function

$$T(g) = \frac{2\pi\sqrt{l}}{\sqrt{g}},$$

where g the acceleration due to gravity, $g \approx 9.81$ m/s^2.

(a) Calculate $T(9.81)$ assuming $l = 1$.

(b) Solve the $T(g)$ function for g to find the function $g(T)$. This function enables one to measure the acceleration due to gravity by measuring the period of a 1-meter-long pendulum! Find $g(2.006)$.

(c) Using the $g(h)$ formula in Exercise 45, calculate $T(g(h))$. This formula predicts how a pendulum's period changes with altitude.

(d) Let $f(h) = T(g(h))$ and $l = 1$. Calculate $f'(0)$ and interpret your result.

47. Speed of Sound The speed of sound s varies according to the temperature of the surrounding air. A reasonable approximation is

$$s(C) = 20.05\sqrt{C + 273.15} \text{ m/s},$$

where C is the air temperature measured in Celsius.

(a) Let $h(F) = s(C(F))$, where F is the temperature in Fahrenheit. Find $h(F)$. (*Hint*: Solve (B.6) of Appendix B for C first.)
(b) Calculate $h(68)$ and compare it to the speed of light c, approximately 300 *million* meters per second. (Your comparison explains why, for example, we *see* fireworks explode in the air before we *hear* them exploding.)
(c) Calculate $h'(68)$ using the Chain Rule and the facts that $C(68) = 20$ and $C'(68) = 5/9$.

48. Calculate $f'(x)$ for $f(x) = |x|$.

49. Calculate $f'(x)$ for $f(x) = \dfrac{x}{|x|}$.

50. Let $h(x) = f(x)(g(x))^{-1}$. Use the Product and Chain rules to derive the Quotient Rule (Theorem 3.7).

51. Find the equations of the lines passing through the origin and tangent to the graph of $f(x) = x^2 + 1$.

52. Let $f(x) = xg(x^2)$. Find $f'(x)$ assuming g' exists (the answer will involve g and g').

Exercises Involving Exponential and Logarithmic Functions

53–60: Calculate the derivative.

53. $f(x) = e^{4x}$

54. $f(x) = 2^{-x^2}$

55. $g(t) = (t^2 + 1)e^{2t}$

56. $h(z) = \dfrac{e^z + e^{-z}}{2}$

57. $f(x) = \ln(x^2 + 5)$

58. $f(z) = e^{-z}\ln(3z)$

59. $h(t) = \ln\dfrac{t}{t^2 + 1}$

60. $g(t) = \ln\dfrac{1 + e^t}{1 - e^t}$

61. Derive the derivative rule $(e^{rx})' = re^{rx}$ directly from the definition of the derivative (3.12) and using Theorem A2.1. (Consulting the calculation in (3.14) may help.)

62. When $f(x) = \ln x$, the limit definition of $f'(x)$, (3.12), yields

$$f'(x) = \lim_{h \to 0} \ln\left(1 + \frac{h}{x}\right)^{1/h}.$$

Show that by letting $t = h/x$ (we're thinking here of x as a fixed positive number), using Theorem A2.1 with $g(h) = h/x$, and using Limit Law 7, we rederive that $f'(x) = \frac{1}{x}$.

63. The Calculus of Cooling Coffee Suppose a cup of coffee at a temperature of T_0 °F is taken off the coffee machine's warming plate and placed on a dining table. Assuming that the ambient temperature of the room is T_a, **Newton's Law of Cooling** yields the following equation for the coffee cup's temperature T as a function of time t (in minutes since the cup was removed from the warming plate):

$$T(t) = T_a + ce^{-bt},$$

where c and b are positive constants.

(a) Two fairly realistic assumptions are that $T_0 = 160$, $T(2) = 120$, and $T_a = 75$. Use this to show that $c = 85$ and $b \approx 0.318$.
(b) Calculate $T'(0)$ for the function obtained in part (a) and interpret your result using the rate of change interpretation of the derivative.
(c) Clalculate $T'(t)$ for the function in part (a).
(d) Find the horizontal asymptote of $T(t)$ (from part (a)) and interpret your results.

64. Ebbinghaus Forgetting Curve In 1885 psychologist Hermann Ebbinghaus conducted an interesting experiment on memory: He memorized nonsense three-letter words (like KAF) and tested himself regularly to see how much of the information he forgot over time. If R denotes what percentage of the information learned initially was retained after t days, Ebbinghaus' results suggested that

$$R(t) = a + (1-a)e^{-bt},$$

where $0 \leq a < 1$ and $b > 0$ are constants.

(a) Calculate $\lim_{t\to\infty} R(t)$ and interpret your result.
(b) Some research suggests that, on average, we forget 70% of what we learned a day ago (assuming no review in the iterim). Use this, along with $a = 0$, to find the associated $R(t)$ function.
(c) Calculate $R'(1)$ for the function you found in part (b) and interpret your result using the rate of change interpretation of the derivative.

65. Wind Power Wind power is a clean, sustainable energy source. But generating power this way requires wind, and ideally high-velocity wind gusts. Luckily, the engineers who design wind turbines have discovered that they can accurately predict the probability of winds of speed v (in m/s) occurring using the function

$$P(v) = ave^{-bv^2},$$

where $a > 0$ and $b > 0$ are parameters that depend in part on the location being studied.

(a) Show that $P'(v) = ae^{-bv^2}[1 - 2bv]$.
(b) Interpret the fact that $P'(0) = a$ using the rate of change interpretation of the derivative.

66. Let $f(x) = x^n$, where n is a real number. Writing $x^n = e^{\ln x^n}$, use the Chain Rule to help show that $f'(x) = nx^{n-1}$. This proves Theorem 3.4 for $x > 0$.

Exercises Involving Trigonometric Functions

67–74: Calculate the derivative.

67. $f(x) = 4x^3 - 3\sin x$

68. $f(x) = \sqrt{x}\cos x$

69. $f(x) = \dfrac{x}{1 - \tan x}$

70. $f(z) = \sin z - z$

71. $g(x) = \cos x + (\cot x)^2$

72. $h(t) = \dfrac{\sin t}{t}$

73. $g(t) = \dfrac{\cos t}{1 + \sin t}$

74. $h(z) = z^4 \sin^2 z$

75. Use (3.12) and (B.24) to show that $(\cos x)' = -\sin x$.

76. Use the Quotient Rule (Theorem 3.7) to show that $(\tan x)' = \sec^2 x$.

77. Use the Chain Rule (Theorem 3.6) to show that $(\csc\ x)' = -\csc\ x \cot\ x$, and that $(\cot\ x)' = -\csc^2\ x$.

78. Exercise 59 in Appendix B relates the slope of a line m to its angle of inclination θ from the x-axis by the equation $m = \tan\theta$. Applying this to the *tangent* line to the graph of a function f at the point $(a, f(a))$ yields

$$f'(a) = \tan\theta,$$

where $-\frac{\pi}{2} < \theta < \frac{\pi}{2}$. Calculate θ for $f(x) = x^3$ at $a = 0$ and $a = \pm 1$. Interpret your results.

79. Let's return to Exercise 60 in Appendix B.

(a) Use the Product Rule (Theorem 3.5) to show that

$$A'(n) = \frac{r^2}{2}\left[\sin\left(\frac{2\pi}{n}\right) - \frac{2\pi}{n}\cos\left(\frac{2\pi}{n}\right)\right].$$

(b) Calculate $\lim_{n\to\infty} A'(n)$ and interpret your result.

80. Measuring Time Using a Pendulum Consider a pendulum of length l, and denote by θ the angle of the pendulum with respect to a vertical line t seconds after it is released from rest at an initial angle of $\theta_0 > 0$ (see diagram below). If the pendulum's oscillations are small and the motion is ideal (for instance, no air resistance is experienced), then

$$\theta(t) = \theta_0 \cos\left(\sqrt{\frac{g}{l}}t\right).$$

θ

(a) Identify the amplitude and period of the trigonometric function $\theta(t)$, and interpret both quantities in terms of the pendulum's motion.

(b) A typical grandfather clock has a pendulum of length 1 meter that swings with an amplitude of 3°. Write down its $\theta(t)$ equation; remember to convert the amplitude to radians, and use $g = 9.81\ \text{m/s}^2$.

(c) Verify that the period of the function you calculated in part (b) is roughly 2 seconds. (30 complete oscillations (two swings) would therefore span 1 minute, making this pendulum a useful time-keeping device.)

(d) A slightly more accurate formula for the period of a pendulum than the one given in Exercise 46 is

$$T(\theta_0) = 2\pi\sqrt{\frac{l}{g}}\left(1 + \frac{1}{16}\theta_0^2\right).$$

Note that this period depends on the amplitude θ_0. Calculate $T(\theta_0)$ for the information given in part (b).

(e) Calculate $T'(\theta_0)$ for the information given in part (b) and interpret your result using the rate of change interpretation of the derivative.

4 Applications of Differentiation

Chapter Preview. *Draw a continuous curve on a piece of paper and you'll notice that your curve has a largest y-value and a smallest y-value. Not a deep insight, I admit. But imagine now that your curve models a company's revenue from selling its products, or the world's population since 2000, or the number of individuals contracting a contagious virus since it was first detected. The extreme values of those curves have important real-world relevance. We'll develop a procedure for calculating those extrema in this chapter. We'll start first with a simpler application of derivatives—the topic of related rates—and then discover a few more facts about what information derivatives give us. We'll then draw on those results to arrive at the grand finale: optimization theory.*

4.1 Related Rates

Relates Rates are calculus problems in which you are asked to relate the instantaneous rates of change of two or more quantities. (Often these are *time* rates of change, i.e., dy/dt.) Your job is then to determine the value of one rate at a particular instant, given the other rate(s).

We've actually already discussed a related rates problem: the expanding square problem in Section 3.10. There we determined that if the side length x of a square was increasing with time, so that x is really $x(t)$, then the time rate of change of the square's area A is

$$\frac{dA}{dt} = 2x\frac{dx}{dt}. \tag{4.1}$$

(This is (3.19), in Section 3.10.) Equation (4.1) is the equation that *relates the rates* in this related rates problem.

We got to (4.1) in Section 3.10 after quite a few calculations. But now that we've learned the Chain Rule, here's the faster way to get to it. (This new derivation will anchor the rest of our calculations in this section.) First, we note that the square's area is $A = x^2$. Then,

$$\begin{aligned}
\frac{dA}{dt} &= \frac{dA}{dx}\frac{dx}{dt} && \text{Leibniz version of Chain Rule, (3.20)}\\
&= \frac{d}{dx}\left(x^2\right)\frac{dx}{dt} && \text{Since } A = x^2\\
&= 2x\frac{dx}{dt}. && \text{Power Rule}
\end{aligned}$$

Let's now use this to solve our first related rates problem.

EXAMPLE 4.1 The side length of a square is increasing at the constant rate of 0.1 feet per second. How fast is the square's area changing when its side length is 1 foot?

Solution Using (4.1): $\frac{dA}{dt} = 2(1)(0.1) = 0.2$ ft/sec. ■

Related rates is one of those topics that's best learned by doing lots and lots of examples. So, let's continue with the examples.

APPLIED EXAMPLE 4.2 You are inflating a balloon that remains spherical as you inflate it. The balloon's volume V and radius r are related by

$$V(r) = \frac{4}{3}\pi r^3.$$

Suppose you inflate the balloon such that r changes at the constant rate of 0.1 inch per second. How fast is the volume of the balloon changing when its radius is 6 inches?

Solution Mimicking what we did in the previous example:

$$\begin{aligned}
\frac{dV}{dt} &= \frac{dV}{dr}\frac{dr}{dt} && \text{Leibniz version of Chain Rule, (3.20)} \\
&= \frac{d}{dr}\left(\frac{4}{3}\pi r^3\right)\frac{dr}{dt} && \text{Since } V = \frac{4}{3}\pi r^3 \\
&= 4\pi r^2 \frac{dr}{dt}. && \text{Constant Multiple and Power Rules}
\end{aligned}$$

Using now the given information:

$$\frac{dV}{dt} = 4\pi(6)^2(0.1) \approx 45.2 \ \text{in}^3/\text{sec}.$$ ■

Both previous examples provided the equation involving the variables whose rates we needed to relate. The next example ups the ante and requires us to come up with that equation ourselves.

APPLIED EXAMPLE 4.3 A traffic camera is tracking a car as it approaches an intersection (see Figure 4.1). Assuming the camera is 300 feet from point A in the intersection, how fast is the distance between the car and the camera changing at the instant the car is 400 feet from point A in the intersection and traveling at a speed of 60 feet per second?

Solution Denote by y the car's distance from point A; we'll measure y in feet and t in seconds. The distance between the car and the traffic camera is the hypotenuse of the triangle in Figure 4.1. Denoting that distance by z (again measured in feet), the Pythagorean Theorem yields:

$$z = \sqrt{(300)^2 + y^2} = \sqrt{9,000 + y^2}.$$

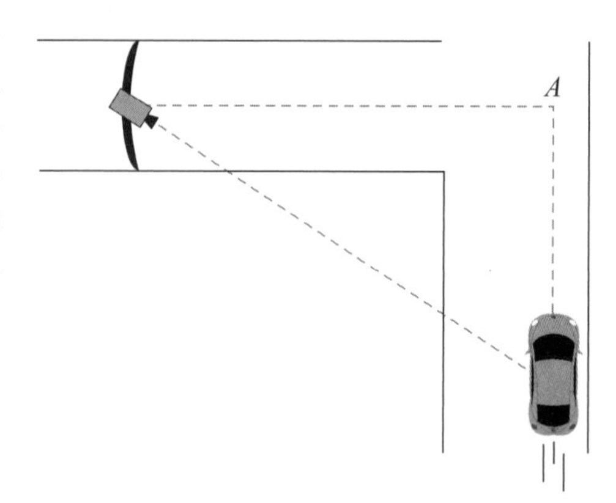

Figure 4.1

We now differentiate z with respect to t:

$$\frac{dz}{dt} = \frac{dz}{dy}\frac{dy}{dt} \qquad \text{Leibniz version of Chain Rule, (3.20)}$$

$$= \left[\frac{1}{2}(9,000 + y^2)^{-1/2}(2y)\right]\frac{dy}{dt} \qquad \text{Differentiating } z = \sqrt{9,000 + y^2}$$

$$= \frac{y}{\sqrt{9,000 + y^2}}\frac{dy}{dt}. \qquad \text{Simplifying} \qquad (4.2)$$

When $y = 400$ we know that $z = \sqrt{9,000 + (400)^2} = \sqrt{25,000} = 500$. And since the car is traveling at 60 feet per second at the instant when $y = 400$, we know that $\frac{dy}{dt} = -60$. (This is negative because the car's distance to the intersection is decreasing.) Substituting these values into (4.2) yields

$$\frac{dz}{dt} = \frac{400}{500}(-60) = -48 \text{ ft/s.} \qquad \blacksquare$$

What set this example apart from the previous two was the **mathematical modeling** part—the portion of the problem that requires translating the given information into math by identifying the relevant variables and coming up with an equation relating them. Intermediate- to advanced-level related rates problems will require that step. In simpler related rates problems—like that of Example 4.2—the main equation and variables are given. The following procedure will help you tackle those more difficult related rates problems.

Box 4.1: How to Set Up a Related Rates Problem

1. Draw a diagram of the situation (if none is given) and label changing quantities.
2. Write down (in mathematical language) what rate the question asks for and what rate(s) are given. (Note: If a quantity is increasing, the rate should be positive; if it's decreasing the rate should be negative.) Tip: Use any units provided to determine what rate(s) are given. (Example: "feet per second" is a rate of the form $\frac{dx}{dt}$, where x measures distance and t time.)
3. Use your content knowledge (e.g., formulas from geometry) along with your diagram to come up with the main equation that relates the variables you've identified.
4. Finally, obtain the related rates equation by differentiating your main equation (often it's with respect to t) using the Leibniz version of the Chain Rule.

APPLIED EXAMPLE 4.4 A fancy coffee maker features a conical water receptacle atop another receptacle containing pre-ground beans. The conical receptacle has a hole at the bottom through which water drips down at a rate of 2 in^3/hour to brew the coffee (Figure 4.2(a)). If the base radius of the conical receptacle is 2 inches and its height is 6 inches, at what rate is the depth of the water in the receptacle changing when the depth of the water remaining is 1 inch?

Solution Following the procedure in Box 4.1, let's first draw a diagram of the situation (Figure 4.2(b)). Water is dripping through the bottom of the inverted cone, so the radius and height of the conical volume of water remaining are changing. Those

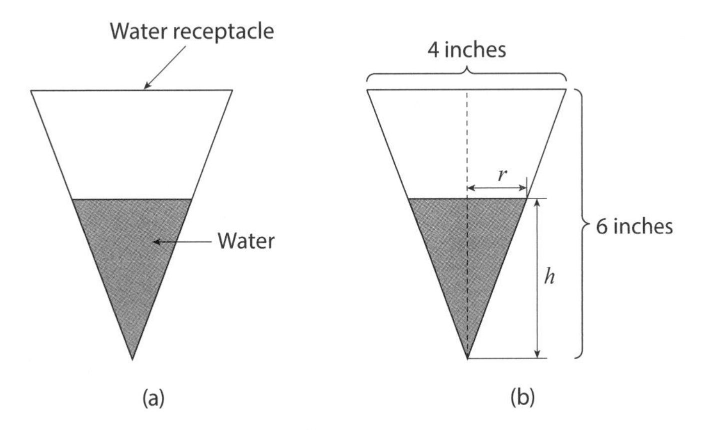

Figure 4.2

quantities are therefore variables, labeled r and h in Figure 4.2(b). Next, let's identify the rates asked for and given in the problem statement.

- **Rate to find:** $\frac{dh}{dt}$ when $h=1$.
- **Rate given:** $\frac{dV}{dt}=-2$ in^3/hour (V for volume, and -2 because the volume is decreasing).

The variables involved in our rates are V and h, so we need an equation relating the volume of a cone to its height. Cue geometry:

$$V=\frac{1}{3}\pi r^2 h, \tag{4.3}$$

where r is the radius of the cone. We aren't given a rate of change for r, so we need to eliminate r somehow. Another glance at Figure 4.2(b) suggests the answer: similar triangles. The water-filled triangle in the figure is similar (in the geometry sense of the word) to the larger conical receptacle's triangle ("similar" because the two triangles' interior angles are all the same). We know from geometry that when two triangles are similar to each other their side lengths are in the same proportion. Applying that here:

$$\frac{2r}{h}=\frac{4}{6} \implies \frac{r}{h}=\frac{1}{3} \implies r=\frac{h}{3}. \tag{4.4}$$

Substituting this into (4.3) yields

$$V=\frac{1}{3}\pi\left(\frac{h}{3}\right)^2 h=\frac{\pi h^3}{27}. \tag{4.5}$$

Now we can differentiate:

$$\frac{dV}{dt}=\left[\frac{\pi}{27}(3h^2)\right]\frac{dh}{dt}=\left(\frac{\pi h^2}{9}\right)\frac{dh}{dt}.$$

Finally, substituting in the given rate and solving for $\frac{dh}{dt}$ when $h=1$ yields

$$-2=\left(\frac{\pi(1)^2}{9}\right)\frac{dh}{dt} \implies \frac{dh}{dt}=-\frac{18}{\pi}\approx -5.7 \text{ inches/hour.}$$ ■

This previous example contained one additional complication you may run into while working through a related rates problem: a "constraint equation." This is what we call (4.4). It forces a specific relationship between variables in the problem (r and h in our case) that you can use to eliminate one variable from the model.

Related Exercises 21–28.

(a)

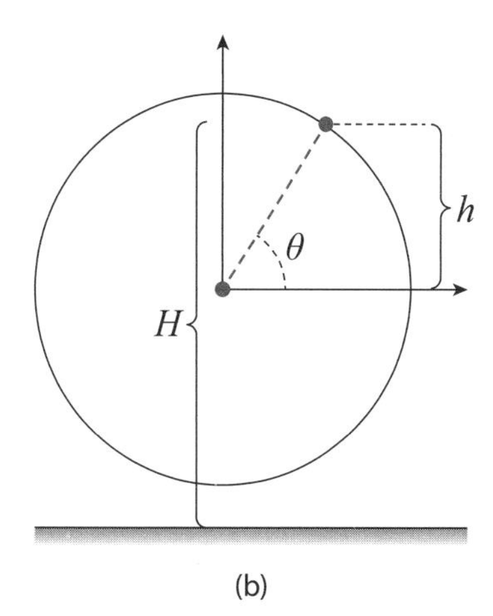

(b)

Figure 4.3

Transcendental Tales

APPLIED EXAMPLE 4.5 You and your friend are aboard a Ferris wheel rotating counterclockwise (Figure 4.3(a)). As the Ferris wheel rotates your height off the ground changes. If you are 502 feet off the ground at the Ferris wheel's highest point, 2 feet off the ground at its lowest point, and the wheel is rotating at the constant rate of $\pi/3$ radians per minute, at what rate is your height off the ground changing at the instant you are 377 feet off the ground and moving upward?

Solution Following the procedure in Box 4.1, let's first draw a diagram of the situation and label the relevant variables (Figure 4.3(b)). Next, let's identify the rates asked for and given in the problem statement.

- **Rate to find:** $\frac{dH}{dt}$ when $H = 377$ feet.
- **Rate given:** $\frac{d\theta}{dt} = \frac{\pi}{3}$. (This rate is positive since the wheel is rotating counterclockwise.)

Our rates involve H and θ. From the diagram in Figure 4.3(b) we see that H can be decomposed as

$$H = 2 + r + h,$$

where r is the Ferris wheel's radius. We calculate that as

$$r = \frac{1}{2}(502 - 2) = 250,$$

half the 500-foot diameter of the wheel. And since

$$\sin\theta = \frac{h}{r}, \qquad \text{so that} \qquad h = r\sin\theta = 250\sin\theta,$$

our H equation becomes

$$H = 252 + 250\sin\theta. \tag{4.6}$$

Differentiating with respect to t:

$$\frac{dH}{dt} = \frac{dH}{d\theta}\frac{d\theta}{dt} = (250\cos\theta)\frac{d\theta}{dt}. \tag{4.7}$$

At the instant $H = 377$, (4.6) tells us that

$$\sin\theta = \frac{377 - 252}{250} = \frac{1}{2}, \qquad \text{so that} \qquad \theta = 30^\circ \text{ or } 120^\circ.$$

However, the prompt wanted us to consider the case in which you are moving upward, so we select $\theta = 30^\circ$. Using this and the given rate in (4.7) then yields

$$\frac{dH}{dt} = (250\cos 30^\circ)\left(\frac{\pi}{3}\right) = \frac{250\pi\sqrt{3}}{6} \approx 227 \text{ ft/min.}$$ ■

Related Exercises 50–51.

Tips, Tricks, and Takeaways

Related rates problems illustrate perfectly calculus' dynamics mindset. Indeed, my first suggestion for tackling such problems—item 1 in Box 4.1—is to label the *changing* quantities. That's easiest to do if you first imagine the action in the problem—water *draining*, a car *speeding* toward an intersection, etc. The world around us is constantly changing, so expect to run into related rates problems in many different disciplines, including the physical, life, and social sciences. Finally, here are three more takeaways.

- The functions in related rates problems are **implicit functions**—we know *that* they depend on a variable (often t) but not *how*. By contrast, functions like $f(x) = x^2$ are **explicit** functions—we know exactly how they depend on their input.
- We differentiate an implicit function through the Leibniz version of the Chain Rule. In general, if $z = f(x)$ and x is an implicit function of t, then,

  $$\frac{dz}{dt} = \frac{dz}{dx}\frac{dx}{dt} = f'(x)\frac{dx}{dt}. \tag{4.8}$$

  The takeaway: $\frac{dz}{dt}$ is the "usual" derivative (that's $z' = f'(x)$) multiplied by $\frac{dx}{dt}$.
- The process of differentiating an implicit function is called **implicit differentiation**.

Next up on our tour of the applications of differentiation is optimization theory. The next section lays the foundation for that by connecting the increase and decrease of a function's graph to its derivative. (We'll later discuss how that helps us maximize and minimize functions.)

4.2 Linearization

The geometric interpretation of the derivative tell us that if the graph of a differentiable function f is increasing then its derivative is positive (being the slope of lines tangent to the graph of f). But is the converse true? That is, if $f'(x) > 0$ on an interval does it follow that the graph of f is increasing on that interval?

Let's make progress on this question by simplifying it. New question: If $f'(a) > 0$, does it follow that the graph of $y = f(x)$ is increasing near $x = a$? Well, recall that $f'(a)$ *equals* the limit as $\Delta x \to 0$ of $\Delta y / \Delta x$ (equation (3.11)). So when Δx is close to zero we expect that $f'(a)$ should be *approximately equal* to $\Delta y / \Delta x$:

$$f'(a) \approx \frac{\Delta y}{\Delta x} \quad \text{when } \Delta x \approx 0. \tag{4.9}$$

Multiplying both sides by Δx yields:

$$\Delta y \approx f'(a)\Delta x \quad \text{when } \Delta x \approx 0. \tag{4.10}$$

This approximation says that a small change in x-values Δx from $x = a$ produces a change in y-values Δy of approximately $f'(a)\Delta x$. Quantifying these changes as

$$\Delta x = x - a, \qquad \Delta y = f(x) - f(a),$$

and substituting these into (4.10) and solving for $f(x)$ yields the following.

Definition 4.1 Linearization. Let f be a differentiable at a. The approximation

$$f(x) \approx f(a) + f'(a)(x - a) \quad \text{for } x \text{ near } a \tag{4.11}$$

is called the **linear approximation** of f at a. The right-hand side linear function,

$$L(x) = f(a) + f'(a)(x - a), \tag{4.12}$$

is called the **linearization** of f at a.

The function $L(x)$ is just the equation of the tangent line at $x = a$.[1] Let me explain why it's called the "linearization" of f. You see, (4.11) says that for x-values near a, *the graph of f is approximately the graph of its tangent line at $x = a$.* In other words, the derivative at $x = a$ *linearizes* the function near $x = a$, and we think of $L(x)$ then as the "linearized" version of $f(x)$.

EXAMPLE 4.6 Calculate the linearization of $f(x) = \sqrt{x}$ at $a = 1$. Then plot your result along with $f(x)$ over the intervals $[0, 2]$, $[0.5, 1.5]$, and $[0.9, 1.1]$. Comment on what you notice.

[1] That line passes through $(a, f(a))$ and has slope $f'(a)$; using these facts in the point–slope equation yields (4.12).

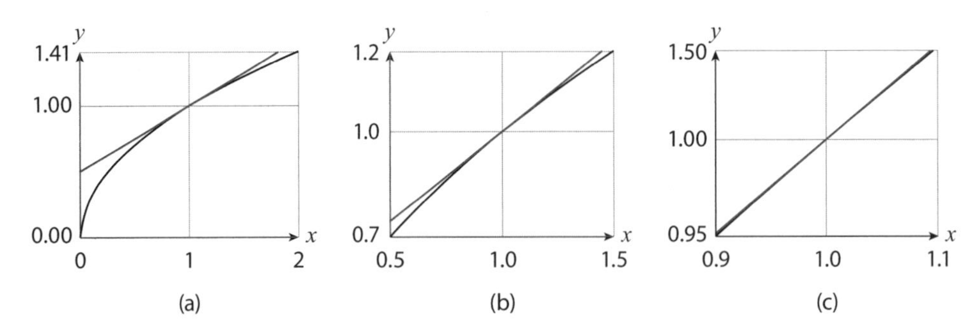

Figure 4.4: From left to right: zooming in on the graph of $f(x)=\sqrt{x}$ and its tangent line $L(x)=\frac{1}{2}(x+1)$ at the point $(1, 1)$.

Solution

$$\begin{aligned} L(x) &= f(1)+f'(1)(x-1) && \text{Equation (4.12) with } a=1 \\ &= 1+\frac{1}{2}(x-1) && \text{Since } f(1)=1,\ f'(x)=\frac{1}{2}x^{-1/2},\ f'(1)=\frac{1}{2} \\ &= \frac{1}{2}(x+1). && \text{Simplifying} \qquad (4.13) \end{aligned}$$

Figures 4.4(a)–(c) plot $L(x)$ and $f(x)$. Notice that as we zoom in to x-values closer to $x=1$ (moving left to right in the figure) the graph of f looks more like its tangent line at $x=1$. ■

Related Exercises 1–4.

Our results tell us that the answer to our earlier question—If $f'(a)>0$, does it follow that the graph of $y=f(x)$ is increasing near $x=a$?—is: *yes.* In the next section we'll return to how this revelation starts us on the path toward optimization theory. But before that, let's discuss two useful applications of linearization. (This is the Applications of Differentiation chapter, after all.)

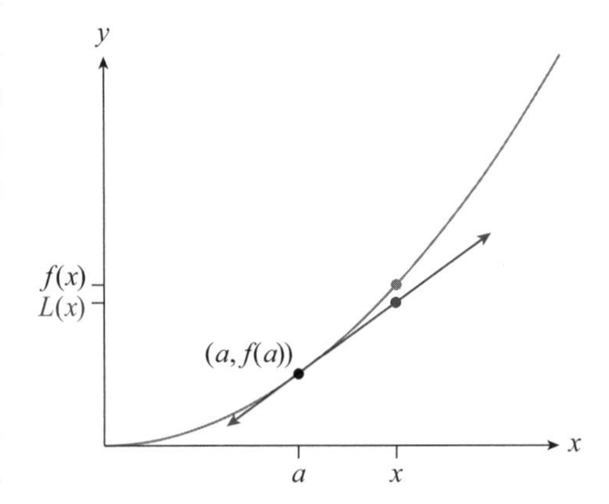

Figure 4.5: When x is near a, the values $f(x)$ are well approximated by the values of the tangent line at $x=a$, $L(x)$.

Approximating Values of Nonlinear Functions

The linear approximation (4.11) is especially useful for approximating values of nonlinear functions. As Figure 4.5 illustrates, (4.11) approximates the actual value of f at x, $f(x)$, with the value of its tangent line at $x = a$, $L(x)$. This quantity is easier to calculate with (being a linear function). And the closer x is to a, the more accurate the approximation $f(x) \approx L(x)$ will be. Here's an example.

EXAMPLE 4.7 Let $f(x) = \sqrt{x}$.

(a) Calculate the linear approximation at $a = 1$.

(b) Use your linear approximation from (a) to estimate $\sqrt{1.05}$. Compare your estimate to the actual value for $\sqrt{1.05}$.

Solution

(a) Using (4.11) and (4.13): $\sqrt{x} \approx \frac{1}{2}(x+1)$ for x near 1.

(b) Substituting $x = 1.05$ into the approximation yields $\sqrt{1.05} \approx \frac{1}{2}(1.05+1) = 1.025$. Since $\sqrt{1.05} = 1.0247\ldots$, our estimate of 1.025 is accurate to two decimal places. Not bad! ■

Related Exercises 5–8.

The Linearization Interpretation of $f'(a)$

Thus far we've considered x to be near a, or equivalently, $\Delta x \approx 0$. However, let's be crude and consider $\Delta x = 1$ (admittedly not a small change in x) in the approximation (4.10). What results is: $\Delta y \approx f'(a)$ for $\Delta x = 1$. This yields the following new interpretation of $f'(a)$.

Box 4.2: The Linearization Interpretation of $f'(a)$

A one-unit increase in the x-value a increases the y-value $f(x)$ by approximately $f'(a)$ (if $f'(a) > 0$), or decreases the y-value by approximately $f'(a)$ (if $f'(a) < 0$).

APPLIED EXAMPLE 4.8 Airlines change the cost of their flights regularly to maximize revenue. Suppose one airline's research team has found that the revenue R (in \$) associated with their Boston to New York flight is modeled by the function

$$R(x) = x(x+90) = x^2 + 90x,$$

where x is the number of tickets sold ($0 \leq x \leq 100$).

(a) Calculate $R'(x)$ and include the units.

(b) If the airline has already sold 50 tickets, calculate $R'(50)$ and interpret your result using the linearization interpretation of the derivative.

(c) Compare your answer to part (b) with the actual revenue increase: $R(51) - R(50)$.

Solution

(a) Using the Power Rule: $R'(x) = 2x + 90$. Since $R(x)$ has units of dollars and x units of number of tickets, $R'(x)$ has units of \$/ticket.

(b) $R'(50) = \$190$ per ticket. Following the linearization interpretation of the derivative, we can say that when the airline has sold 50 tickets, selling one more ticket increases revenue by approximately \$190.

(c) $R(51) = \$7,191$ and $R(50) = \$7,000$, so $R(51) - R(50) = \$191$. This is just \$1 more than the estimate obtained via the interpretation in part (b). ■

Related Exercises 33.

Transcendental Tales

Linearization is especially useful for approximating values of transcendental functions.

EXAMPLE 4.9 Let $f(x) = e^x$.

(a) Calculate the linear approximation of f at $a = 0$. Plot your result.

(b) Use your approximation to estimate $e^{0.1}$. Compare your result to the actual value of $e^{0.1}$.

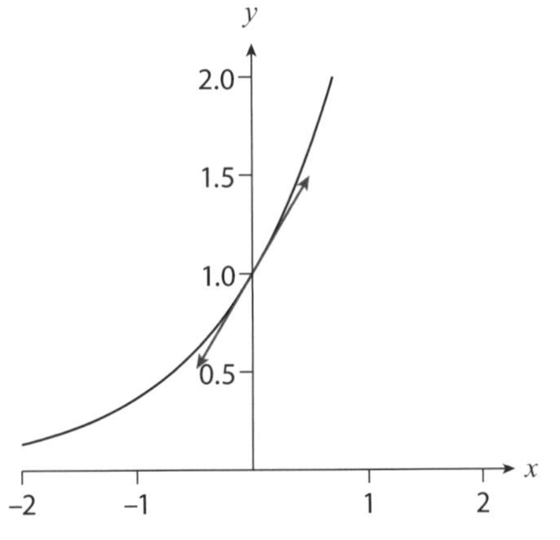

Figure 4.6: The linear approximation of $f(x) = e^x$ by $L(x) = 1 + x$ near $x = 0$.

Solution

(a) Since $f(0) = 1$, $f'(x) = e^x$ (Theorem 3.9), and $f'(0) = 1$, (4.11) yields $e^x \approx 1 + x$ for x near 0. Figure 4.6 plots both functions.

(b) The actual value is: $e^{0.1} = 1.105\ldots$; our linear approximation yields $e^{0.1} \approx 1.1$ (two decimal place accuracy). ■

EXAMPLE 4.10 Show that

$$\sin x \approx x \quad \text{and} \quad \cos x \approx 1, \quad \text{for } x \text{ near } 0. \tag{4.14}$$

Solution Let $f(x) = \sin x$ and $g(x) = \cos x$. Applying (4.11) with $a = 0$ yields

$$\sin x \approx f(0) + f'(0)x,$$
$$\cos x \approx g(0) + g'(0)x.$$

Since $f(0) = \sin 0 = 0$ and $g(0) = \cos 0 = 1$, and since (3.15) implies that $f'(0) = \cos 0 = 1$ and $g'(0) = -\sin 0 = 0$, we get that

$$\sin x \approx 0 + 1 \cdot x = x,$$
$$\cos x \approx 1 + 0 \cdot x = 1.$$

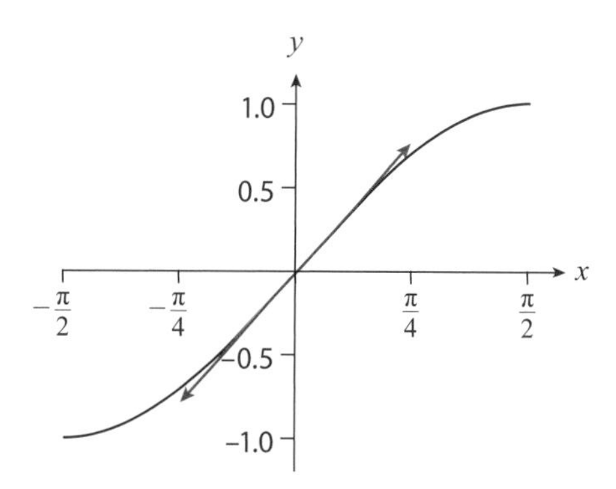

Figure 4.7: The linear approximation of $f(x) = \sin x$ by $L(x) = x$ near $x = 0$.

Figure 4.7 illustrates the $\sin x \approx x$ approximation. ■

Related Exercises 49(a) and 62–63.

4.3 The Increasing/Decreasing Test

Linearization marks our first foray into optimization theory. With it, we can detect whether the graph of f is increasing or decreasing near a *particular* x-value. The following theorem extends these results to statements about *intervals* of x-values.[2]

Theorem 4.1 Let f be differentiable on (a, b). Then,

(a) If $f'(x) > 0$ for all x in (a, b), then f is increasing on that interval.

(b) If $f'(x) < 0$ for all x in (a, b), then f is decreasing on that interval.

As an example, consider $f(x) = x^2$. Since $f'(x) = 2x$, the theorem tells us that the graph of f is decreasing for $x < 0$ (since $f'(x) = 2x < 0$ for $x < 0$) and increasing for $x > 0$ (see Figure 4.8(a)). You can see how this helps us build our theory of optimization: If a function's graph switches from decreasing to increasing as we cross a particular x-value (like $x = 0$ in the figure) then that could be the location of the minimum value of the function. Note my usage of "could"—the graph could switch direction again, later on, possibly yielding a lower y-value. Figure 4.8(b) illustrates the more general twists and turns a function's graph could take. Notice how the *sign* of f' helps us detect—via Theorem 4.1—the increasing and decreasing portions of the graph. Notice also that the graph switches from increasing to decreasing (or vice versa) at x-values for which $f'(x) = 0$ (e.g., $x = c$) or $f'(x)$ DNE (e.g., $x = b$). These

[2]The proof uses the **Mean Value Theorem**; see Section A4.1 in the online appendix to this chapter.

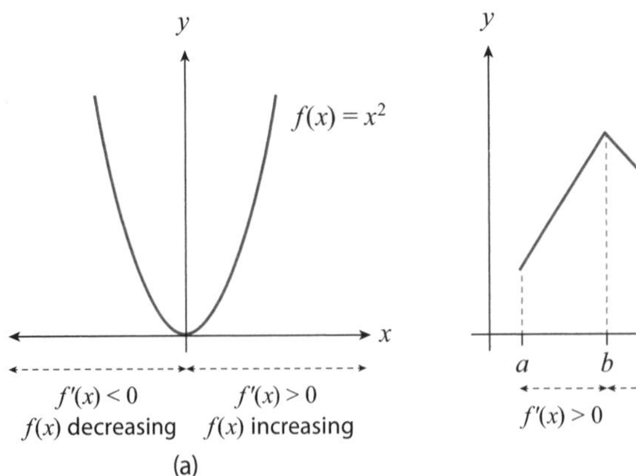

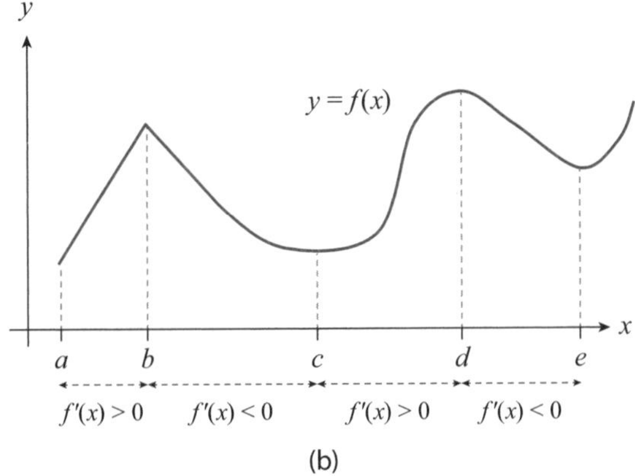

Figure 4.8

x-values seem *critical* to determining the maximum and minimum values of the graph of f. That explains the following definition.

Definition 4.2 Critical Numbers, Values, and Points. Let f be a function and c be in the domain of f, where c is not an endpoint of the domain of f. We then say

(a) c is a **critical number of** f if $f'(c)=0$ or $f'(c)$ does not exist.

(b) $f(c)$ is a **critical value of** f if c is a critical number of f.

(c) $(c, f(c))$ is a **critical point of** f if c is a critical number of f.

For example, $x=0$ is a critical number of $f(x)=x^2$ (since $f'(0)=0$), and b, c, and d are all critical numbers of the function graphed in Figure 4.8(b). With this new terminology and the insights we've now acquired we can state the following procedure for determining where a function f is increasing or decreasing.

Box 4.3: The Increasing/Decreasing Test

To determine the interval(s) on which a function f is increasing or decreasing:

1. Find the critical numbers and plot them on a number line.
2. Evaluate $f'(x)$ for x-values in the intervals between those critical numbers (and in the domain of f).
3. Use Theorem 4.1 to determine the intervals of increase and decrease of f.

I'll spend the rest of this section helping you get acquainted with this procedure. We'll return to how it's another stepping stone to optimization theory in the next section.

EXAMPLE 4.11 Find the intervals of increase and decrease for $f(x) = x^3 - 3x$.

Solution Let's follow the procedure in Box 4.3.

1. We start with $f'(x) = 3x^2 - 3$. There are no x-values where $f'(x)$ DNE. Setting $f'(x) = 0$ yields $3x^2 - 3 = 0$, whose solutions are $x = \pm 1$. Thus, $x = -1$ and $x = 1$ are the only critical numbers.

2. We now plot the critical numbers on a number line:

   Next, we select any number in the three intervals this number line divides the real line into, and substitute those values into $f'(x)$. Choosing $x = -2$, $x = 0$, and $x = 2$ yields

$$f'(-2) = 9 > 0, \qquad f'(0) = -3 < 0, \qquad f'(2) = 9 > 0.$$

   Let me update our number line (I'll call the diagram below a "sign chart" for future reference):

$f'(x)$: + + +   − − −   + + +
−1   1

3. Theorem 4.1 then implies $f(x)$ is (a) increasing on $(-\infty, -1)$ and $(1, \infty)$, and (b) decreasing on $(-1, 1)$. ■

The function in this previous example is the same one from Example 3.11. Our results now help us better understand the associated Figure 3.9—whenever the graph of $f'(x)$ (the bottom graphs) is below the x-axis the graph of f (the top graphs) is decreasing, with a similar conclusion holding when "below" and "decreasing" are replaced with "above" and "increasing."

EXAMPLE 4.12 Find the intervals of increase and decrease for $f(x) = \sqrt[3]{x^2} - x$.

Solution Employing again the procedure in Box 4.3:

1. First we rewrite f as $f(x) = x^{2/3} - x$. Then,

$$f'(x) = \frac{2}{3}x^{-1/3} - 1 = \frac{2 - 3x^{1/3}}{3x^{1/3}}.$$

We see that $f'(0)$ DNE, so $x = 0$ is one critical number. Setting $f'(x) = 0$ yields $2 - 3x^{1/3} = 0$, whose solution is $x = 8/27$. Thus, $x = 0$ and $x = 8/27$ are the only critical numbers.

2. We now plot the critical numbers on a number line and select x-values in the resulting intervals we'll substitute into $f'(x)$. Selecting $x = -1$, $x = 1/27$, and $x = 1$ yields

$$f'(-1) = -\frac{5}{3} < 0, \qquad f'\left(\frac{1}{27}\right) = 1 > 0, \qquad f'(1) = -\frac{1}{3} < 0,$$

which results in the following sign chart:

$$f'(x): \quad - - - \quad \underset{0}{|} \quad + + + \quad \underset{\frac{8}{27}}{|} \quad - - -$$

3. Theorem 4.1 then implies $f(x)$ is decreasing on $(-\infty, 0)$ and $(8/27, \infty)$, and increasing on $(0, 8/27)$. ■

Related Exercises 9–12 (only (a)–(c)), 31, and 38(a).

Figure 4.9 illustrates the results of the previous example.

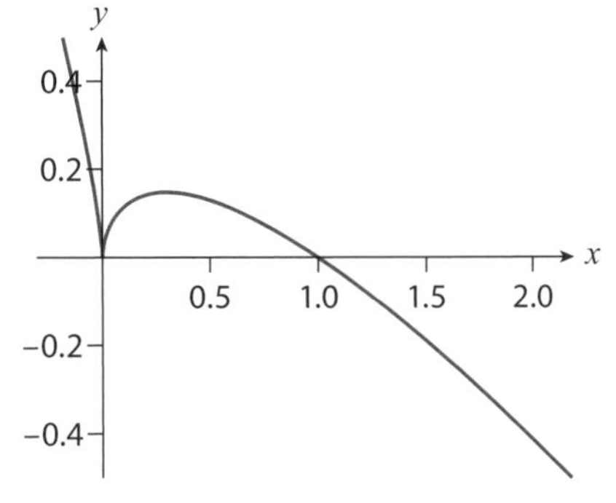

Figure 4.9: $f(x) = x^{2/3} - x$.

Transcendental Tales

One can show that an exponential function b^x is increasing for all x if $b > 1$, and decreasing for all x if $0 < b < 1$. (Exercise 43(a) guides you through the proof; just skip the "no local extrema" part of the question, for now.) Similar statements are true for logarithmic functions $\log_b x$ (Exercise 44(a) guides you through the proof; again, skip the "no local extrema" part). Let's now explore more complicated examples.

EXAMPLE 4.13 Find the intervals of increase and decrease for $f(x) = e^{2x} + e^{-x}$.

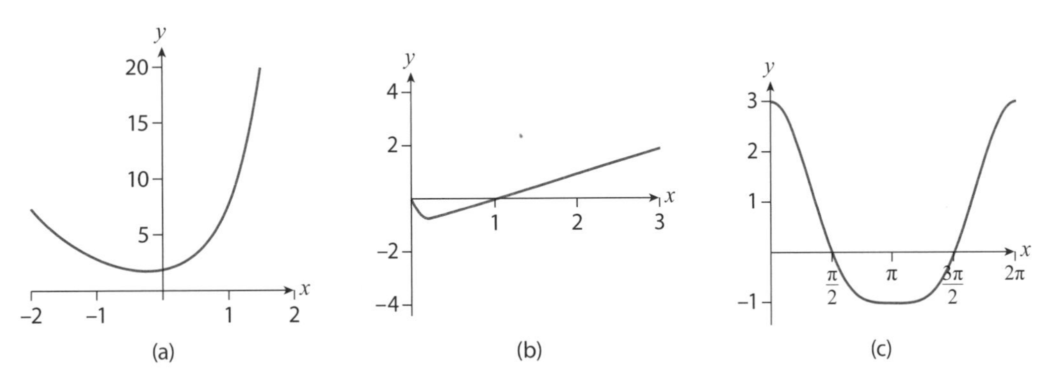

Figure 4.10: Portions of the graphs of (a) $f(x)=e^{2x}+e^{-x}$, (b) $g(x)=\sqrt{x}\ln x$, and (c) $f(x)=2\cos x+\cos^2 x$.

Solution Note that

$$f'(x)=2e^{2x}-e^{-x}=\frac{2e^{3x}-1}{e^x}.$$

There are no x-values for which $f'(x)$ DNE (the denominator is never zero). Setting $f'(x)=0$ yields

$$e^{3x}=\frac{1}{2} \quad\Longrightarrow\quad x=-\frac{\ln 2}{3}\approx -0.23.$$

This is the only critical number. Choosing the test points $x=-1$ and $x=0$ yields the following sign chart.

$$f'(x):\quad - - - - - \quad\big|\quad + + + + + \qquad \text{(at } -\tfrac{\ln 2}{3}\text{)}$$

We conclude from Theorem 4.1 that f is decreasing on $\left(-\infty,-\frac{\ln 2}{3}\right)$ and increasing on $\left(-\frac{\ln 2}{3},\infty\right)$. Figure 4.10(a) shows a portion of the graph of f. ■

EXAMPLE 4.14 Find the intervals of increase and decrease for $g(x)=\sqrt{x}\ln x$.

Solution Let's first note that the domain of this function is $(0,\infty)$. Next, we calculate the derivative using the Product Rule:

$$g'(x)=\left(\frac{1}{2}x^{-1/2}\right)\ln x+\frac{\sqrt{x}}{x}=\frac{2+\ln x}{2\sqrt{x}}.$$

Since $x>0$ the denominator is again always positive. Setting $g'(x)=0$ yields $x=e^{-2}$ (obtained by solving $2+\ln x=0$). This is our only critical number. Choosing the test points $x=e^{-3}$ and $x=1$ yields the sign chart below.

$$g'(x): \quad - - - - - \quad + + + + +$$
$$e^{-3}$$

We conclude from Theorem 4.1 that f is decreasing on $\left(0, e^{-3}\right)$ and increasing on $\left(e^{-3}, \infty\right)$. Figure 4.10(b) shows a portion of the graph of g. ■

Related Exercises 39–42 (only (a)–(b)), 45(a), and 48(a)–(c).

Let's now apply what we've learned to trigonometric functions. Returning now to Figures B.20 (a) and (b) it's clear that the graphs of $\sin x$ and $\cos x$ increase and decrease over many intervals. But due to the periodicity of sine and cosine, these features end up being copy cats of features of the functions for x-values in the interval $[0, 2\pi)$. (See Exercise 57 for the details.) So let's work through a slightly more complicated example.

EXAMPLE 4.15 Let $f(x) = 2\cos x + \cos^2 x$, where $0 \leq x \leq 2\pi$. Find the (sub)intervals of increase and decrease.

Solution Note that

$$f'(x) = -2\sin x - 2(\cos x)(\sin x) = -2(\sin x)(1 + \cos x).$$

The only critical numbers occur when $f'(x) = 0$, which yields

$$\sin x = 0 \qquad \text{or} \qquad \cos x = -1.$$

The former equation yields $x = 0$ and $x = 2\pi$ (recall that our interval of interest is $[0, 2\pi]$); the latter equation yields $x = \pi$. Thus, our critical numbers are: 0, π, 2π. Choosing the test points $x = \pi/2$ and $x = 3\pi/2$ yields the following sign chart.

$$f'(x): \quad - - - - - \quad + + + + +$$
$$0 \qquad \pi \qquad 2\pi$$

We conclude from Theorem 4.1 that f is decreasing on $(0, \pi)$ and increasing on $(\pi, 2\pi)$. Figure 4.10(c) shows the graph of f. ■

Related Exercises 53–56 (only part (a)–(b)), 58(a), 59(a)–(b), 60(a)–(b), and 61(a)–(b).

4.4 Optimization Theory: Local Extrema

Our work in the previous section allows us to identify the "hills" and "valleys" of a function's graph using calculus. Let me introduce the math jargon for that.

Definition 4.3 Let f be a function and c in the domain of f. We then say:

(a) f has a **local maximum at** c if $f(c) \geq f(x)$ for x near c.

(b) f has a **local minimum at** c if $f(c) \leq f(x)$ for x near c.

In either case, we refer to $f(c)$ as "the" local extremum.

For example, returning to Figure 4.8(b), that function has local maxima at $x=b$ and $x=d$ (the graph looks like a "hill" near the y-values $f(b)$ and $f(d)$). Also, that function has local minima at $x=a$, $x=c$, and $x=e$ (the graph looks like a "valley" near the y-values $f(a), f(c)$, and $f(e)$). Notice that these x-values consist of the critical numbers plus the endpoints of the interval. That's no surprise, because *we can use the Increasing/Decreasing Test to help us locate local extrema.* The following theorem provides the details.

Theorem 4.2 The First Derivative Test. Let f be a function and c a critical number of f.

(a) If f' changes sign from positive to negative as we cross $x=c$, then f has a local maximum at c.

(b) If f' changes sign from negative to positive as we cross $x=c$, then f has a local minimum at c.

(c) If f' does not change sign as we cross $x=c$, then there is no local extremum at c.

Figure 4.11 illustrates the theorem. Because a local maximum is a good candidate for *the* maximum of a function f (and a local minimum a good candidate for *the* minimum), we'll spend the rest of this section hunting for local extrema. (The next section then returns to the quest for the "absolute" extrema.)

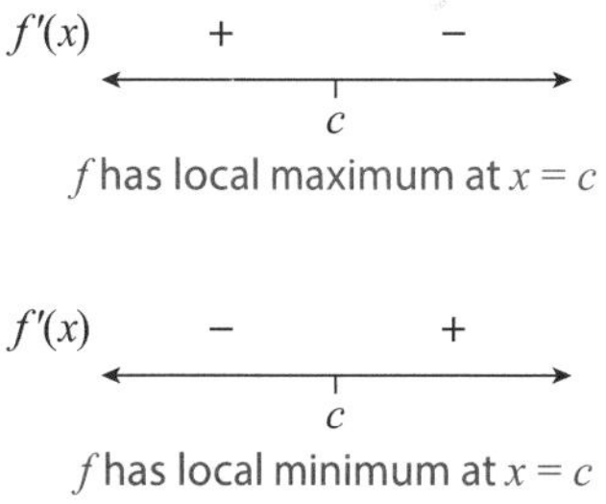

$f'(x)$ + −

c

f has local maximum at $x=c$

$f'(x)$ − +

c

f has local minimum at $x=c$

Figure 4.11

EXAMPLE 4.16 Find the local extrema of $f(x) = x^4 - 8x^3$.

Solution Since $f'(x) = 4x^3 - 24x^2 = 4x^2(x-6)$, the only critical numbers are $x=0$ and $x=6$. Choosing the test points $x=-1$, $x=1$, and $x=7$, we get the following sign chart:

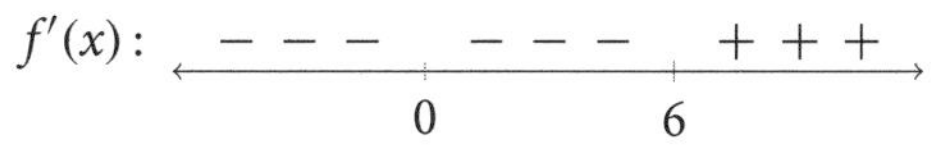

$f'(x)$: − − −   − − −   + + +

0   6

Since f' changes sign from negative to positive as we cross $x = 6$, Theorem 4.2 tells us that $f(6)$ is a local minimum. Since f' doesn't change sign as we cross $x = 0$, however, Theorem 4.2 tells us that there is no local extremum at $x = 0$. ■

Related Exercises 9–12 (only (d)), and 37(a).

Transcendental Tales

EXAMPLE 4.17 Find the local extrema for the functions f and g from Examples 4.13 and 4.14.

Solution

(a) Applying Theorem 4.2 to the $f'(x)$ sign chart we already calculated in the example tells us that f has a local minimum at $x = -\ln 2/3$. Since no endpoints are given, we conclude that this is the only local extremum of f. (Figure 4.10(a) confirms this.)

(b) By similar reasoning, Theorem 4.2 applied to the $g'(x)$ sign chart we already calculated in the example tells us that g has a local minimum at $x = e^{-3}$. Since the interval of interest $(0, \infty)$ does not include the endpoints, we conclude that $x = e^{-3}$ is the only local extremum of f on that interval. (Figure 4.10(b) confirms these results.) ■

Related Exercises 39–42 (only part (c)), 43(a), and 44(a).

EXAMPLE 4.18 Find the local extrema for the function f and interval given in Example 4.15.

Solution Applying Theorem 4.2 to the $f'(x)$ sign chart we already calculated in the example tells us that f has a local minimum at $x = \pi$. Focusing now on $x = 0$ and $x = 2\pi$, since they *are* included in the interval of interest let's analyze them next. Our sign chart tells us that the function's graph decreases on the interval $(0, \pi)$. Thus, $x = 0$ must be a local maximum of f. Similarly, since the sign chart tells us that the function's graph increases on $(\pi, 2\pi)$, we conclude that $x = 2\pi$ must be a local maximum of f. (Figure 4.10(c) confirms these results.) ■

Related Exercises 53–56 (only part (c)).

Tips, Tricks, and Takeaways

The First Derivative Test (Theorem 4.2) helps us classify critical numbers as local extrema or not. You may wonder, however, if there are other x-values that would be good candidates for local extrema. The following theorem addresses that question.

Theorem 4.3 Fermat's Theorem. Suppose f has a local maximum or minimum at c, where c is not an endpoint of the domain of f. Then c is a critical number of f.

Here are some important takeaways from this theorem.

- The Theorem *does not say that if c is a critical number then f has a local extremum at c.* (Note that $x = 0$ was a critical number in Example 4.16 yet it was not a local extremum of the function.)
- The "contrapositive" of Fermat's Theorem—the statement that if c is *not* a critical number of f then f *does not* have a local extremum at c—is true. Takeaway: *Searching for local extrema at non-critical numbers is futile.*
- Note the assumption that c is not an endpoint of the domain. Takeaway: We need to investigate the endpoints separately.

Fermat's Theorem basically tells us that the "hills" and "valleys" of a function's graph must occur at either the critical numbers or the endpoints of the interval given. This sets us up nicely for the last stop on this tour of optimization theory: finding the absolute highest "hill" and absolute lowest "valley."

4.5 Optimization Theory: Absolute Extrema

We're now ready to investigate the largest and smallest y-values of a function (the extrema). In analogy with Definition 4.3, here's the formal definition of the concept.

Definition 4.4 Let f be a function defined on an interval I, and let c be a number in I. We then say:

(a) f has an **absolute maximum at** c if $f(c) \geq f(x)$ for all x in I. We then call $f(c)$ the absolute maximum on I.

(b) f has an **absolute minimum at** c if $f(c) \leq f(x)$ for all x in I. We then call $f(c)$ the absolute minimum on I.

When the interval I is all real numbers, we simply say $f(c)$ is the absolute minimum (or absolute maximum).

Note the "for all x in I" here, versus the "for x near c" language in Definition 4.3. This means we're looking at *all* the x-values in the interval of interest I and picking out the one(s) with the absolute largest (and smallest) $f(x)$-values. For example, referring back to Figure 4.8 again, that function's absolute minimum in the interval $[a, e]$ it's graphed on is $f(c)$, and its absolute maximum on that interval is $f(d)$.

Let's now use the theory we've developed for *local* extrema to help us find *absolute* extrema. When f has only one critical number inside the interval of interest, here's the first result: local extremum = absolute extremum.

Theorem 4.4 Suppose f is continuous on an interval I and has only one critical number c inside I. If c is a local maximum, then $f(c)$ is the absolute maximum of f on I. Similarly, if c is a local minimum, then $f(c)$ is the absolute minimum of f on I.

Great. But what if f has *multiple* critical numbers inside I? Well, we know from the previous section that the "hills" and "valleys" of a function's graph occur at either the critical numbers or the endpoints of the given interval. So, we need only identify the highest "hill" and lowest "valley" to determine the absolute extrema. The procedure in Box 4.4 does exactly that.

Box 4.4: How to Find the Absolute Extrema of a Continuous Function Defined on a Closed Interval $[a, b]$

1. Find the critical numbers in the interval (a, b).
2. Calculate the associated critical values, and also calculate $f(a)$ and $f(b)$.
3. The absolute maximum is the largest of the numbers in Step 2; the absolute minimum is the smallest of the numbers in Step 2.

EXAMPLE 4.19 Find the absolute extrema of the function $f(x) = (x^2 - 1)^3$ on the interval $[-\sqrt{2}, \sqrt{2}]$.

Solution First, note that $f(x)$ is a polynomial; we showed in Chapter 2 that such functions are continuous. And since $[-\sqrt{2}, \sqrt{2}]$ is a closed interval, the procedure from Box 4.4 applies.

1. To find the critical numbers we first differentiate f using the Chain Rule:

$$\begin{aligned} f'(x) &= 3(x^2 - 1)^2(2x) = 6x(x^2 - 1)^2 = 6x[(x-1)(x+1)]^2 \\ &= 6x(x-1)^2(x+1)^2. \end{aligned}$$

   It follows that the only critical numbers are $x = -1$, $x = 0$, and $x = 1$.

2. The associated critical values, as well as $f(-\sqrt{2})$ and $f(\sqrt{2})$ (the y-values at the endpoints of the interval $[-\sqrt{2}, \sqrt{2}]$) are

$$f(-1) = 0, \quad f(0) = -1, \quad f(1) = 0, \quad f(-\sqrt{2}) = 1, \quad f(\sqrt{2}) = 1.$$

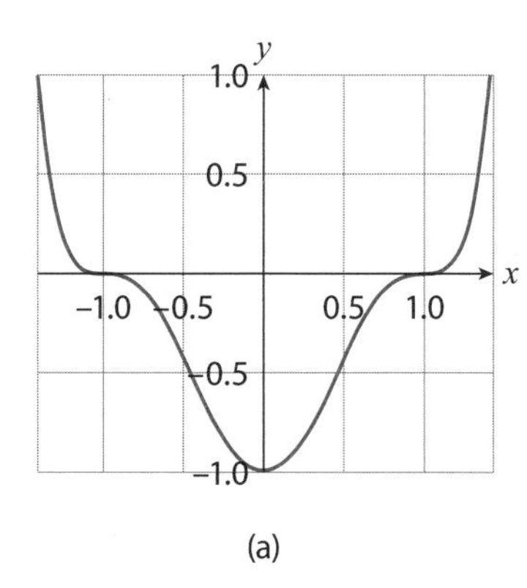

(a)

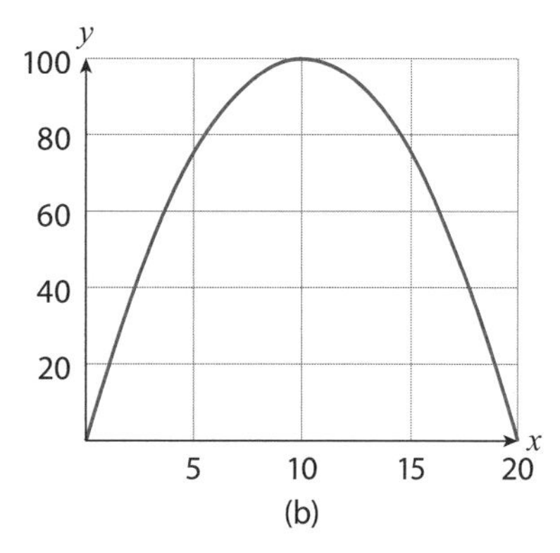

(b)

Figure 4.12: The graphs of (a) $f(x) = (x^2 - 1)^3$ on $[-\sqrt{2}, \sqrt{2}]$, and (b) $A(x) = 20x - x^2$ for $0 < x < 20$.

3. Finally, comparing these we see that the largest values are $f(-\sqrt{2}) = f(\sqrt{2})$; these are the absolute maxima of f on $[-\sqrt{2}, \sqrt{2}]$. (As this example shows, multiple absolute maxima are possible.) Since $f(0)$ is the smallest y-value it is the absolute minimum of f on $[-\sqrt{2}, \sqrt{2}]$. Figure 4.12(a) shows the graph of f. ■

Related Exercises 13–16.

APPLIED EXAMPLE 4.20 You may have noticed that the bedrooms in the house you live in are nearly square in dimensions. Let's work out why. To begin, suppose you are adding a new rectangular bedroom to your house and only need to add two walls. You can afford to build up to 20 feet worth of walls.

(a) Find the living area A of the bedroom as a function of its width x (in feet).

(b) On what intervals is $A(x)$ increasing, decreasing?

(c) Find the critical numbers of $A(x)$ and classify them as local maxima, local minima, or neither.

(d) What bedroom dimensions maximize the living area?

Solution

(a) Let y be the length (in feet) of the new rectangular bedroom. We're given that $x + y = 20$. Note that this implies that $0 < x < 20$ and $0 < y < 20$. (Since x and y are distances they cannot be negative. Also, if $x = 0$ or $x = 20$ there is no bedroom, only a line of length 20 feet in which no one could live.) Since the area of the bedroom is $A = xy$, then substituting in $y = 20 - x$ yields the function

$$A(x) = x(20 - x) = 20x - x^2, \quad 0 < x < 20.$$

(b) We have $A'(x) = 20 - 2x$. Moreover, $A'(x) = 0$ yields only $x = 10$. Therefore, choosing, say, $x = 1$ and $x = 15$ yields the following sign chart:

$$A'(x): \quad + + + + + \quad \underset{10}{|} \quad - - - - -$$

We conclude from Theorem 4.1 that $A(x)$ is increasing on the interval $(0, 10)$ and decreasing on the interval $(10, 20)$.

(c) There are no x-values for which $A'(x)$ does not exist. And since $A'(x) = 0$ only when $x = 10$, the only critical number is $x = 10$. Finally, since A' changes sign from positive to negative as we cross $x = 10$, Theorem 4.2 tells us that $x = 10$ is a local maximum for $A(x)$.

(d) Since $A(x)$ is increasing for all $x < 10$ and decreasing for all $x > 10$, at $x = 10$ we don't just have a local maximum, we have the largest y-value of $A(x)$. (Figure 4.12(b) on the previous page shows the graph of $A(x)$.) That critical *value* is $A(10) = 10(20 - 10) = (10)^2 = 100$ ft^2. Thus, the dimensions of the bedroom with the largest living area are $10' \times 10'$. ■

Transcendental Tales

EXAMPLE 4.21 Find the absolute extrema of $f(x) = xe^{-x}$ on the interval $[0, 3]$.

Solution f is a continuous function and the given interval is closed, so let's follow the procedure in Box 4.4. We first find $f'(x)$:

$$f'(x) = e^{-x} - xe^{-x} = \frac{1 - x}{e^x}.$$

We see that the only critical numbers of f occur when $f'(x) = 0$, which yields $x = 1$. The associated critical value is $f(1) = e^{-1}$. The y-values at the endpoints of the interval are $f(0) = 0$ and $f(3) = 3e^{-3}$. We conclude that f has an absolute maximum at $x = 1$ and an absolute minimum at $x = 0$; Figure 4.13(a) shows the relevant portion of the graph of f. ■

EXAMPLE 4.22 Find the absolute extrema of $g(x) = x - 2\ln(x^2 + 1)$ on the interval $[0, 10]$.

Solution Since g is continuous on $[0, 10]$, let's again follow the procedure in Box 4.4. We first calculate that

$$g'(x) = 1 - 2\left(\frac{2x}{x^2 + 1}\right) = 1 - \frac{4x}{x^2 + 1}.$$

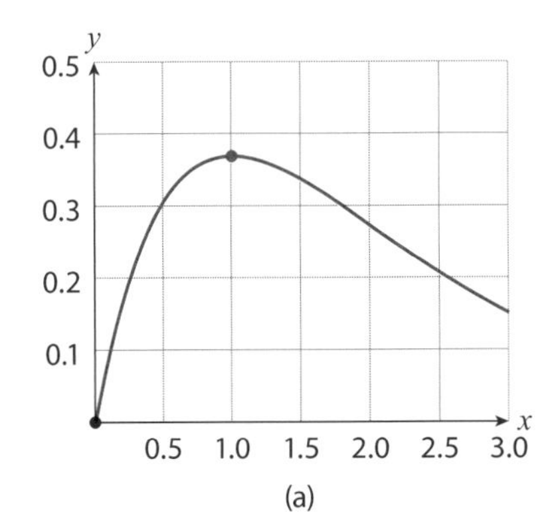

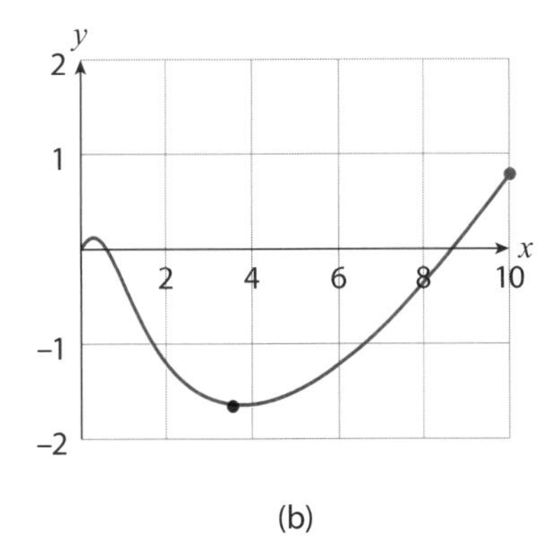

Figure 4.13: Portions of the graphs of: (a) $f(x) = xe^{-x}$ for $0 \leq x \leq 3$, and (b) $g(x) = x - 2\ln(x^2 + 1)$ for $0 \leq x \leq 10$. The black dots are the locations of the absolute minima, the blue dots the absolute maxima.

The only critical numbers of g occur when $g'(x) = 0$, which yields

$$\frac{4x}{x^2+1} = 1 \quad \Longrightarrow \quad x^2 - 4x + 1 = 0.$$

The quadratic formula then yields the two solutions $x = 2 - \sqrt{3}$ and $x = 2 + \sqrt{3}$. Since

$$g(0) = 0, \quad g(2-\sqrt{3}) \approx 0.1, \quad g(2+\sqrt{3}) \approx -1.7, \quad g(10) \approx 0.8,$$

we conclude that g has an absolute minimum at $x = 2 + \sqrt{3} \approx 3.7$ and an absolute maximum at $x = 10$; Figure 4.13(b) shows the relevant portion of the graph of g. ■

Related Exercises 39–42 (only part (d)), 45(b), 46(a), and 53–56 (only part (d)).

Tips, Tricks, and Takeaways

We've introduced two optimization results in this section: the procedure in Box 4.4 and Theorem 4.4. The procedure in Box 4.4 is based on the intuition that the graphs of continuous functions can be drawn without lifting your pen (discussed in Chapter 2), and therefore if the endpoints of the interval are included, such graphs should have absolute extrema. The following theorem confirms this gut feeling.

Theorem 4.5 The Extreme Value Theorem. Suppose f is a function continuous on $[a, b]$. Then f has both an absolute maximum and an absolute minimum on $[a, b]$.

Great! But what if f is continuous on a nonclosed interval? That's where Theorem 4.4 may help. If there's only one critical number (inside I) and we're still dealing with a continuous function, then local = absolute when it comes to classifying extrema. This is what we used in Applied Example 4.20.[3] That example was also special because, like the Related Rates problems earlier in this chapter, we had to do some mathematical modeling to determine the appropriate function and interval to maximize. This is a typical step in a real-world optimization problem. The next section focuses on such applications of optimization theory.

4.6 Applications of Optimization

This section is intended to showcase the real-world applications of the optimization theory developed in the previous section. As such, let me do that by working through several examples with you.

APPLIED EXAMPLE 4.23 When a person catches a cold his or her body's immune system reacts to the virus and eventually produces a thick mucus that accumulates at the back of the throat. Coughing helps expel this mucus by pushing air through the trachea. If we imagine the tracheal tube as a cylinder (which is roughly true), the velocity v of the air rushing through the trachea during a cough is well approximated by

$$v(r) = k(r_0 - r)r^2, \qquad \frac{r_0}{2} \leq r \leq r_0,$$

where $k > 0$ is a constant and $r_0 > 0$ the original radius of the tracheal tube. Supposing $r_0 = 1$ cm (roughly the radius of a typical adult's trachea), find the absolute maximum of $v(r)$. Then, given that experiments show that $r \approx \frac{2}{3}r_0$ during a cough, interpret this r-value in the context of your answer.

Solution Since $v(r) = k(1-r)r^2$ is a continuous function (it's a polynomial) defined on a closed interval, let's use our procedure from Box 4.4. We first find the critical numbers inside the interval. By the Product Rule:

$$v'(r) = k(-1)r^2 + k(1-r)(2r) = kr(2-3r).$$

Thus, $v'(r) = 0$ yields $r = 0$ or $r = \frac{2}{3}$. We reject the $r = 0$ critical number since it is not inside the given interval. Next, we calculate the critical values as well as $v(1/2)$ and $v(1)$:

$$v\left(\frac{1}{2}\right) = \frac{k}{8}, \qquad v\left(\frac{2}{3}\right) = \frac{4k}{27}, \qquad v(1) = 0.$$

Finally, we look for the largest of these y-values (since we're looking for the absolute maximum). Since $\frac{4}{27} > \frac{1}{8}$, we conclude that on the interval of interest the absolute maximum of $v(r)$ occurs at $r = \frac{2}{3} = \frac{2}{3}r_0$. This matches what experiments have found,

[3]There exist yet other procedures that handle the cases skipped by the two methods we've developed. We'll discuss some of those in the next section.

meaning that our trachea seems to contract to the radius that maximizes the velocity of air rushing through it during a cough. ■

Optimization is also useful for designing products and minimizing commuting times. Here are examples of that.

APPLIED EXAMPLE 4.24 Consider the cylindrical can of soda pictured in Figure 4.14. If the can is made of aluminum and is to hold 21.65 cubic inches of volume (about 12 fluid ounces), what dimensions minimize the amount of aluminum used to build the can?

Solution This time around we need to come up with the function to be minimized (and its interval). One key piece of insight: the amount of aluminum used depends on the surface area of the can. Thus, to minimize the aluminum used, we should minimize the surface area of the cylindrical can. Referring to Figure 4.14, the surface area S of that can is the sum of the areas of the top, side, and bottom:

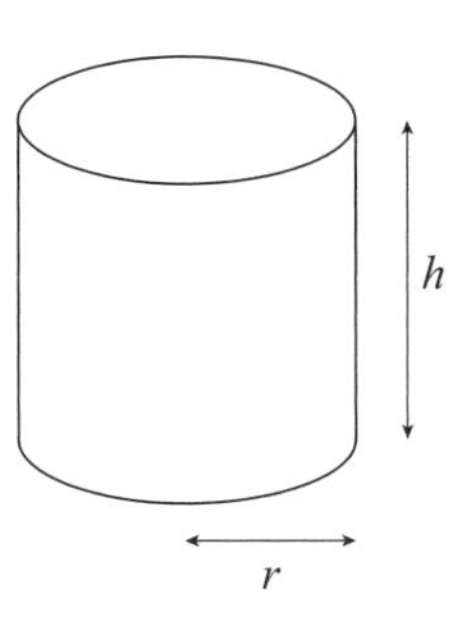

Figure 4.14

$$S = \pi r^2 + 2\pi rh + \pi r^2 = 2\pi r^2 + 2\pi rh, \tag{4.15}$$

where we'll measure r and h in inches. Now, since the volume of the can needs to be 21.65 cubic inches, and the volume of our cylinder is $V = \pi r^2 h$, then,

$$\pi r^2 h = 21.65. \tag{4.16}$$

Solving this equation for h and substituting the result into (4.15) yields

$$S(r) = 2\pi r^2 + 2\pi r \left(\frac{21.65}{\pi r^2}\right) = 2\pi r^2 + \frac{43.3}{r}.$$

We now need the interval for r. Anything less than half an inch would be hard to hold in your hand, so we can safely assume $r \geq 0.5$. Similarly, no one is likely to buy a soda can whose height is less than 1 inch. From equation (4.16), when $h \geq 1$ we have $r \leq 2.63$. Let's be generous and assume $r \leq 3$. We then have our function and interval:

$$S(r) = 2\pi r^2 + \frac{43.3}{r}, \qquad 0.5 \leq r \leq 3.$$

Let's now apply our procedure from Box 4.4. We first find the critical numbers inside the interval:

$$S'(r) = 4\pi r - \frac{43.3}{r^2}, \quad \text{and} \quad S'(r) = 0 \Longrightarrow r^3 = \frac{43.3}{4\pi}.$$

Thus, $r = \sqrt[3]{43.3/(4\pi)} \approx 1.5$ inches is the only critical number. Moving on to the second step in our procedure, since

$$S\left(\sqrt[3]{\frac{43.3}{4\pi}}\right) \approx 43, \qquad S(0.5) \approx 88, \qquad S(3) \approx 71,$$

and the first number is the smallest of the three, we conclude that the absolute minimum of $S(r)$ occurs at the critical number, $r \approx 1.5$ inches. The corresponding height of the can, using (4.16), is:

$$h = \frac{21.65}{\pi r^2} \approx 3.$$

Thus, a 12 oz can about 3 inches in diameter and about 3 inches in height minimizes the aluminum used to construct it. ■

APPLIED EXAMPLE 4.25 Suppose you leave home (point A in Figure 4.15) to drive 5 miles down a straight road to get to work (point C). One day you notice that there's another straight road that runs from your house to point B, 3 miles down the road, and eventually connects to your place of employment (point C), 4 miles down that road. You, having read this book, get an idea: calculate the route that minimizes the amount of fuel your car uses. (This will save you money.) Supposing your car achieves 20 miles/gallon fuel economy on road C (it's a popular route) and 30 miles/gallon on the side roads from AB to C, how far down the side roads AB should you drive to minimize the fuel cost of the trip?

Solution The number of gallons g used up in driving the x miles down road AB and then the y miles to point C afterwards is

$$g = \frac{x}{30} + \frac{y}{20}.$$

Now, notice from the geometry of the situation that

$$y^2 = 4^2 + (3 - x)^2.$$

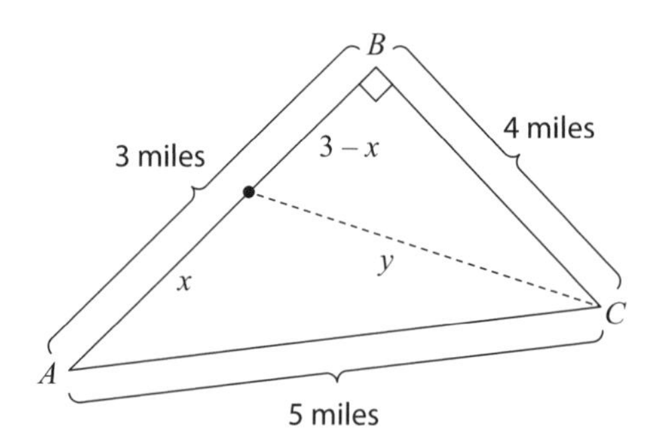

Figure 4.15: An illustration of the possible routes associated with Example 4.25.

(This follows from the Pythagorean Theorem; see (A.2) in Appendix A.) Thus, we can express g as a function of x:

$$g(x) = \frac{x}{30} + \frac{\sqrt{16 + (3 - x)^2}}{20}. \tag{4.17}$$

We now need an interval for x. As Figure 4.15 suggests, $0 \leq x \leq 3$. We now apply our procedure from Box 4.4. Using the Chain Rule:

$$\begin{aligned} g'(x) &= \frac{1}{30} + \frac{1}{20}\left[\frac{1}{2}[16+(3-x)^2]^{-1/2}(2(3-x))(-1)\right] \\ &= \frac{1}{30} + \frac{x-3}{20\sqrt{16+(3-x)^2}}. \end{aligned} \tag{4.18}$$

Setting $g'(x) = 0$ and simplifying yields

$$\frac{x-3}{20\sqrt{16+(3-x)^2}} = -\frac{1}{30} \implies 3-x = \frac{2}{3}\sqrt{16+(3-x)^2}.$$

Squaring both sides and simplifying yields

$$(3-x)^2 = \frac{64}{5} \implies x = 3 \pm \frac{8}{\sqrt{5}}.$$

But both $3+8/\sqrt{5} \approx 6.6$ and $3-8/\sqrt{5} \approx -0.6$ are outside our interval of $[0, 3]$, so we reject those critical numbers. All that's left to do from our procedure then is to calculate $g(0)$ and $g(3)$:

$$g(0) = 0.25, \qquad g(3) = 0.3.$$

Since $g(0)$ is the smaller number, the fuel-minimizing route is to take road AC all the way to work; that will use up 0.25 gallon of fuel. ■

The Calculus of Fairness Applied Example C.3 in Appendix C uses optimization theory to describe how to optimally split a divisible quantity (like pizza) between two parties in a fair manner.

We've now done a few optimization problems. And as we've seen, some are harder than others. The previous two examples, in particular, contain a few common characteristics of tough optimization problems:

- *You have to determine the equation to be optimized.* This is called the **objective function**.
- *There are relationships between the variables in the objective function.* We call these relationships **constraints**.
- *The interval of interest isn't given.* You need to determine it based on the particular physical situation modeled by the problem.

Figuring out the objective function, the constraints (if any), and the interval are the hardest parts of any optimization problem. These are similar to the mathematical modeling steps we had to go through for some of the Related Rates problems from Section 4.1. In analogy with Box 4.1, here's a relatively straightforward procedure to help you tackle optimization problems.

Box 4.5: How to Solve an Optimization Problem with a Closed Interval

1. *If possible, draw a diagram and identify the variables in the problem.* This can help you visualize the problem.
2. *Look for keywords for clues about the objective function.* For instance, "minimize the area" suggests the objective function is an area, and the next step should be to identify the areas present in the problem.
3. *Identify the constraints (if any).* These may come from your diagram, or by reading closely the description of a variable (e.g., "at most" to indicate the largest possible value).
4. *Use the constraint(s), if any, along with your diagram to help figure out the interval.*
5. *If the interval is closed and the objective function continuous, use our procedure from Box 4.4 to find the absolute maximum or minimum.*
6. *Make sure you finish by answering the question.* Some exercises ask for the actual minimum value, others for the dimensions of the minimum solution. Make sure you give the appropriate answer (including the units, if applicable).

I encourage you to employ this procedure while working through the suggested exercises below.

Related Exercises 29–30, 32, and 34–36.

Transcendental Tales

Exercise 19 of Appendix B discusses the *height* of a football as a function of time (neglecting air resistance). Let me show you how to find the maximum *range* of the football (which is of great interest to the average NFL quarterback).

APPLIED EXAMPLE 4.26 Suppose we throw a sufficiently heavy object (i.e., not a feather) up in the air with an initial velocity v_0 (in ft/s) and at an angle θ with respect to the ground (where $0 \leq \theta \leq \frac{\pi}{2}$). Denote by R the horizontal distance (in feet) traveled by the object before it returns to its initial height. (R is called the **range** of the projectile.) Neglecting air resistance,

$$R(\theta) = \frac{v_0^2}{g}\sin(2\theta), \quad 0 \le \theta \le \frac{\pi}{2}.$$

(Here $g \approx 32$ ft/s^2 denotes the acceleration of gravity.)

(a) Find the critical numbers of R.

(b) Use calculus to find the throwing angle θ that produces maximum range; what is the maximum range?

Solution

(a) Since

$$R'(\theta) = \frac{v_0^2}{g}\cos(2\theta)(2) = \frac{2v_0^2}{g}\cos(2\theta)$$

is a continuous function, the only critical numbers occur where $R'(\theta) = 0$. This yields $\cos(2\theta) = 0$. Since cosine is zero only once in the interval $[0, \pi/2]$ (at $\pi/2$), the only critical number is $\theta = \pi/4$.

(b) Following the procedure in Box 4.4 we calculate:

$$R(0) = 0, \qquad R\left(\frac{\pi}{4}\right) = \frac{v_0^2}{g}, \qquad R\left(\frac{\pi}{2}\right) = 0.$$

We conclude that R has an absolute maximum at $\theta = \pi/4$. Thus, a throwing angle of 45° produces maximum range. (Again, we have neglected air resistance.) Moreover, the maximum range is v_0^2/g. Note that this is a quadratic function of the initial throwing velocity v_0. (So, a quarterback who increases his throwing velocity by a factor of x will increase the throw's range by a factor of x^2.) ■

Optimal Holding Time of an Asset Applied Example C.4 in Appendix C uses optimization theory to determine the optimal holding time of an asset of value (the math involved uses exponential functions).

Optimal Branching of Blood Vessels Applied Example C.5 in Appendix C uses optimization theory to determine the angle that minimizes resistance to blood flow at the branching point of a blood vessel.

Related Exercises 47, 49, and 58–60.

Tips, Tricks, and Takeaways

First, let me say it out loud: *Optimization problems are the hardest calculus problems.* After working through the previous examples, you can see why: *They require knowledge of nearly everything we've done thus far.* In addition, many of them include a mathematical modeling step in which you need to determine the objective function and interval yourself. Takeaway: *Don't despair.* Instead, practice, practice, and practice.

Optimization theory is a milestone in our study of calculus. Now that we've discussed it, I'll begin transitioning us into the next chapter. I'll do that by returning to the setting of the last section in the previous chapter: higher-order derivatives. It turns out that the second derivative, in particular, can help us out when it comes to finding local extrema. Moreover, the way in which it helps also leads to a new insight: f'' measures the curvature of a graph.

4.7 What the Second Derivative Tells Us About the Function

Recall that $f'(a)$ is the slope of the line tangent to the graph of f at the point $(a, f(a))$. We've yet to develop an analogous interpretation for $f''(a)$. To do so, let's return to its origin—helping us describe how $f'(x)$ changes. Insight is provided by replacing f with f' everywhere in Theorem 4.1 to get the following theorem.

Theorem 4.6 Let f be twice differentiable on (a, b). Then,

(a) If $f''(x) > 0$ for all x in (a, b), then f' is increasing on that interval.

(b) If $f''(x) < 0$ for all x in (a, b), then f' is decreasing on that interval.

Figure 4.16 illustrates this theorem. In (a), $f''(x) = 2$, which is always positive. The theorem then implies $f'(x)$ is increasing (i.e., the slopes of the tangent

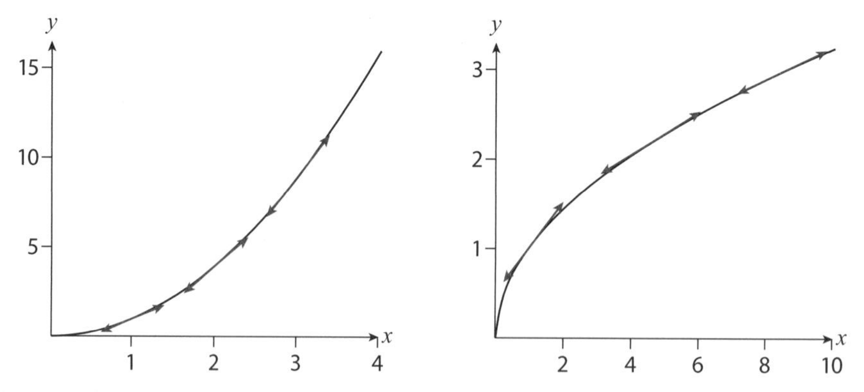

Figure 4.16: (a) $f(x) = x^2$ (concave up) and (b) $g(x) = \sqrt{x}$ (concave down), along with a few of their tangent lines.

lines to f are getting *steeper*) as the x-values increase. In (b), $f''(x) = -(1/4)x^{-3/2}$, which is negative when $x > 0$. The theorem then implies $f'(x)$ is decreasing (i.e., the slopes of its tangent lines to f are getting *less steep*) as the x-values increase.

Now, recalling our linearization insight—that the graph of f near $x = a$ is indistinguishable from the graph of the tangent line at $x = a$—we can replace "f' is increasing on that interval" in Theorem 4.6 with "the graph of f *curve upwards* on that interval" (as in Figure 4.16(a)). This yields the following new terminology and rewording of Theorem 4.6.

Definition 4.5 Concavity. Let f be a function defined on an interval I. We say:

(a) f is **concave up on** I if the graph of f is above the graph of its tangent lines on I.

(b) f is **concave down on** I if the graph of f is below the graph of its tangent lines on I.

Theorem 4.7 The Concavity Test. Let f be a function defined on an interval I. Then,

(a) If $f''(x) > 0$ for all x in I, then f is concave up on I.

(b) If $f''(x) < 0$ for all x in I, then f is concave down on I.

Like Figures 4.16(a) and (b) suggest, *concave up portions of a graph look "U"-ish*, while *concave down portions of the graph look "∩"-ish*. So, our first takeaway is that *f'' measures the curvature of the graph*: curving up and "U" shaped when $f''(x) > 0$, and curving down and "∩" shaped when $f''(x) < 0$.

In addition to the *sign* of f'' telling us information about how the graph of f curves, the *numerical value* of f'' is a measure of the curvature itself. This is illustrated in Figure 4.17. There I've graphed three parabolas. Note that $f''(x) = \frac{1}{2}$, $g''(x) = 2$, and $h''(x) = 4$, and that the graphs of the parabolas become more curved as we move from f to g to h. So, our second takeaway: *A larger $f''(x)$ value implies a curvier graph.* (Compare this information to what we get from f': the *slope* (or steepness) of the graph of f.)

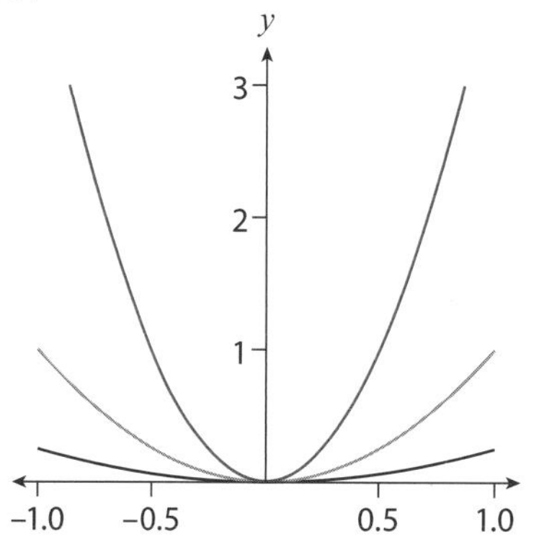

Figure 4.17: $f(x) = \frac{1}{4}x^2$, $g(x) = x^2$, $h(x) = 4x^2$.

Interactive Figure

Taking this analogy with f' one final step further, we learned in this chapter that the x-values at which $f'(x)$ changes sign as we cross them have meaning (they are local extrema on the graph of f). Similarly, the x-values where $f''(x)$ changes sign as we cross them also have meaning: they are the locations where the graph of f changes concavity. This concept is also important and has a special name.

Definition 4.6 Inflection Point. We say f has an **inflection point** at $x = c$ if the graph of f changes concavity as we cross $x = c$.

We can use the same sign chart technique we used for f' to find inflection points. First, we find $f''(x)$ and determine where it's equal to zero or does not exist. Let's call these x-values "candidate inflection points." (These x-values are essentially the critical numbers of f'.) Then, we use a sign chart to test points to see if $f''(x)$ changes sign as we cross each candidate inflection point. If so, Theorem 4.7 tells us there's a change in concavity, and that candidate inflection point is promoted to an inflection point.

EXAMPLE 4.27 Consider the function $f(x) = x^3 - 3x$ from Example 3.13.

(a) On what intervals is f concave up/down?

(b) Find the inflection point(s).

Solution

(a) By the Power Rule, $f'(x) = 3x^2 - 3$, and $f''(x) = 6x$. Since $f''(x) > 0$ when $x > 0$, Theorem 4.7 tells us that the graph of f is concave up for $x > 0$. Similarly, since $f''(x) < 0$ when $x < 0$, we conclude that the graph of f is concave down for $x < 0$.

(c) The only "candidate inflection point" is $x = 0$ (it is the only x-value at which $f''(x) = 6x$ is zero or does not exist). Since $f''(x) < 0$ for $x < 0$ and $f''(x) > 0$ for $x > 0$, $f''(x)$ changes sign as we cross $x = 0$. Thus, $x = 0$ is the only inflection point. ■

Figure 4.18(a) illustrates our results. Note how the graph of f switches from "∩"-shaped to "U"-shaped as we cross $x = 0$. This switch corresponds to when f'' changes sign (Figure (b)).

Notice too from Figure 4.18(a) that f has a local maximum at $x = -1$ and a local minimum at $x = 1$. At those points, $f''(-1) < 0$ and $f''(1) > 0$. This hints to a possible connection between the second derivative and local extrema. The following theorem gives the details.

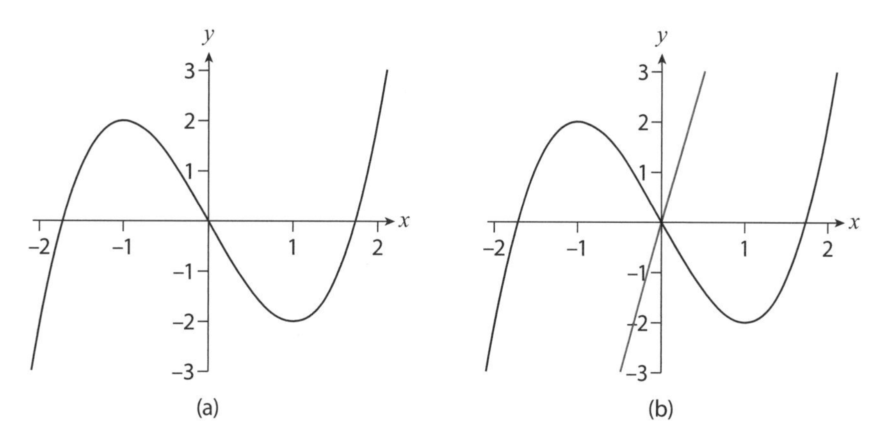

Figure 4.18: (a) $f(x)=x^3-3x$, and (b) $f(x)=x^3-3x$, and $f''(x)=6x$ (blue line).

> **Theorem 4.8 The Second Derivative Test.** Suppose f'' is continuous on an interval including c, and that $f'(c)=0$. Then,
>
> (a) If $f''(c)>0$ then f has a local minimum at $x=c$.
>
> (b) If $f''(c)<0$ then f has a local maximum at $x=c$.

The utility of this theorem is that it dispenses with the need for a sign chart to investigate local extrema (as was required by the First Derivative Test, Theorem 4.2); Theorem 4.8 instead requires you calculate $f''(x)$ and substitute in the critical number(s) obtained from setting $f'(x)=0$. The catch, however, is in the theorem's assumptions: it requires f'' to be continuous near $x=c$ and $f'(c)=0$. The First Derivative Test is more broadly applicable than that. Takeaway: If it's easy to calculate the second derivative (e.g., f is a polynomial) and the hypotheses of the Second Derivative Test are satisfied, use that; otherwise, the First Derivative Test is your go-to theorem for investigating local extrema.

> **The Cube Rule in Political Science** Applied Example C.6 in Appendix C applies our results about the second derivative to describe the "cube rule" in political science, which relates the proportion of seats in the U.S. House of Representatives won by the president's party to his or her percentage of the popular vote in the presidential election.

Related Exercises 17–20, 37(b), and 38(b).

Transcendental Tales

EXAMPLE 4.28 Find the intervals of concavity and the inflection points for the functions f and g from Example 4.13: (a) $f(x)=e^{2x}+e^{-x}$ (b) $g(x)=\sqrt{x}\ln x$

Solution

(a) We already calculated $f'(x)$: $f'(x) = 2e^{2x} - e^{-x}$. To explore the concavity of f, note that

$$f''(x) = 4e^{2x} + e^{-x} = \frac{4e^{3x}+1}{e^x} > 0$$

for all x. It follows from Theorem 4.7 that f is concave up for all x, and thus it has no inflection points. This agrees with Figure 4.10(a).

(b) We had previously calculated that $g'(x) = \frac{2+\ln x}{2\sqrt{x}}$. To explore the concavity of g, let's first calculate $g''(x)$ using the Quotient Rule:

$$g''(x) = \frac{\frac{1}{x}(2\sqrt{x}) - \frac{\ln x+2}{\sqrt{x}}}{4x} = -\frac{\ln x}{4x^{3/2}}.$$

Since we're only considering $x > 0$ (due to the domain of g), the only possible candidates for inflection points occur when $g''(x) = 0$. This yields $\ln x = 0$, whose solution is $x = 1$. Using the test points $x = 0.5$ and $x = 2$ yields the following sign chart.

$$g''(x):\quad + + + + + \quad \underset{1}{|} \quad - - - - -$$

It follows from Theorem 4.7 that g is concave up on the interval $(0, 1)$ and concave down on $(1, \infty)$. Thus, $x = 1$ is the (only) inflection point. This agrees with Figure 4.10(b). ■

The Spread of Infectious Diseases Applied Example C.7 in Appendix C explores the **logistic equation**, a mathematical model of growth that has various applications, including to the spread of infectious diseases such as the common cold.

Related Exercises 39–42 (part (e) only), 43(b), 44(b), 46(b), 48(d).

EXAMPLE 4.29 Let $f(x) = 2\cos x + \cos^2 x$, where $0 \le x \le 2\pi$. (This is the same function and domain from Example 4.15.) Find the intervals of concavity and the inflection points.

Solution We had previously calculated that $f'(x) = -2(\sin x + \sin x \cos x)$. From this we calculate that

$$f''(x) = -2(\cos x + \cos^2 x - \sin^2 x).$$

There are no x-values for which $f''(x)$ is undefined, so the only candidate inflection points occur where $f''(x)=0$. Setting $f''(x)=0$ yields

$$\cos x+\cos^2 x-\sin^2 x=0.$$

Substituting in $\sin^2 x=1-\cos^2 x$ (from (B.22) in Appendix B) yields

$$2\cos^2 x+\cos x-1=0.$$

This is a quadratic equation, in disguise. To see that we can let $z=\cos x$. This transforms the equation into $2z^2+z-1=0$, which factors into $(2z-1)(z+1)=0$. The $z=-1$ and $z=1/2$ solutions then become $\cos x=-1$ and $\cos x=1/2$. The solutions to these equations (inside the interval $[0,2\pi]$) are

$$x=\frac{\pi}{3}, \qquad x=\pi, \qquad x=\frac{5\pi}{3}.$$

Choosing the test points $x=\pi/4$, $x=\pi/2$, $x=3\pi/2$ yields the following sign chart.

$f''(x)$: $--$ | $++$ | $++$ | $--$

0 $\quad \frac{\pi}{3}$ $\quad \pi$ $\quad \frac{5\pi}{3}$ $\quad 2\pi$

It follows from Theorem 4.7 that f is concave down on the intervals $\left(0,\frac{\pi}{3}\right)$ and $\left(\frac{5\pi}{3},2\pi\right)$, and concave up on the intervals $\left(\frac{\pi}{3},\pi\right)$ and $\left(\pi,\frac{5\pi}{3}\right)$. Thus, $x=\pi/3$ and $x=5\pi/3$ are the only inflection points. This agrees with Figure 4.10(c). ■

Related Exercises 52–56 (only part (e)) and 61.

4.8 Parting Thoughts

We have now completed our study of differentiation. While there's always more that can be said, what we've already discussed is more than enough to give you a good understanding of derivatives and their applications. This chapter, in particular, has illustrated well the dynamics mindset of calculus as well as the previous chapter's title (that derivatives measure change).

In the next chapter we'll return to the last of the three Big Problems discussed in Chapter 1: the "area under the curve" problem. Solving that problem will lead to the last big character in the calculus story: the integral. And despite arising from a purely geometric problem—calculating the area under a curve—having nothing to do with derivatives, we will discover that the integral is intimately related to derivatives. The result relating the two—the Fundamental Theorem of Calculus—is the crowning achievement of calculus.

CHAPTER 4 EXERCISES

1–4: Find the linearization $L(x)$ of the function at the indicated a-value.

1. $f(x) = (x-1)^2,\ a=1$ **2.** $f(x) = \sqrt{x},\ a=1$

3. $f(x) = \frac{1}{x},\ a=1$ **4.** $f(x) = x^3,\ a=2$

5–8: Approximate the number using linearization; compare your answer to the actual value.

5. $\sqrt{10}$ **6.** $(1.01)^6$ **7.** $\frac{1}{\sqrt{3}}$ **8.** $\sqrt[3]{2}$

9–12: Determine the intervals on which the function is (a) increasing and (b) decreasing. Then, find the (c) the critical numbers and (d) the local extrema of the function.

9. $f(x) = 2x^3 + 3x^2 - 36x$

10. $f(x) = x + \frac{1}{x}$

11. $f(x) = x^4 - 2x^3 - x^2 + 2x$

12. $f(x) = \frac{x^2}{x+3}$

13–16: Find the (a) absolute maximum and (b) absolute minimum x-values of the function on the given interval.

13. $f(x) = x^3 - 3x + 1, \quad [0,3]$

14. $f(x) = x^4 - 2x^2 + 3, \quad [-2,3]$

15. $f(x) = (x^2-1)^3, \quad [0,1]$

16. $f(x) = \frac{x}{x^2+1}, \quad [0,2]$

17–20: Determine the intervals on which f is (a) concave up, (b) concave down. Finally, (c) identify any inflection points of f.

17. $f(x) = 2x^3 - 3x^2 - 12x$

18. $f(x) = 2 + 3x - x^3$

19. $f(x) = 2 + 2x^2 - x^4$

20. $f(x) = x\sqrt{x+3}$

21. An ice cube in your drink begins to melt. Supposing that the cube's side length is decreasing at the constant rate of 2 inches/minute, how fast is the cube's volume decreasing at the instant the cube's side length is 1/3 inch?

22. Imagine a cylindrical tank of radius 20 cm that is filled with water. A small hole is now drilled at the bottom of its circular base and the water begins to drain at 25 cm^3/sec. How fast is the water level in the tank dropping?

23. Repeat problem 22 above, but instead assume that the tank has radius 1 meter and that the water is drained at the rate of 3 liters per second.

24. A baseball player is at first base. The batter hits the ball and the player at first base starts running toward second base. Supposing the player runs at the rate of 15 ft/sec, and using the fact that a baseball diamond is actually a square with sides 90 ft, find the rate at which the player's distance from third base is decreasing when she's halfway from first to second base.

25. A granary is preparing to transport its grain. A conveyor belt at the facility is pouring the grain into a truck at the rate of 15 cm^3/sec. Assuming the pile of grain is a cone whose base diameter is always equal to its altitude, how fast is the altitude of the pile changing when the pile is 3 cm high?

26. A child at a park holding a balloon lets it go. The balloon rises at the constant rate of 5 m/sec. When it's 50 m up in the air, a dog runs underneath it, traveling in a straight line at a speed of 10 m/sec. How fast is the distance between the dog and the balloon changing 2 seconds later?

27. You and a sibling have just finished attending a family reunion. You get in your cars and leave for your home from the same position and at the same time. You travel north at 30 mph, while your sibling travels east at 40 mph. Calculate the rate at which the distance between you two is changing after 1 hour.

28. Imagine an 18 ft tall street lamp. A 6 ft tall person walks under and away from the lamp, on her way to a restaurant. If she's walking at a speed of 5 ft/sec, how fast is her shadow lengthening?

29. Aquathlon An aquathlon is a race consisting of swimming followed by running. Suppose Maria takes part in such an event. She starts at the north bank of a straight river that is 2 miles wide, and the finish line is 6 miles east on the south bank of the river. She can swim at a speed of 1 mile per hour and run at a speed of 3 miles per hour. To what point on the south bank of the river should Maria swim to in order to minimize the time taken to complete the aquathlon?

30. Maximizing baseball ticket revenue Fenway Park—the oldest baseball stadium in the country—holds about 38,000 spectators. Suppose that the average ticket costs \$100 (there are different tiers to the ticket prices), and that at that price, the average attendance throughout the season is 25,000. Now suppose the Boston Red Sox conduct a poll and find that for each drop in average ticket price of \$10, average attendance would increase by 1,000.

(a) Find the average price p as a linear function of average attendance x.

(b) The revenue generated by selling x tickets is $R(x) = xp(x)$. Use your answer to part (a) to help you determine the average ticket price that will yield maximum revenue.

31. Average revenue generated by Amazon.com Amazon.com sells many products. Let's denote the revenue it generates by selling x units of a certain product (e.g., shampoo) by $R(x)$. Companies like Amazon adjust their prices often to maximize revenue, and one popular metric they look at is the *average revenue*, $\overline{R}(x) = R(x)/x$, generated by the products they sell.

(a) Calculate $\overline{R}'(x)$.

(b) Show that the critical numbers of $\overline{R}(x)$ are $x = 0$, and x-values that satisfy $R'(x) = \overline{R}(x)$; interpret this last condition.

32. Maximizing blood velocity The velocity v of blood flowing through a nearly cylindrical section of an artery is well approximated by

$$v(r) = k(R^2 - r^2),$$

where k is a constant, R the radius of the artery, and r the distance from the central axis of the artery. (The v equation is known as *Poiseuille's Law.*) Show that the maximum blood velocity occurs along the central axis.

33. The acceleration due to gravity Return to Exercise 45 in Chapter 3 and interpret $g'(0)$ using the linearization interpretation of the derivative.

34. Let x and y be two numbers whose sum is 100. Find the absolute maximum of their product, xy. Does the absolute minimum exist? Briefly explain.

35. Returning to Example 4.24, suppose the cylindrical can's volume is V. Show that the minimum aluminum is used when $h = 2r$ (i.e., the can's height equals its diameter).

36. A wire of length 10 ft is cut into two pieces. One piece is used to make a square, and the other a triangle whose side lengths are all equal (i.e., an *equilateral* triangle). Letting A denote the sum of the resulting shape's areas, find the absolute minimum of A.

37. Consider a general quadratic polynomial $f(x)=ax^2+bx+c$ (where $a\neq 0$).

(a) Prove that $x=-\frac{b}{2a}$ is a local minimum if $a>0$, and a local maximum if $a<0$.

(b) Prove that f is concave down if $a<0$ and concave up if $a>0$.

38. Consider a general cubic polynomial $g(x)=ax^3+bx^2+cx+d$ (where $a\neq 0$), and let $D=b^2-3ac$.

(a) Prove that: (1) if $D>0$ then g has two critical numbers, (2) if $D=0$ then g has one critical number, and (3) if $D<0$ then g has no critical numbers.

(b) Prove that the only possible inflection point of g is $x=-\frac{b}{3a}$.

Exercises Involving Exponential and Logarithmic Functions

39–42: Find (a) the interval(s) of increase/decrease, (b) the critical number(s), (c) the local extrema (if any), (d) the absolute extrema on the interval [1, 2], and (e) the intervals of concavity and inflection points (if any) inside that same interval.

39. $f(x)=xe^{-x}$ **40.** $g(x)=e^x-x$

41. $h(t)=t^2-8\ln t$ **42.** $f(z)=\dfrac{2\ln z}{z^2}$

43. Let $f(x)=b^x$ be an exponential function.

(a) Given that $f'(x)=(\ln b)b^x$, explain why f is increasing for all x if $b>1$, and decreasing for all x if $0<b<1$. Explain why it follows that f has no local extrema.

(b) Given that $f''(x)=(\ln b)^2b^x$, explain why f is concave up for all x. Explain why it follows that f has no inflections points.

44. Let $g(x)=\log_b x$ be a logarithmic function.

(a) Given that $g'(x)=\frac{1}{x(\ln b)}$, find the intervals of increase/decrease in the cases $0<b<1$ and $b>1$. Explain why g has no local extrema.

(b) Given that $g''(x)=-\frac{1}{x^2(\ln b)}$, find the intervals of concavity in the cases $0<b<1$ and $b>1$. Explain why g has no inflection points.

45. Consider the function $f(x)=x^ne^{-x}$, where n is a positive integer. Show that: (a) the only nonzero critical number is $x=n$, and (b) that f has an absolute maximum at that critical number.

46. The Bell Curve Functions of the form

$$f(x)=\frac{1}{b\sqrt{2\pi}}e^{-(x-a)^2/(2b^2)},$$

where $b>0$ and a are constants, are called **normal distributions**. They are widely used in statistics to describe the distribution of human heights, students' exam scores, and even IQ scores.

(a) Show that f has an absolute maximum at $x=a$ (called the **mean** of the distribution).

(b) Show that f has inflection points at $x=a-b$ and $x=a+b$. (The number b is called the **standard deviation**.)

(c) Use your results thus far to sketch a graph of f for $a>0$. You will then see why the graph of f is often referred to as a "bell curve."

47. The Origin of the Universe The prevailing explanation for the origin of our universe is the **Big Bang Theory**, which posits that our universe was once a tiny high-density and high-temperature "singularity," and following a "big bang," rapidly inflated and expanded to become the universe we know today. This expansion spread out the initially high-temperature universe and created a cooler environment that enabled atoms to form. Today the leftover heat from the Big Bang is called the **Cosmic Microwave Background Radiation (CMB)**. The temperature of the CMB radiation is nearly constant at about 2.7 Kelvin, but there are very tiny fluctuations. Below is a 2012 all-sky image of the CMB from NASA's WMAP probe (the shades of gray correspond to temperature fluctuations).

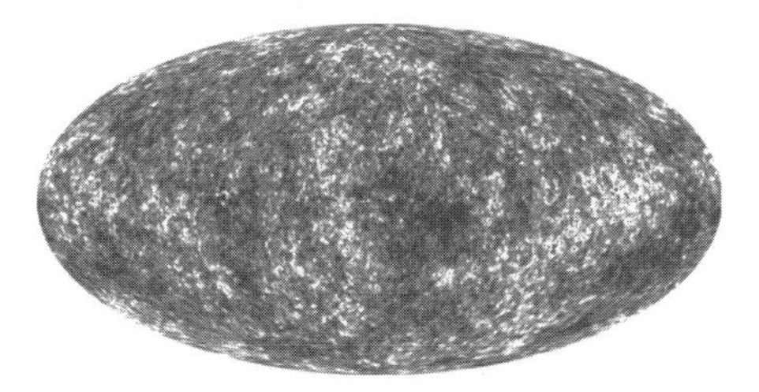

The distribution of the radiation energy R given off by the CMB varies with the wavelength λ of the light it emits, and is very accurately modeled by the function

$$R(\lambda) = \frac{a}{\lambda^5} \frac{1}{e^{b/(2.725\lambda)} - 1},$$

where $\lambda > 0$, and a and b are known constants.

(a) Let's make life easier and pretend that $a = 1$ and $b = 5.45$. Calculate $R'(\lambda)$.

(b) The only critical number of R is ≈ 0.4. Use this to show that R has an absolute maximum at $x \approx 0.4$.

48. Gompertz Curves The graphs of the functions

$$G(t) = e^{\frac{a}{b}(1-e^{bt})},$$

where a and b are positive real numbers and t a nonnegative real number, are called **Gompertz survival curves**. They are used to model the probability of surviving to age t (measured in years) after a successful birth at age 0. Let's suppose that $a = b$ for simplicity, and that $b = 0.085$ (this comes from empirical data for certain populations).

(a) Write out the $G(t)$ function that results. Then evaluate $G(0)$ and interpret your result.

(b) Evaluate $\lim_{t\to\infty} G(t)$ and interpret your result.

(c) Show that $G'(t) < 0$ for all t (here $t \geq 0$) and interpret this result.

(d) Show that $G''(t) > 0$ for all t (here $t \geq 0$) and interpret your result.

49. Wind Power Wind power is a clean, sustainable energy source. But generating power this way requires wind, and ideally high-velocity wind gusts. Luckily, the engineers that design wind turbines have discovered that they can accurately predict the probability of winds of speed v (in m/s) occurring using the function

$$P(v) = ave^{-bv^2},$$

where $a > 0$ and $b > 0$ are parameters that depend in part on the location being studied. The derivatives $P'(v)$ and $P'(0)$ were calculated in Exercise 65 in Chapter 3.

(a) Use linearization to show that $P(v) \approx av$ for v near 0.

(b) Determine the most probable wind velocity for a region in which $b = 1/2$.

Exercises Involving Trigonometric Functions

50. You and a friend go to a park to launch a toy rocket. After lighting the rocket, you move 20 ft away. The rocket launches, and at the instant its angle of elevation is 45°, that angle is increasing at the rate of 3° per second. How fast is the rocket's altitude changing at that instant?

51. A lighthouse 1 mile from a straight shoreline turns on at night. Its beacon shines a spot of light on the shoreline, and revolves at the rate of 5 revolutions per minute. How fast is the spot of light moving when the angle between the ray of light and the line connecting the shoreline and lighthouse is 30°?

52. Let $f(x) = \sin x$ and $g(x) = \cos x$. We know that $f'(x) = g(x)$ and $g'(x) = -f(x)$. Show that $f''(x) = -f(x)$ and $g''(x) = -g(x)$. (Thus, the sine and cosine functions both satisfy the **differential equation** $y'' + y = 0$.)

53–56: Find (a) the interval(s) of increase/decrease inside the given interval, (b) the critical number(s), (c) the local extrema (if any) inside the given interval, (d) the absolute extrema inside the given interval (if any), and (e) the intervals of concavity and inflection points (if any).

53. $f(x) = 2\cos x + \sin^2 x, \quad [0, \pi]$

54. $g(x) = 4x - \tan x, \quad [-\pi/3, \pi/3]$

55. $h(t) = 2\cos t + \sin(2t), \quad [0, \pi/2]$

56. $g(s) = s + \cot(s/2), \quad [\pi/4, 7\pi/4]$

57. Recall that $f(x) = \sin x$ and $g(x) = \cos x$ are both 2π-periodic: $f(x) = f(x + 2\pi)$ and $g(x) = g(x + 2\pi)$ for all x. Use this to help prove that f', f'', g', and g'' are all also 2π-periodic. It follows that all features of f and g that would be discerned from calculus (e.g., critical numbers, local extrema, etc.) need be determined for x-values only in the interval $[0, 2\pi)$.

58. Moving Boxes Efficiently via Calculus Picture a heavy box of mass m kg on the floor with a rope attached to it. Suppose you pull on the rope to attempt to drag the box. A simple model for the force F (in Newtons) required, assuming the rope makes an angle θ with respect to the floor, is

$$F(\theta) = \frac{\mu m g}{\cos\theta + \mu \sin\theta}, \qquad 0 \le \theta \le \frac{\pi}{2},$$

where $g \approx 9.8\ \text{m/s}^2$ is the acceleration of gravity and $0 \le \mu \le 1$ is the **coefficient of static friction**.

(a) Show that the only critical number of F in the interval given occurs when $\tan\theta = \mu$.

(b) Show that when $\tan\theta = \mu$,

$$F(\theta) = \frac{\mu m g}{\sqrt{1 + \mu^2}}.$$

(c) Explain why

$$\frac{\mu m g}{\sqrt{1 + \mu^2}} \le \mu m g \le m g,$$

and use this, along with part (a), to help you conclude that F is minimized when $\tan\theta = \mu$. (If the box is made of cardboard and the floor is wood, $\mu \approx 0.5$ and $\theta \approx 27°$.)

59. The Shape of Planetary Orbits One of the triumphs of Newton's Law of Gravity (Exercise 35 in Chapter 2) was to explain why planets in our solar system orbit the sun in elliptical orbits. Using Newton's result, one can show that if we put the Sun at the origin of the plane (see accompanying diagram), a planet's distance r to the Sun is very nearly modeled by the angular version of the ellipse equation (i.e., the ellipse equation in "polar coordinates"):

$$r(\theta) = \frac{a(1 - e^2)}{1 + e\cos\theta},$$

where $0 \le \theta \le 2\pi$ is the angle of the planet from the x-axis, $0 \le e < 1$ is the **eccentricity** of the orbit, and $a > 0$ is the ellipse's semi-major axis length.

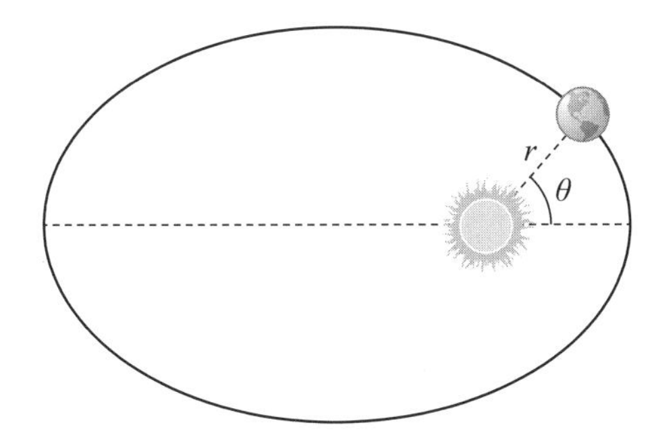

(a) Calculate $r(0)$ and $r(\pi)$, and explain why $r(\pi) > r(0)$.

(b) Show that $\theta = \pi$ is the only critical number of r inside the interval $[0, 2\pi]$.

(c) For Earth's orbit, $e \approx 0.017$ and $a \approx 9.3 \times 10^7$ miles. Use your result from part (b) to calculate the closest and furthest Earth gets from the Sun (the **perihelion** and **aphelion**, respectively).

Here is some context for Problems 60–61.

Fermat—that's the same Fermat for which Theorem 4.3 is named after—in 1662 formulated his **Principle of Least Time**, which states that light rays travel along the path that minimizes the travel time. Using this principle, Fermat was able to explain the *law of reflection*—which states that the image distance inside a flat mirror is the same as the object's distance in front of the mirror—as well as the *law of refraction*, which helps explain why a straw inside a glass of water appears bent at the surface of the water. These two laws are derived in the next two exercises using optimization theory.

60. The Law of Reflection A light ray emanates from point A, reaches point P on the mirror at an incident angle θ_i, then reflects off the mirror at a reflection angle θ_r, and eventually reaches point B (see diagram below) where A and B are a units above P.

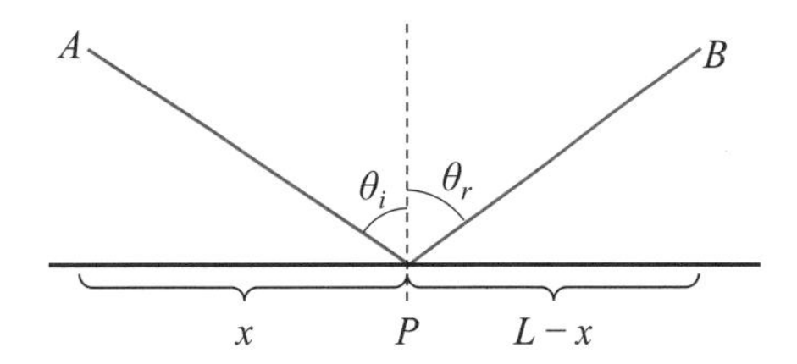

(a) Let t_1 and t_2 denote the time it takes the light ray to traverse the distances AP and PB, respectively. If c denotes the speed of light, show that

$$t_1(x) = \frac{\sqrt{a^2 + x^2}}{c},$$

$$t_2(x) = \frac{\sqrt{a^2 + (L - x)^2}}{c}.$$

(b) Let $t(x) = t_1(x) + t_2(x)$ be the light's total travel time. Show that the only critical number of $t(x)$ in the interval $0 \leq x \leq L$ is $x = L/2$.

(c) Using your result form part (b) and the fact that

$$\sqrt{2a^2 + L^2} < a + \sqrt{a^2 + L^2},$$

show that $t(x)$ has an absolute minimum at $x = L/2$.

(d) Use the diagram above to show that the equation that produced $x = L/2$ is equivalent to $\sin\theta_i = \sin\theta_r$. It follows from the symmetry of the triangles in the diagram that $\theta_i = \theta_r$.

61. The Law of Refraction (Snell's Law) The diagram below shows a light ray emanating from point A in a medium in which light travels at velocity v_1, reaching the interface with another medium at a distance x away from A, and terminating at point B in a medium in which light travels at velocity v_2.

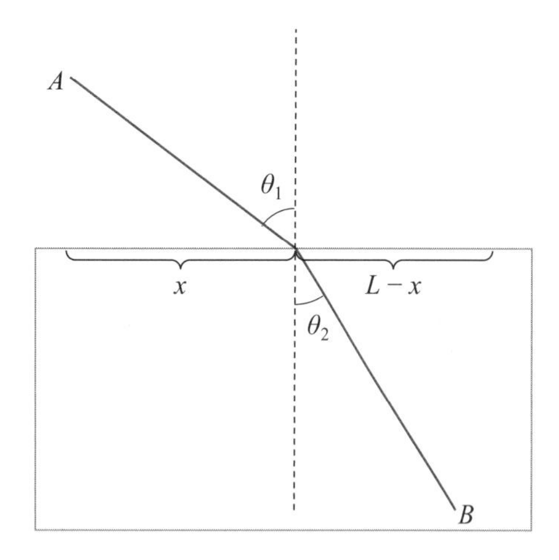

The incident angle θ_1 is related to the refracted angle θ_2 via **Snell's Law**:

$$\frac{\sin\theta_1}{\sin\theta_2} = \frac{v_1}{v_2}.$$

Let's derive this law using optimization theory.

(a) If A is a units above the interface and B is b units below it, show that the total time it takes the light ray to travel from A to B is

$$t(x) = \frac{\sqrt{x^2+a^2}}{v_1} + \frac{\sqrt{(L-x)^2+b^2}}{v_2},$$
$$0 \le x \le L.$$

(b) Show that the only critical number $x = x_c$ in the interval of interest occurs when

$$\frac{x_c}{v_1\sqrt{x_c^2+a^2}} = \frac{L-x_c}{v_2\sqrt{(L-x_c)^2+b^2}}.$$

(c) The second derivative of $t(x)$ is

$$t''(x) = \frac{a^2}{v_1(x^2+a^2)^{3/2}} + \frac{b^2}{v_2[(L-x)^2+b^2]^{3/2}}.$$

Explain why it follows that $t(x)$ is concave up on $[0, L]$.

(d) Use parts (b) and (c), along with a few theorems from the chapter, to help you conclude that $x = x_c$ is the absolute minimum of $t(x)$.

(e) Finally, rewrite the equation in part (b) in terms of $\sin\theta_1$ and $\sin\theta_2$ to derive Snell's Law.

62. Show that $\tan x \approx x$ for x near zero.

63. Let's return briefly to Exercise 60 in Appendix B.

(a) Explain why $\sin\left(\frac{2\pi}{n}\right) \approx \frac{2\pi}{n}$ for large n.

(b) Use part (a) and the $A(n)$ formula from the exercise to conclude that $A(n) \approx \pi r^2$ for large n.

5 Integration: Adding Up Change

Chapter Preview. *In 1666, roughly a year after Isaac Newton started working on what would become calculus, a German gentleman named Gottfried Leibniz had just acquired his law license. Yet Leibniz quickly grew frustrated with law and developed an interest in mathematics instead. He focused his work on the third Big Problem from Chapter 1: the Area under the Curve Problem. Leibniz's work led to the notion of the* definite integral, *the third pillar of calculus. In 1693 he made a breakthrough—Leibniz formulated and proved what we today call the Fundamental Theorem of Calculus. This theorem relates integration to differentiation, connecting the Area under the Curve Problem to the Slope of the Tangent Line Problem and unifying all of calculus. We'll build up to that theorem in this chapter by returning first to where our calculus story started: the Instantaneous Speed Problem.*

5.1 Distance as Area

We started Chapter 3 by trying to make sense of a falling apple's instantaneous speed. A few pages later, we had the answer: $s(t) = d'(t)$ (i.e., the apple's instantaneous speed is the derivative of its distance function). We can now calculate derivatives quickly, so if we're given $d(t)$ we can easily calculate $s(t)$. But how do we go backwards? That is: *How can we calculate the distance function for an object given its instantaneous speed function*? Let's employ a tried and true strategy for tackling tough math problems like this one: *Simplify the problem.*

Let's make things simpler by imagining a car driving down a highway at a *constant* speed (that's the simplification) of $s(t) = 60$ miles per hour. Using "distance = rate × time," we know the car travels 60 miles in 1 hour, 120 miles in 2 hours, and in general, $60t$ miles in t hours. That's the car's distance function: $d(t) = 60t$. We solved the problem!

Okay, but what if $s(t)$ *isn't* constant? Well then "distance = rate × time" won't help, because the rate ($s(t)$) is changing with time. A good example of this is our old friend from Chapter 3—the falling apple. As the apple falls, it picks up speed due to gravitational acceleration. Galileo's famous Leaning Tower of Pisa experiment—where he dropped balls of different masses to see if they'd hit the ground at the same time—suggested to him that gravity accelerates objects at the constant rate of 32 ft/s^2 (independent of the object's mass). In our notation, this means that $s'(t) = 32$. Following the same reasoning as in the car example above, we conclude that $s(t) = 32t$ is the instantaneous speed of the apple t seconds into its fall. (I'm assuming the apple is dropped from rest.) Question: How do we calculate $d(t)$ from *this* $s(t)$

function? Some 200 years *before* Galileo was even born (i.e., circa 1350s), a Parisian scholar named Nicole Oresme already had the answer: *Calculate the area under the graph of $s(x)$ and between $x=0$ and $x=t$.*

Figure 5.1(a) illustrates Oresme's approach. The shaded (triangular) region is the aforementioned area. Denoting that area by $A(t)$—the "area under the graph up to x-coordinate t"—we see that $A(t)=\frac{1}{2}(t)(32t)=16t^2$. That's the distance function Galileo deduced using experiments! (Recall (3.2).)

Oresme's realization of distance as area (the title of this section, by the way) works, but we don't yet understand *why*. Not knowing that, it's unclear if we could use the same approach to calculate $d(t)$ from other $s(t)$ functions. But let's not give up on Oresme's approach just yet. As usual, the problem is the static mindset inherent in Figure 5.1(a). Figure 5.1(b) shows a more dynamic mindset; there I've imagined the apple falling for an additional time Δt beyond the time t it's already been falling for. The change in area—the lighter shaded region—is

$$
\begin{aligned}
\Delta A &= A(t+\Delta)-A(t)\\
&= \frac{1}{2}[32t+32(t+\Delta t)](\Delta t)\\
&= 32t(\Delta t)+16(\Delta t)^2,
\end{aligned}
$$

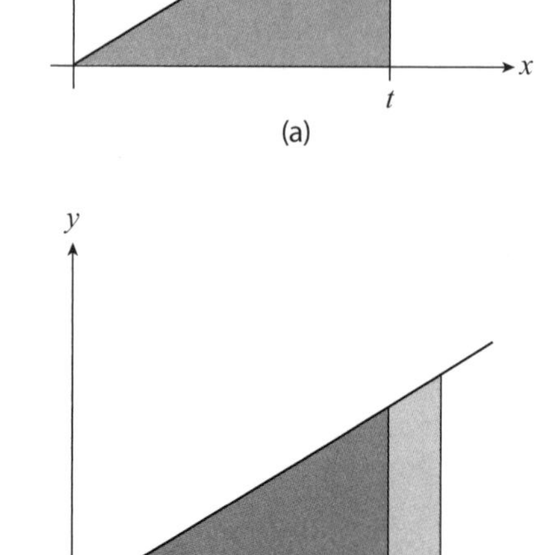

Figure 5.1: The graph of $s(x)=32x$, along with (a) the (shaded) area under s and between $x=0$ and $x=t$ (denoted by $A(t)$ in the text) and (b) the lighter shaded trapezoidal region (equal to $A(t+\Delta t)-A(t)$).

where I've used the formula for the area of a trapezoid: $A=\frac{1}{2}(h_1+h_2)b$, where h_1 and h_2 are the two heights of the trapezoid and b its base length. It follows that

$$
\frac{\Delta A}{\Delta t}=32t+16(\Delta t), \qquad \text{so that} \qquad A'(t)=\lim_{\Delta t\to 0}\frac{\Delta A}{\Delta t}=32t.
$$

But wait! That's $s(t)$! Thus

$$
A'(t)=s(t). \tag{5.1}
$$

And since $s(t)=d'(t)$, we can also say that $A'(t)=d'(t)$. The two functions A and d, therefore, have the same tangent line slopes at every point. These functions must either be equal or shifted by a constant: $A(t)=d(t)+C$, where C is a real number.

(Exercise 14 guides you through the proof.) Using the fact that $A(0) = d(0) = 0$ (no distance is traversed by the apple zero seconds into its fall), we arrive at the final result: $A(t) = d(t)$. Since we've already calculated that $A(t) = 16t^2$ (using Figure 5.1(a)), we've just derived the apple's distance function $d(t) = 16t^2$ by calculating an area under a curve!

APPLIED EXAMPLE 5.1 The gravitational acceleration near the surface of the Moon is about 5.4 ft/s^2. Suppose an astronaut on the Moon drops an apple from rest. Calculate the apple's distance function.

Solution We're given that $s'(t) = 5.4$; this implies that $s(t) = 5.4t$. Emulating the analysis from when s was equal to $32t$ yields $d(t) = \frac{1}{2}(t)(5.4t) = 2.7t^2$. ■

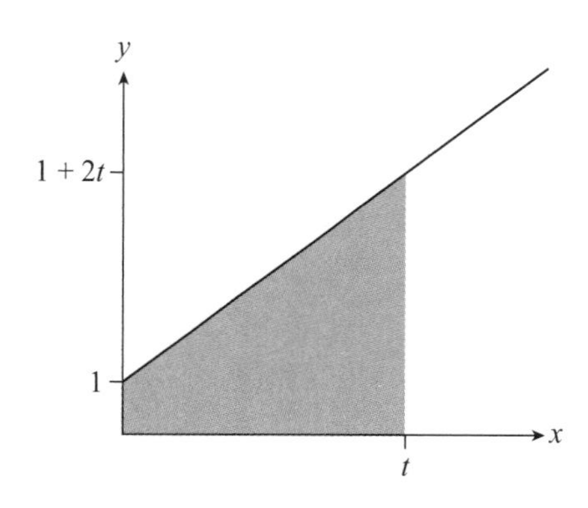

Figure 5.2: $s(x) = 1 + 2x$.

EXAMPLE 5.2 An object's instantaneous speed function is $s(x) = 1 + 2x$ (graphed in black in Figure 5.2).

(a) Calculate $A(t)$, the area under the graph of $s(x)$ over the interval $[0, t]$.

(b) Calculate $A'(t)$, then verify (5.1) by repeating the calculation that led up to it.

(c) Calculate the object's distance function.

Solution

(a) $A(t)$ is the area of the trapezoidal shaded region in Figure 5.2. Thus: $A(t) = \frac{t}{2}[1 + (1 + 2t)] = t(1 + t) = t + t^2$.

(b) Mimicking what we did before, if we imagine a small increase Δt in t in Figure 5.2, the additional area added is a trapezoid of width Δt and heights $s(t) = 1 + 2t$ and $s(t + \Delta t) = 1 + 2(t + \Delta t)$. Therefore

$$\Delta A = \frac{1}{2}[(1 + 2t) + (1 + 2(t + \Delta t))](\Delta t) = (1 + 2t)(\Delta t) + (\Delta t)^2.$$

It follows that

$$A'(t) = \lim_{\Delta t \to 0} \frac{\Delta A}{\Delta t} = \lim_{\Delta t \to 0} \frac{(1 + 2t)(\Delta t) + (\Delta t)^2}{\Delta t}$$
$$= \lim_{\Delta t \to 0} [1 + 2t + \Delta t] = 1 + 2t = s(t).$$

(c) We now know that $A'(t) = s(t)$. The same argument from before implies that $A(t) = d(t)$. Using the result of part (a), we conclude that $d(t) = t + t^2$. ■

> **Runway Lengths** Applied Example C.8 in Appendix C uses areas of triangles to help estimate the runway length needed for a jet airplane to take off safely.

Related Exercises 1–3.

Tips, Tricks, and Takeaways

The main takeaway from the previous examples—as well as Exercise 3 at the end of this chapter—is this: *The distance function $d(t)$ for an object with piecewise linear $\mathscr{s}(x)$ is the area under that graph bounded by $x=0$ and $x=t$* (assuming $d(0)=0$). This is an improvement over what we could do at the start of this section—calculate the distance traveled only for a *constant* $\mathscr{s}$ function (remember the car example?). But it's still a far cry from being able to calculate $d(t)$ for *any* $\mathscr{s}(x)$ function. We'll start solving that problem two sections from now. But first, let's introduce some new notation for the area under a curve and discuss its new insights.

5.2 Leibniz's Notation for the Integral

Mathematicians are lazy creatures in that we prefer not to write too much. So let's introduce better notation that is shorthand for "the area under the graph of $\mathscr{s}(x)$ and bounded between $x=0$ and $x=t$." We've denoted that by $A(t)$, but this notation makes no reference to $\mathscr{s}$ or $x=0$ (the left-hand boundary of the region whose area $A(t)$ refers to). Let's make progress by returning to (5.1). Employing our linearization result (4.10), (5.1) implies that

$$\Delta A \approx \mathscr{s}(x)\,\Delta x \quad \text{when } \Delta x \approx 0. \tag{5.2}$$

When $\Delta x \to 0$ we're considering an infinitesimal change in x (like we discussed in Chapter 1). Recall that Leibniz introduced the notation dx to denote that infinitesimal change (see the discussion that follows (3.16)). Finally, recalling our discussion of how linear approximations arising from linearization get better as Δx gets closer to zero, we expect that the infinitesimal change dx in the right boundary of the region under the graph of $\mathscr{s}$ should result in an infinitesimal change

$$dA = \mathscr{s}(x)\,dx \tag{5.3}$$

in the area of that region.[1] We visualize (5.3) as the area of a rectangle with infinitesimal width dx and height $\mathscr{s}(x)$ (Figure 5.3(a)). To Leibniz, the area $A(t)$ under the graph of $\mathscr{s}$ was then the sum of these infinitesimally small areas dA as x ranged from

[1] This is yet another manifestation of the calculus workflow (Figure 1.3)—the finite changes in (5.2) become the infinitesimal changes in (5.3) as $\Delta x \to 0$.

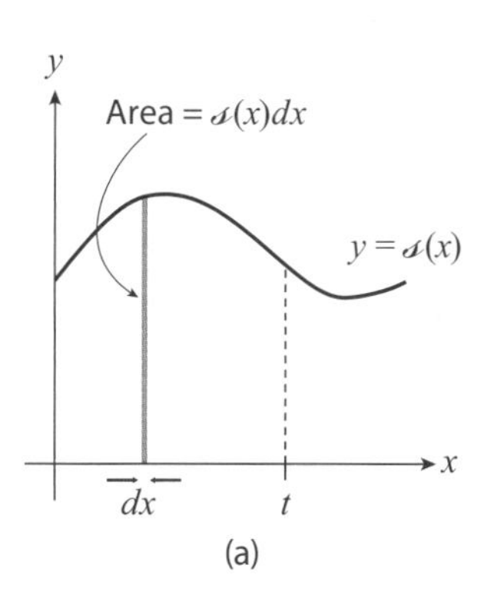

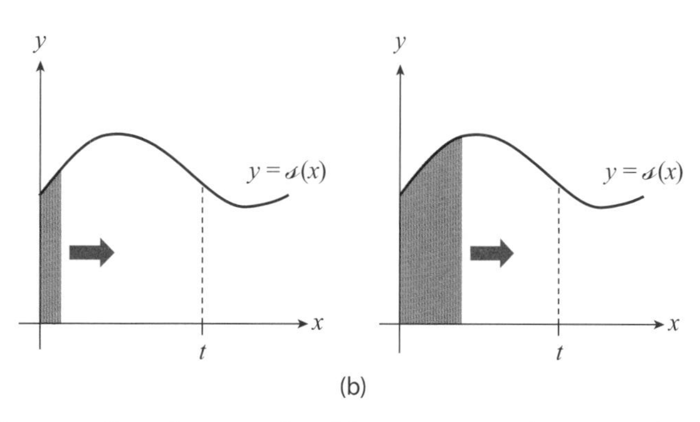

Figure 5.3: (a) One of Leibniz's infinitesimally wide rectangles. (b) Sweeping out the area under the curve by adding the areas of Leibniz's rectangles.

0 to t (Figure 5.3(b)):

$$\begin{aligned} A(t) &= \text{sum of } dA \text{ from } x=0 \text{ to} x=t \\ &= \text{sum of } s(x)\,dx \text{ from } x=0 \text{ to } x=t. \end{aligned}$$

The lazy mathematicians soon replaced "sum of" with "S", and over time that symbol morphed into $\int$:

$$\begin{aligned} A(t) &= \text{S } s(x)\,dx \text{ from } x=0 \text{ to } x=t \\ &= \int s(x)\,dx \text{ from } x=0 \text{ to } x=t. \end{aligned}$$

Today we append the bounds $x=0$ to $x=t$ to the **integral sign** $\int$ to arrive at our new notation:

$$A(t) = \int_0^t s(x)\,dx. \tag{5.4}$$

The right-hand side is called the **definite integral** of $s(x)$, that function is called the **integrand**, and 0 and t are called the **lower** and **upper limits of integration**, respectively.

EXAMPLE 5.3 Express the area of the shaded region in Figure 5.1(a) as a definite integral.

Solution

$$\int_0^t 32x\,dx.$$ ■

EXAMPLE 5.4 Express the area of the shaded region in Figure 5.2 as a definite integral.

Solution

$$\int_0^t (1+2x)\,dx.$$ ■

Tips, Tricks, and Takeaways

In addition to being yet another perfect illustration of calculus' dynamics mindset, what I've just explained has one key takeaway: *Integration is about adding up small changes* (*infinitesimal* changes, actually), hence this chapter's title. The progression from (5.2) to (5.4) was meant to illustrate that at every step of the way. Keep this in mind when you look at the right-hand side of (5.4).

One last finer takeaway: *Definite integrals are sums of tiny areas of rectangles.* Figure 5.3 further illustrates this second takeaway.

5.3 The Fundamental Theorem of Calculus

Alright, let's now return to generalizing (5.1). Our objective: Calculate the area under the graph of a function $f(x)$ and between $x = a$ and $x = t$. In the notation of (5.4), we're looking for

$$A(t) = \int_a^t f(x)\,dx.$$

Let's retain two core properties of $s(x)$: it's a continuous, nonnegative function. (We'll generalize our results even further in a later section.) Visually, then, we're looking for the area of a region like that of the blue-colored region in Figure 5.4. As we did before, let's investigate the effect on $A(t)$ of a small change Δt in t:

$$A(t+\Delta t) - A(t) = \int_a^{t+\Delta t} f(x)\,dx - \int_a^t f(x)\,dx = \int_t^{t+\Delta t} f(x)\,dx. \tag{5.5}$$

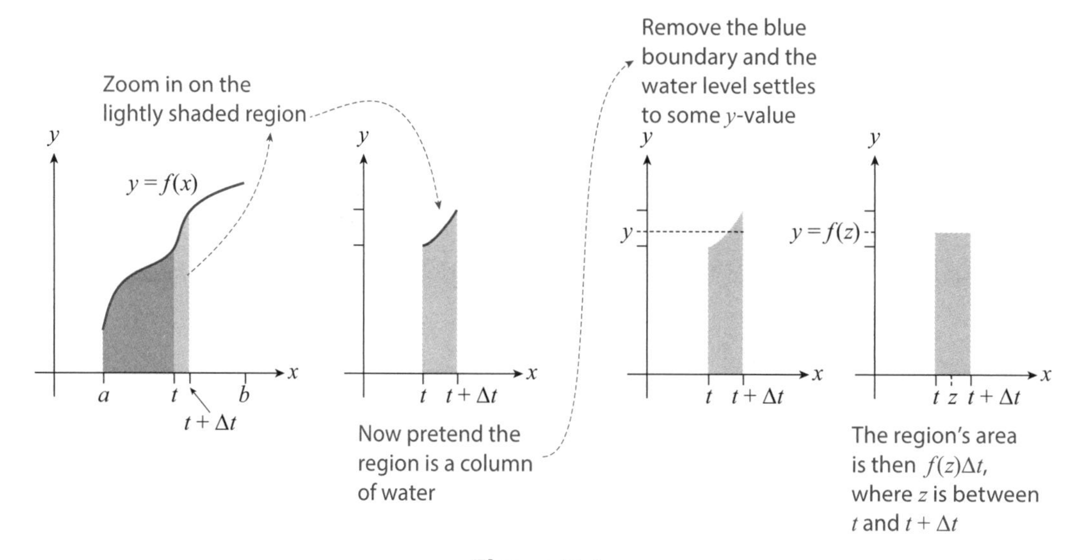

Figure 5.4

The integral on the far right of this equation is the area of the lightly shaded region in Figure 5.4 (leftmost plot). At this point in our earlier analysis we exploited the trapezoidal nature of that shaded region (recall Figure 5.1(b)). But that is no longer the case in Figure 5.4. No problem though—dynamics mindset to the rescue!

Let's pretend that the lightly shaded region is a column of water (Figure 5.4, second plot). When we remove the top "lid" (the graph of f), the water settles down to some y-value. As I've illustrated in the figure that y-value is the output of some x-value z: $y=f(z)$, where $t \leq z \leq t+\Delta t$.[2] Conclusion: the area of the lightly shaded region is the area of the *rectangle* of base length Δt and height $f(z)$ (Figure 5.4, last plot). So, (5.5) becomes

$$A(t+\Delta t)-A(t)=f(z)\Delta t. \tag{5.6}$$

Dividing by Δt and taking the limit of both sides yields

$$\lim_{\Delta t\to 0}\frac{A(t+\Delta t)-A(t)}{\Delta t}=\lim_{\Delta t\to 0}f(z). \tag{5.7}$$

We recognize the left-hand side as $A'(t)$; to calculate the right-hand side, we recall that $t \leq z \leq t+\Delta t$. Thus, as $\Delta t \to 0$, z approaches t. Conclusion:

$$A'(t)=f(t). \tag{5.8}$$

We've generalized (5.1)! (We used an intuitive argument to get from (5.5) to (5.7); Exercise 36 guides you through a more formal argument.) In a later section I'll indicate how our argument can be modified to account for the possibility that $f(x)<0$ for some x-values. So let me time travel a bit and give you this extended result now. It's what we today call the **Fundamental Theorem of Calculus**. Here is the formal statement of the Theorem, published by Leibniz in 1693.

Theorem 5.1 The Fundamental Theorem of Calculus. Suppose $f(x)$ is continuous on $[a,b]$, and define the function $A(t)$ by

$$A(t)=\int_a^t f(x)\,dx, \tag{5.9}$$

where $a \leq t \leq b$. Then $A(t)$ is continuous on $[a,b]$, differentiable on (a,b), and $A'(t)=f(t)$.

You might be thinking: "*That's* the Fundamental Theorem of Calculus?! It doesn't look very *fundamental* to me." I'll get back to why it is in the Takeaways subsection below. But right now, let's get more comfortable with the theorem itself.

[2] This follows from the fact that f is continuous and the Intermediate Value Theorem (Section A2.4 in the online appendix to Chapter 2).

EXAMPLE 5.5 Consider $f(x)=1$ on the interval $[0,5]$ and let t be inside this interval.

(a) Show that

$$\int_0^t 1\,dx = t. \tag{5.10}$$

(b) Verify Theorem 5.1 in this setting.

Solution

(a) The integral $\int_0^t 1\,dx$ is the area of a rectangle with width t and height 1, which is $(t)(1)=t$.

(b) We know that $f(x)=1$ is a continuous function (in particular, continuous on $[0,5]$). We just calculated that $A(t)=t$; this too is a continuous function (in particular, continuous on $[0,5]$). Moreover, since $A'(t)=1$ (by the Power Rule), A is differentiable (in particular, differentiable on $(0,5)$), and we see that $A'(t)=f(t)$. ■

EXAMPLE 5.6 Consider $f(x)=x$ on the interval $[0,5]$ and let t be inside this interval.

(a) Show that

$$\int_0^t x\,dx = \frac{t^2}{2}. \tag{5.11}$$

(b) Verify Theorem 5.1 in this setting.

Solution

(a) The integral $\int_0^t x\,dx$ is the area of a triangle with base length t and height t (similar to the shaded region in Figure 5.1). That area is $\frac{1}{2}(t)(t)=\frac{t^2}{2}$, verifying (5.11).

(b) We know that $f(x)=x$ is a continuous function (in particular, continuous on $[0,t]$). We just calculated that $A(t)=\frac{t^2}{2}$; this too is a continuous function (in particular, continuous on $[0,t]$). Moreover, since $A'(t)=t$ (by the Power Rule), A is differentiable (in particular, differentiable on $(0,t)$), and we see that $A'(t)=f(t)$. ■

Related Exercises 6–9.

Tips, Tricks, and Takeaways

Figure 5.5 helps you begin to appreciate why Theorem 5.1 is so fundamental. Assuming that f is continuous, the theorem's workflow is (1) integrate $f(x)$ to get $A(t)$;

(2) differentiate $A(t)$ to get $f(t)$. (Note that $f(t)$ and $f(x)$ are the same function.) The first big revelation:

Differentiation and integration undo each other!

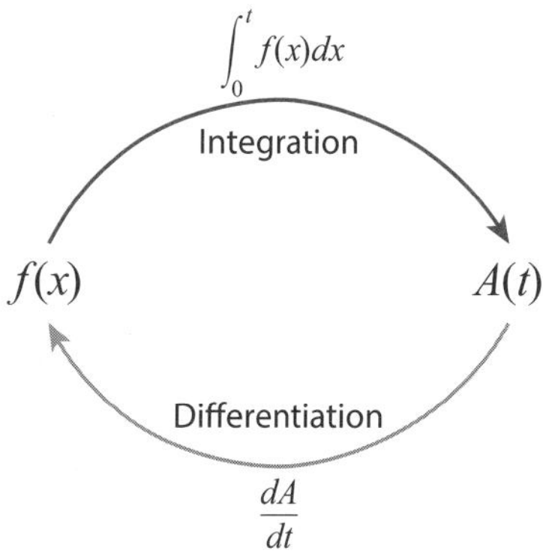

Figure 5.5: Differentiation and integration are inverse processes.

Therefore, Theorem 5.1 ties together, in one simple equation, two of the pillars of calculus: differentiation and integration.

Beyond the specific revelation that differentiation and integration undo each other, Theorem 5.1 shows us that our third and final Big Problem from Chapter 1—the "area under the curve" problem which has now become the "calculate the definite integral" problem—is related to derivatives, a topic we've spent two chapters mastering! This suggests an interesting idea: perhaps we can use *derivatives* to help us calculate definite *integrals.* (What a plot twist!) F.T.C., as we call Theorem 5.1, indeed hints that's the case. We'll explore that in the next section. Before then, however, one final insight—there is another route to the definite integral you'll likely encounter in a calculus textbook: the **Riemann sums** route. We'll have no need for that concept in this book, but see Section A5.1 (in the online appendix to this chapter) if you're interested in Riemann sums. Okay, back to F.T.C.

5.4 Antiderivatives and The Evaluation Theorem

To calculate $A(t)$ we've been graphing f and then using geometry formulas to find the area under that curve. But F.T.C. (Theorem 5.1) suggests another way. It says that if f is continuous, then $f(t)=A'(t)$. In other words, given f we automatically know the *derivative* of $A(t)$. All we have to do to find $A(t)$, then, is to "undo" that derivative. That process is aptly called **antidifferentiation.**

> **Definition 5.1** Suppose $F'(x)=f(x)$. We then call F the **antiderivative** of f.

In general, then, F is the function which *differentiates to* $f(x)$. Example: if $f(x)=2x$ then one antiderivative is $F(x)=x^2$ (since $F'(x)=2x=f(x)$). Another antiderivative is $F(x)=x^2+5$. The most general antiderivative is $F(x)=x^2+C$, where C is any real number.

Let's now employ the antiderivatives viewpoint to extract from Theorem 5.1 a faster way to calculate $A(t)$. To begin, suppose F is an antiderivative of the function f in the Theorem, so that $F'(t)=f(t)$. We also know that $A'(t)=f(t)$ (where A is defined by (5.9)). Therefore, $F'(t)=A'(t)$. It follows from Exercise 14 (with d replaced by F) that $A(t) = F(t) + C$. Substituting this in (5.9) yields

$$\int_a^t f(x)\,dx = F(t) + C. \tag{5.12}$$

When $t = a$ we get $\int_a^a f(t)\,dt = F(a) + C$. But the integral on the left-hand side is the area under the graph of f and bounded by $x = a$ and $x = a$, which is zero. Thus, $0 = F(a) + C$, so that $C = -F(a)$. Using this in (5.12) and writing $t = b$ yields the following corollary of Theorem 5.1.

Theorem 5.2 The Evaluation Theorem. Suppose f is continuous on $[a, b]$ and that F is an antiderivative of f (i.e., $F'(x) = f(x)$). Then,

$$\int_a^b f(x)\,dx = F(b) - F(a). \tag{5.13}$$

EXAMPLE 5.7 The Power Rule tells us that $\left(x^3\right)' = 3x^2$. Use this result to help you calculate

$$\int_0^1 (3x^2)\,dx.$$

Solution Here $f(x) = 3x^2$, which is continuous on $[0, 1]$. Moreover, $\left(x^3\right)' = 3x^2$ tells us that $F(x) = x^3$ is an antiderivative of f. It follows from Theorem 5.2 that

$$\int_0^1 (3x^2)\,dx = F(1) - F(0) = 1^3 - 0^3 = 1. \quad \blacksquare$$

EXAMPLE 5.8 The Power Rule tells us that $\left(\sqrt{x}\right)' = \frac{1}{2\sqrt{x}}$. Use this result to help you calculate

$$\int_1^4 \left(\frac{1}{2\sqrt{x}}\right) dx.$$

Solution Here $f(x) = \frac{1}{2\sqrt{x}}$, which is continuous on $[1, 4]$. Moreover, $\left(\sqrt{x}\right)' = \frac{1}{2\sqrt{x}}$ tells us that $F(x) = \sqrt{x}$ is an antiderivative of f. It follows from Theorem 5.2 that

$$\int_1^4 \left(\frac{1}{2\sqrt{x}}\right) dx = F(4) - F(1) = \sqrt{4} - \sqrt{1} = 1. \quad \blacksquare$$

Related Exercises 10–12 and 27.

Tips, Tricks, and Takeaways

First, some things you should know:

- Often we use the shorthand $F(x)|_a^b$ for $F(b) - F(a)$, so that (5.13) becomes

$$\int_a^b f(x)\,dx = F(x)|_a^b.$$

- Replacing x with t (or any other letter) in (5.13) changes nothing. For that reason we refer to x as a **dummy variable**.
- Theorem 5.2 is also called the "Fundamental Theorem of Calculus, Part 2" in some textbooks.

Now for the main takeaway—*The Evaluation Theorem converts the Area under the Curve Problem to a new problem: the Find the Antiderivative of f Problem.* Indeed, if you know an antiderivative F of f (and f is continuous on $[a, b]$), then the Evaluation Theorem says that the area under the graph of $f(x)$ and bounded by $x = a$ and $x = b$—the left-hand side in (5.13)—is simply $F(b) - F(a)$. For this reason we will spend much of the remainder of the chapter discussing antiderivatives. The next section discusses their properties and begins to build a repository of antiderivative formulas.

5.5 Indefinite Integrals

The antiderivatives viewpoint is very useful. But writing things like "$F(x) = x^2$ is an antiderivative of $f(x) = 2x$" takes too long (remember the lazy mathematicians?). Let's first take care of the pervasive "an" in these sentences.

Theorem 5.3 Suppose F is an antiderivative of f (i.e., $F' = f$). Then $F(x) + C$, where C is any constant, is also an antiderivative of f.

(The proof is very simple: $(F(x) + C)' = F'(x) = f(x)$.) We can now abuse the English language a bit and say things like "$F(x) = x^2 + C$ is the antiderivative of $f(x) = 2x$." (It's an abuse of the English language because C could be any real number, so there are many different formulas for F, clashing with the usage of "the" in the sentence.) Finally, let's condense this statement via the following notation.

Definition 5.2 The Indefinite Integral. Let F be an antiderivative of f, so that $F' = f$. We then write

$$\int f(x)\,dx = F(x) + C \tag{5.14}$$

and call the left-hand side the **indefinite integral of f**.

Note the use of the symbol $\int$ here again. But be careful in reading too much into this: the definite integral produces a *number* (the area under the graph of f)

while the indefinite integral produces a *function* (the most general antiderivative of f).

Since indefinite integrals are just new notation for antidifferentiation, the indefinite integral is just the reverse process of differentiation:

$$F'(x) = f(x) \iff \int f(x)\,dx = F(x) + C. \tag{5.15}$$

For example:

$$(x^2)' = 2x \iff \int 2x\,dx = x^2 + C.$$

This finally condenses the more wordy "$F(x) = x^2 + C$ is the antiderivative of $f(x) = 2x$."

The equivalence (5.15) gives us a wealth of antiderivatives; *simply take the differentiation results we've already worked out in the past two chapters, read them in right-to-left order, and add in the indefinite integral sign and the "+C" in the right places.* For example, returning to the first sentences in Examples 5.7 and 5.8:

$$\int x^3\,dx = 3x^2 + C, \qquad \int \left(\frac{1}{2\sqrt{x}}\right) dx = \sqrt{x} + C.$$

These results come from the Power Rule (Theorem 3.4), so we can follow our prescription (5.15) to write down the integral version of the Power Rule:

$$(x^m)' = mx^{m-1} \iff \int mx^{m-1}\,dx = x^m + C.$$

A more user-friendly formula is obtained by substituting $m - 1 = n$ in and solving for the indefinite integral of x^n. This yields the following theorem.

■ Theorem 5.4 The Integral Version of the Power Rule.

$$\int x^n\,dx = \frac{x^{n+1}}{n+1} + C, \quad n \neq -1. \tag{5.16}$$

Note the requirement that $n \neq -1$. (The integral of $\frac{1}{x}$ turns out to be a logarithm, as discussed near the end of this section.) One particularly tricky instance of this theorem is when $n = 0$. In that case, (5.16) yields

$$\int 1\,dx = x + C. \tag{5.17}$$

EXAMPLE 5.9 Calculate $\int x^2\,dx$.

Solution Setting $n=2$ in (5.16) yields: $\int x^2\,dx=\dfrac{x^3}{3}+C.$

EXAMPLE 5.10 Calculate $\displaystyle\int_0^1 x^2\,dx.$

Solution We just calculated a family of antiderivatives for x^2 ($\frac{x^3}{3}+C$). We can choose any of these to use in the Evaluation Theorem. Choosing $C=0$, we get that $F(x)=\frac{x^3}{3}$ is an antiderivative of $f(x)=x^2$. Therefore, according to the Evaluation Theorem:

$$\int_0^1 x^2\,dx=\frac{x^3}{3}\bigg|_0^1=\frac{1}{3}.$$

■

EXAMPLE 5.11 Calculate $\displaystyle\int \frac{1}{x^2}\,dx.$

Solution Since $\frac{1}{x^2}=x^{-2}$, using (5.16) with $n=-2$ yields

$$\int \frac{1}{x^2}\,dx=\frac{x^{-1}}{-1}+C=-\frac{1}{x}+C.$$

■

EXAMPLE 5.12 Calculate $\displaystyle\int \sqrt{x}\,dx.$

Solution Writing $\sqrt{x}=x^{1/2}$ and using (5.16) with $n=1/2$ yields

$$\int \sqrt{x}\,dx=\frac{x^{3/2}}{\frac{3}{2}}+C=\frac{2x^{3/2}}{3}+C.$$

■

Related Exercises 17–19, 41, and 50.

Tips, Tricks, and Takeaways

Example 5.10 illustrates the fact that F can be *any* antiderivative of f when it comes to using the Evaluation Theorem; it need not be the most general one ($F(x)+C$). For this reason we will always select the $C=0$ antiderivative of f when using the Evaluation Theorem.[3]

We've only one learned how to integrate one function at a time thus far. In the next section we'll learn how to integrate combinations of functions (e.g., a sum or difference of two functions).

[3]Using any other antiderivative, like $F(x)+7$, won't change the result in the Evaluation Theorem, since $[F(b)+7]-[F(a)+7]=F(b)-F(a)$, the same result as using $F(x)$ (i.e., with $C=0$).

5.6 Properties of Integrals

Like the Limit Laws from Chapter 3, the indefinite and definite integrals satisfy various properties that help us calculate them. The first few mimic the first few derivative rules we discussed: the Sum, Difference, and Constant Multiple Rules (Theorem 3.3).

Theorem 5.5 Properties of the Integral. Suppose f and g are continuous on $[a, b]$, and let c be a real number. Then,

1. **The Sum Rule:** $\int_a^b [f(x)+g(x)]\,dx = \int_a^b f(x)\,dx + \int_a^b g(x)\,dx$
2. **The Difference Rule:** $\int_a^b [f(x)-g(x)]\,dx = \int_a^b f(x)\,dx - \int_a^b g(x)\,dx$
3. **The Constant Multiple Rule:** $\int_a^b [cf(x)]\,dx = c\int_a^b f(x)\,dx$

Moreover, these rules also hold if the definite integral is replaced by an indefinite integral.

These rules can be proven using antiderivatives and Theorem 3.3. Exercise 13 guides you through one of those proofs.

In addition to the rules above, the following additional rules hold for definite integrals.

Theorem 5.6 Additional Properties of the Definite Integral. Suppose f and g are continuous on $[a, b]$, and let c be a real number. Then,

1. $\int_a^c f(x)\,dx = \int_a^b f(x)\,dx + \int_b^c f(x)\,dx$
2. $\int_a^b f(x)\,dx = -\int_b^a f(x)\,dx$
3. $\int_a^a f(x)\,dx = 0$
4. If $f(x) \leq g(x)$ for every x in $[a, b]$ then $\int_a^b f(x)\,dx \leq \int_a^b g(x)\,dx$

Property 1 tells us that we can split the calculation of the area under a curve into a sum of two different area calculations. (Importantly, while we think of b as being between a and c in that property, it need not be.) Property 2 tells us that swapping the limits of integration multiplies the original value of the definite integral by -1. Property 3 merely reflects the fact that the area under the graph of f between $x=a$ and $x=a$ is zero (a fact we've already used). Finally, Property 4 says that if the graph of f is at or below the graph of g, then the area under the graph of f will be less than or equal to the area under the graph of g. Let's now illustrate these properties through a couple of examples.

EXAMPLE 5.13 Calculate $\int (x^2 - x)\, dx$.

Solution

$$\int (x^2 - x)\, dx = \int x^2\, dx - \int x\, dx \quad \text{Indefinite integral version of 2, Theorem 5.5}$$

$$= \frac{x^3}{3} - \frac{x^2}{2} + C. \quad \text{Using (5.16)}$$

■

EXAMPLE 5.14 Calculate $\int_0^9 (3\sqrt{x} + 9x^2)\, dx$.

Solution First, let's find the antiderivative of $f(x) = 3\sqrt{x} + 9x^2$:

$$\int (3\sqrt{x} + 9x^2)\, dx = 3\int x^{1/2}\, dx + 9\int x^2\, dx \quad \text{Sum and Constant Multiple Rules, Theorem 5.5}$$

$$= 3\left(\frac{2}{3}x^{3/2}\right) + 9\left(\frac{x^3}{3}\right) + C \quad \text{Using (5.16)}$$

$$= 2x^{3/2} + 3x^3 + C. \quad \text{Simplifying}$$

Selecting $C=0$ and using that result in the Evaluation Theorem:

$$\int_0^9 (3\sqrt{x} + 9x^2)\, dx = \left[2x^{3/2} + 3x^3\right]_0^9 = 2,241.$$

■

Related Exercises 20–23 (Hint: Simplify first), 28–29, and 49.

We can now integrate may of the common combinations of functions we'll encounter—but not all; we'll return to this point in Section 5.9—but as we'll see in the next section, the Difference Rule in Theorem 5.5 will force us to reinterpret what quantity the definite integral yields.

5.7 Net Signed Area

Consider the integral $\int_0^1 (-1)\, dx$. Our understanding of the definite integral as yielding the area under the graph of $f(x)$ does not apply here, because the x-axis—which has been the *bottom* boundary of the area defined by all the integrals we've calculated thus far—is actually *above* the function $f(x) = -1$. So, we need to reinterpret what the definite integral means when the graph of f dips below the x-axis. That's where Theorem 5.5 comes in—the Constant Multiple Rule implies that

$$\int_0^1 (-1)\, dx = (-1) \int_0^1 1\, dx = -1,$$

since the second definite integral is 1. The first equation here literally tells us that $\int_0^1 (-1)\, dx$ is -1 times the area under the graph of $f(x) = 1$. This is why some interpret $\int_0^1 (-1)\, dx$ as "negative area." But such a thing doesn't exist, so I'll interpret $\int_0^1 (-1)\, dx = -1$ as "there's 1 unit of area below the x-axis," as illustrated in Figure 5.6.

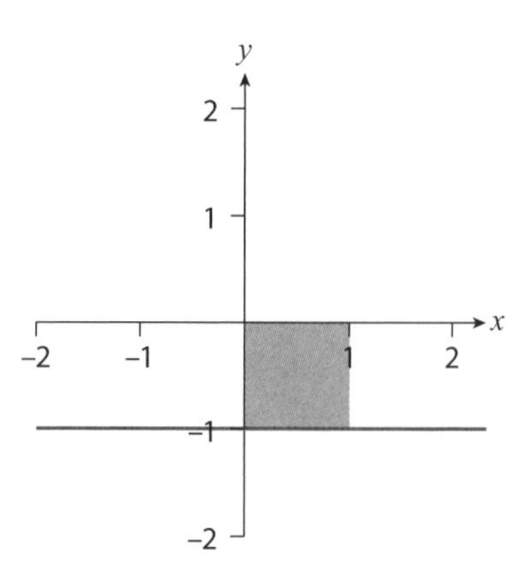

Figure 5.6: $f(x) = -1$ and $\int_0^1 (-1)\, dx$ (the negative of the area of the shaded region).

What we've just done generalizes rather easily. To wit: If $f(x)$ has both positive and negative values inside the interval $[a, b]$, then,

$$\int_a^b f(x)\, dx = A_+ - A_-, \tag{5.18}$$

where A_+ denotes the sum of all areas above the x-axis and A_- the sum of all areas below the x-axis. Thus, *in general the definite integral yields a* ***net signed area***. The "net" part of the phrase describes the subtraction present in (5.18); the "signed area" describes the possibility that the resulting number—which we previously thought of as the area under the curve—may be negative.

EXAMPLE 5.15 Calculate $\displaystyle\int_0^2 (x-1)\, dx$ using the Evaluation Theorem and also (5.18).

Solution Using the Difference Rule (from Theorem 5.5), (5.16), and the Evaluation Theorem:

$$\int_0^2 (x-1)\, dx = \int_0^2 x\, dx - \int_0^2 1\, dx = \left.\frac{x^2}{2}\right|_0^2 - x\big|_0^2 = 2 - 2 = 0.$$

Figure 5.7 illustrates our answer. The region below the x-axis (the darker shaded region) has area $\frac{1}{2}$, so that $A_- = \frac{1}{2}$. The lighter shaded region above the x-axis has area $\frac{1}{2}$ too, so that $A_+ = \frac{1}{2}$. Therefore, from (5.18)

$$\int_0^2 (x-1)\, dx = A_+ - A_- = \frac{1}{2} - \frac{1}{2} = 0. \quad ■$$

We've now covered everything you need to know about the basics of integration. The next section applies all we've learned to transcendental functions. But if you're skipping those, skip the next section and head right to Section 5.9; there we'll discuss a very useful integration technique that follows from the Chain Rule, called u-substitution.

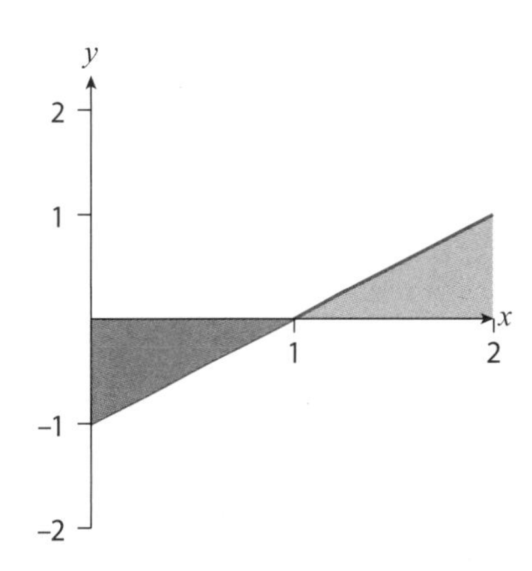

Figure 5.7: The function $f(x) = x-1$, along with the shaded region between f, the x-axis, and bounded by $[0,1]$ (the darker blue-colored region), and the similar region bounded by $[1,2]$ (the lighter blue-colored region).

5.8 (Optional) Integrating Transcendental Functions

Let's start with the rules for integrating exponential functions. Returning to Theorem 3.8, in view of (5.15) we immediately get the following integration rules.

Theorem 5.7

$$\int b^x\,dx = \frac{b^x}{\ln b} + C, \qquad \int e^x\,dx = e^x + C. \tag{5.19}$$

Let's now apply the same approach to finally calculate the integral of $1/x$ (which is not covered by (5.16)). We'll need the following slight generalization of Theorem 3.9 (see Exercise 48 for the derivation):

$$\frac{d}{dx}(\ln|x|) = \frac{1}{x}. \tag{5.20}$$

Using this in (5.15) yields the following integration rule.

Theorem 5.8

$$\int \frac{1}{x}\,dx = \ln|x| + C. \tag{5.21}$$

Let me illustrate the preceding two theorems through a few examples.

EXAMPLE 5.16 Calculate $\int 3e^x\,dx$.

Solution

$$\begin{aligned} \int 3e^x\,dx &= 3\int e^x\,dx && \text{Constant Multiple Rule, Theorem 5.5} \\ &= 3e^x + C. && \text{Using (5.19)} \end{aligned}$$

■

EXAMPLE 5.17 Calculate $\int (x^2 + 2^x)\, dx$.

Solution

$$\begin{aligned}\int (x^2 + 2^x)\, dx &= \int x^2\, dx + \int 2^x\, dx && \text{Sum Rule, Theorem 5.5}\\ &= \frac{1}{3}x^3 + \frac{2^x}{\ln 2} + C. && \text{Using (5.16) and (5.19)}\end{aligned}$$

■

EXAMPLE 5.18 Calculate $\int \frac{x^2+1}{2x}\, dx$.

Solution We first simplify: $f(x) = \frac{x^2+1}{2x} = \frac{x}{2} + \frac{1}{2x}$. Then,

$$\begin{aligned}\int \frac{x^2+1}{2x}\, dx &= \frac{1}{2}\int x\, dx + \frac{1}{2}\int \frac{1}{x}\, dx && \text{Sum and Constant Multiple Rules, Theorem 5.5}\\ &= \frac{x^2}{2} + \frac{1}{2}\ln|x| + C. && \text{Using (5.16) and (5.21)}\end{aligned}$$

■

Related Exercises 42–43, and 52.

Let's now pivot to discussing the integration of trigonometric functions. Applying (5.15) to (3.15) and (3.23) yields the following theorem.

Theorem 5.9

$$\int \cos x\, dx = \sin x + C, \qquad \int \sin x\, dx = -\cos x + C, \qquad \int \sec^2 x\, dx = \tan x + C.$$

Additionally, the results of Example 3.41 and Exercise 77 of Chapter 3, applied to (5.15), yield the following rules pertaining to the reciprocal trigonometric functions.

Theorem 5.10

$$\int \sec x \tan x\, dx = \sec x + C, \qquad \int \csc^2 x\, dx = -\cot x + C,$$

$$\int \csc x \cot x\, dx = -\csc x + C.$$

EXAMPLE 5.19 Calculate $\int_0^{\pi} (\sin x + \cos x)\, dx$.

Solution Let's first find the antiderivative of $f(x) = \sin x + \cos x$:

$$\int (\sin x + \cos x)\, dx = \int \sin x\, dx + \int \cos x\, dx \qquad \text{Sum Rule, Theorem 5.5}$$

$$= -\cos x + \sin x + C \qquad \text{Using Theorem 5.9}$$

Then, using the Evaluation Theorem:

$$\int_0^{\pi} (\sin x + \cos x)\, dx = [-\cos x + \sin x]_0^{\pi} = (-\cos \pi + \sin \pi) - (-\cos 0 + \sin 0)$$

$$= 1 - (-1) = 2. \qquad \blacksquare$$

Related Exercises 54–55, and 59–60.

5.9 The Substitution Rule

Thus far we've developed the rules for integrating sums, differences, and constant multiples of functions (along with a few specialized rules for integrating particular functions, like (5.16)). This has paralleled the development of the differentiation rules in Chapter 3. Continuing along this path, we'd develop the integration rule analogue of the Product Rule. That's called **integration by parts**, but because it's covered in a second semester calculus course, we'll skip instead to the integral rule analogue of the Chain Rule.

To get there, let's apply the equivalence (5.15) to the Chain Rule (Theorem 3.6):

$$\frac{d}{dx}[F(g(x))] = F'(g(x))g'(x) \quad \Longleftrightarrow \quad \int F'(g(x))g'(x)\, dx = F(g(x)) + C. \tag{5.22}$$

(I'll explain why I used F and not f soon.) The integrand in the rightmost equation is messy. Let's make it look simpler by introducing $u = g(x)$. This yields:

$$\int F'(g(x))g'(x)\, dx = \int F'(u)g'(x)\, dx. \tag{5.23}$$

Recalling now (5.3), the same argument used to get to that equation tells us that

$$du = g'(x)\, dx.$$

Using this in (5.23) then yields

$$\int F'(g(x))g'(x)\, dx = \int F'(u)\, du.$$

Finally, letting $F' = f$ yields the following theorem.

Theorem 5.11 The Substitution Rule. Suppose f is continuous on an interval I, and $g(x)$ is differentiable and has range I. Then for $u = g(x)$,

$$\int f(g(x))g'(x)\,dx = \int f(u)\,du. \tag{5.24}$$

You'll often hear this technique referred to as "u-substitution." Now on to the examples.

EXAMPLE 5.20 Calculate $\int 2x(x^2+1)^{100}\,dx$.

Solution The integrand contains the composite function $(x^2+1)^{100}$; it's "inner" function is x^2+1. So, let's try setting $u = g(x)$:

$$u = x^2 + 1 \quad \Longrightarrow \quad du = 2x\,dx.$$

Substituting these into the integral yields

$$\int 2x(x^2+1)^{100}\,dx = \int u^{100}\,du.$$

Using (5.16) with $n = 100$, this integrates to $\frac{u^{101}}{101} + C$. That's not the end of the calculation though, since we should end up with a function in the same variable we started with. So, we substitute $u = x^2 + 1$ back in to get

$$\int 2x(x^2+1)^{100}\,dx = \frac{(x^2+1)^{101}}{101} + C. \qquad \blacksquare$$

EXAMPLE 5.21 Calculate $\int_0^2 x(x^2+4)^3\,dx$.

Solution Here $u = x^2 + 4$ seems to be the logical choice (it's the "inside" function of the composition $(x^2+4)^3$). Letting $u = x^2 + 4$, $du = 2x\,dx$. Dividing both sides by 2 yields $\frac{1}{2}du = x\,dx$. Then, (5.24) and (5.16) yield

$$\int x(x^2+4)^3\,dx = \int u^3\left(\frac{1}{2}du\right) = \frac{1}{2}\int u^3\,du = \frac{u^4}{8} + C = \frac{(x^2+4)^4}{8} + C. \tag{5.25}$$

We've now found an antiderivative of $x(x^2+4)^3$. The Evaluation Theorem (Theorem 5.2) then implies

$$\int_0^2 x(x^2+4)^3\,dx = \left.\frac{(x^2+4)^4}{8}\right|_0^2 = 480. \qquad \blacksquare$$

EXAMPLE 5.22 Calculate $\int \frac{x}{\sqrt{1+x^2}}\,dx$.

Solution For reasons similar to those in the previous example, the logical choice here is $u = x^2 + 1$. Then $du = 2x\,dx$. Dividing both sides by 2 yields $\frac{1}{2}du = x\,dx$. Then, (5.24) and (5.16) yields

$$\int \frac{x}{\sqrt{1+x^2}}\,dx = \frac{1}{2}\int u^{-1/2}\,du = u^{1/2} + C = \sqrt{1+x^2} + C. \quad ■$$

EXAMPLE 5.23 Calculate $\int \sqrt{x+1}\,dx$.

Solution The only viable choice for u is $x+1$. Letting $u = x+1$, we have $du = 1\,dx$; employing (5.24) then yields

$$\int \sqrt{x+1}\,dx = \int 1\cdot\sqrt{x+1}\,dx = \int \sqrt{u}\,dx = \frac{2u^{3/2}}{3} + C = \frac{2(x+1)^{3/2}}{3} + C. \quad ■$$

EXAMPLE 5.24 Calculate $\int x^5\sqrt{1+x^2}\,dx$.

Solution This is the most challenging example yet. But hopefully your gut should tell you to choose $u = 1 + x^2$. Then $du = 2x\,dx$ and $\frac{1}{2}du = x\,dx$. Substituting in what we know thus far yields

$$\frac{1}{2}\int x^4\sqrt{u}\,du.$$

(I siphoned off one x from x^5 to use $x\,dx = \frac{1}{2}du$.) We now need to relate x to u to complete the substitution. But since $u = 1 + x^2$, then $x^2 = u - 1$, and so $x^4 = (u-1)^2$. Thus,

$$\begin{aligned}\frac{1}{2}\int x^4\sqrt{u}\,du &= \frac{1}{2}\int \sqrt{u}(u-1)^2\,du = \frac{1}{2}\int \sqrt{u}(u^2 - 2u + 1)\,du \\ &= \frac{1}{2}\int \left[u^{5/2} - 2u^{3/2} + u^{1/2}\right]\,du.\end{aligned}$$

The properties of integrals (from Theorem 5.6) and (5.16) yield

$$\frac{1}{2}\int \left[u^{5/2} - 2u^{3/2} + u^{1/2}\right]\,du = \frac{1}{2}\left(\frac{2u^{7/2}}{7} - \frac{4u^{5/2}}{5} + \frac{2u^{3/2}}{3}\right) + C.$$

(Each integral generates its own arbitrary constant, but these can be added together to form another arbitrary constant, which is the C in the equation.) Substituting back in $u = 1 + x^2$ finally yields

$$\int x^5\sqrt{1+x^2}\,dx = \frac{(1+x^2)^{7/2}}{7} - \frac{2(1+x^2)^{5/2}}{5} + \frac{(1+x^2)^{3/2}}{3} + C. \quad ■$$

Related Exercises 23–26, and 30–31.

Transcendental Tales

EXAMPLE 5.25 Calculate the integrals below.

(a) $\int_0^1 2xe^{x^2}\,dx$ (b) $\int \frac{1}{x+1}\,dx$ (c) $\int \frac{x^2+2x+1}{x^2+1}\,dx$

Solution

(a) Let $u = x^2$, so that $du = 2x\,dx$. Equations (5.24) and (5.19) then yield

$$\int 2xe^{x^2}\,dx = \int e^u\,du = e^u + C = e^{x^2} + C.$$

Therefore,

$$\int_0^1 2xe^{x^2}\,dx = e^{x^2}\Big|_0^1 = e - 1.$$

(b) Let $u = x+1$, so that $du = dx$. Equations (5.24) and (5.21) then yield

$$\int \frac{1}{x+1}\,dx = \int \frac{1}{u}\,du = \ln|u| + C = \ln|x+1| + C.$$

(c) Let's first simplify the function:

$$\frac{x^2+2x+1}{x^2+1} = 1 + \frac{2x}{x^2+1}.$$

Then, by part 1 of Theorem 5.6:

$$\int \frac{x^2+2x+1}{x^2+1}\,dx = \int 1\,dx + \int \frac{2x}{x^2+1}\,dx.$$

The first integral yields $x + C_1$ (from (5.17)). To calculate the second, let $u = x^2+1$, so that $du = 2x\,dx$. Equations (5.24) and (5.21) then yield

$$\int \frac{2x}{x^2+1}\,dx = \int \frac{1}{u}\,du = \ln|u| + C_2 = \ln|x^2+1| + C_2.$$

We conclude that

$$\int \frac{x^2+2x+1}{x^2+1}\,dx = x + \ln(x^2+1) + C.$$

(We don't need the absolute value around x^2+1 since that quantity is always positive. Also, I added C_1 and C_2 to produce C.) ■

EXAMPLE 5.26 Calculate the integrals: (a) $\int \tan x\,dx$ (b) $\int \cot x\,dx$

Solution

(a) Since $\tan x = \frac{\sin x}{\cos x}$, letting $u = \cos x$ we have $du = -\sin x\,dx$, so that (5.24) yields

$$\int \tan x\,dx = \int \frac{\sin x}{\cos x}\,dx = -\int \frac{1}{u}\,du.$$

Here we need (5.21); we conclude that

$$\int \tan x\,dx = -\ln|\cos x| + C = \ln|\sec x| + C.$$

(b) Since $\cot x = \frac{\cos x}{\sin x}$, letting $u = \sin x$ we have $du = \cos x\,dx$, so that (5.24) yields

$$\int \cot x\,dx = \int \frac{\cos x}{\sin x}\,dx = \int \frac{1}{u}\,du.$$

Here we need (5.21) again; we conclude that

$$\int \cot x\,dx = \ln|\sin x| + C.$$ ■

Related Exercises 44–47, 51, and 53.

EXAMPLE 5.27 Calculate the integrals below.

(a) $\int x^2 \cos(x^3)\,dx$ (b) $\int \sec^2(2x)\,dx$ (c) $\int_0^{\pi/4} \cos(2x)\,dx$

Solution

(a) Let $u = x^3$. Then $du = 3x^2\,dx$, and (5.24) along with Theorem 5.9 yields

$$\int x^2 \cos(x^3)\,dx = \frac{1}{3}\int \cos u\,du = \frac{1}{3}\sin u + C = \frac{1}{3}\sin(x^3) + C.$$

(b) Letting $u = 2x$, we have $du = 2\,dx$. Then, (5.24) along with Theorem 5.10 yields

$$\int \sec^2(2x)\,dx = \frac{1}{2}\int \sec^2 u\,du = \frac{1}{2}\tan u + C = \frac{1}{2}\tan(2x) + C.$$

(c) Using the substitution $u = 2x$, $du = 2\,dx$ yields

$$\int \cos(2x)\,dx = \frac{1}{2}\sin(2x) + C.$$

Setting $C = 0$ and applying (5.13) then yields

$$\int_0^{\pi/4} \cos(2x)\,dx = \frac{1}{2}\sin(2x)\Big|_0^{\pi/4} = \frac{1}{2}\left(\sin\frac{\pi}{2} - 0\right) = \frac{1}{2}.$$ ■

Related Exercises 56–58, and 62–63.

Tips, Tricks, and Takeaways

- The u-substitution technique is useful only when the integrand is of the form $f(g(x))g'(x)$. Such integrands contain a composite function, $f(g(x))$, multiplied by the derivative of the "inside" function, $g'(x)$. So, the first takeaway: *u-substitution should be used only when the integrand is a composite function.* (This reflects the technique's origin in the Chain Rule, which should only be used to differentiate composite functions.) You should then try setting $u = g(x)$, where $g(x)$ is the "inner" function in the composition.
- The substitution $u = g(x)$ converts $f(g(x))$ to $f(u)$; nothing difficult there. However, the remaining part of the integral, namely $g'(x)\,dx$, also gets transformed—into du. Therefore, the complete substitution is:

$$u = g(x), \quad du = g'(x)\,dx.$$

- Once you've transformed the integral to one involving u's and (hopefully) calculated the resulting integral, don't forget to transform variables back to x (using $u = g(x)$).

One last bit about u-substitution. Though we've used the technique thus far only to help us calculate indefinite integrals, it works just as well for calculating definite integrals. Let me illustrate what I mean by returning to Example 5.21. Since $u = x^2 + 4$ in that example, the upper limit of integration $x = 2$ becomes $u = 8$; the lower limit of integration $x = 0$ becomes $u = 4$. Thus

$$\int_0^2 x(x^2+4)^3\,dx = \frac{1}{2}\int_4^8 u^3\,du = \frac{1}{2}\left[\frac{u^4}{4}\right]_4^8 = 480,$$

the same answer we obtained. The suggested exercises below further explore this usage of u-substitution.

Related Exercises 37–40, and 61.

We've now learned a lot about integration. Additional integration techniques and theory are typically covered in a second-semester college calculus course. Because we started this chapter by exploring the real-world context for integration, let's return to that and learn more about the real-world applications of integration.

5.10 Applications of Integration

Let me end the chapter by discussing two brief applications of integration. (A second-semester calculus course is where the vast majority of the applications of integration are discussed.) I will introduce these via two examples.

APPLIED EXAMPLE 5.28 In a simplified version of the **Andersen Fitness Test**, a person runs back and forth between point A and point B a set distance apart for 2 minutes, pausing momentarily at each end to touch the floor. (The aim of the test is to cover the greatest distance.) Suppose Emilia's velocity as she runs is given by the function

$$v(t) = 80(t-1)^3 - 80(t-1),$$

measured in ft/s, and where $0 \le t \le 2$ (see Figure 5.8).

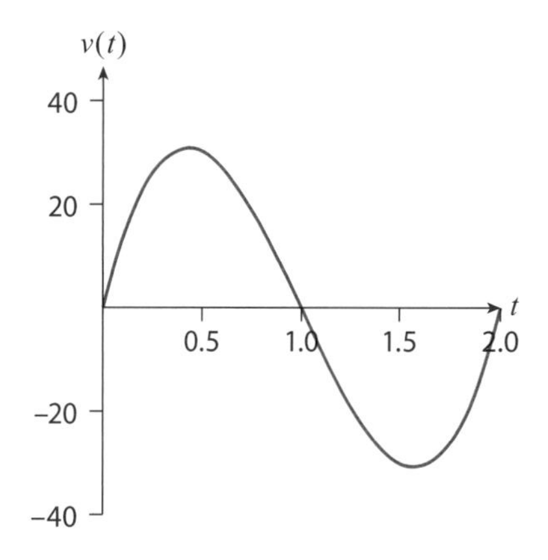

Figure 5.8: The graph of $v(t) = 80(t-1)^3 - 80(t-1)$ on the interval $[0, 2]$.

(a) At what time(s) is Emilia not moving? When is Emilia moving right, and when is she moving left?

(b) Assume Emilia starts at point A. Calculate Emilia's **position function** $s(t)$ from point A.

(c) How far away is point B?

(d) Calculate $\int_0^2 v(t)\,dt$. How does this relate to Emilia's **displacement** over the 2 minutes, the quantity $s(2) - s(0)$? (Here $v = s'$.)

Solution

(a) We need to solve $v(t) = 0$. Factoring yields

$$80(t-1)[(t-1)^2 - 1] = 0 \quad \Longrightarrow \quad t = 0, 1, 2.$$

Thus, Emilia is not moving at the start of the test $(t = 0)$, the end of the test $(t = 2)$, and at one time in the middle $(t = 1)$.

When Emilia is moving right, her position function $s(t)$ is increasing (she is getting farther away from point A), and so $s'(t) = v(t)$ is positive. As Figure 5.8 shows, $v(t) > 0$ on the interval $(0, 1)$. We conclude that Emilia is moving right during the first minute of her 2-minute run. Since $v(t) < 0$ on $(1, 2)$, she is moving left during the second minute of her 2-minute run.

(b) Since $s'(t) = v(t)$, we have that

$$\begin{aligned} s(t) &= \int v(t)\,dt = \int [80(t-1)^3 - 80(t-1)]\,dt \\ &= 80\int (t-1)^3\,dt - 80\int t\,dt + 80\int 1\,dt, \end{aligned}$$

where I've used a few properties of integrals to break up the calculation. We can use u-substitution to calculate the first integral; with $u = t - 1$ and $du = dt$, we have

$$\int (t-1)^3\,dt = \int u^3\,du = \frac{u^4}{4} + C = \frac{(t-1)^4}{4} + C.$$

The second and third integrals in the $s(t)$ equation are easy to do via (5.16) and (5.17). Therefore

$$\begin{aligned} s(t) &= 80\left(\frac{(t-1)^4}{4}\right) - 80\left(\frac{t^2}{2}\right) + 80t + C \\ &= 20(t-1)^4 - 40t^2 + 80t + C. \end{aligned}$$

Since Emilia starts at Point A, we know that $s(0) = 0$, which tells us that $C = -20$. Thus,

$$s(t) = 20(t-1)^4 - 40t^2 + 80t - 20.$$

(c) According to the rules, Emilia must stop at point B to touch the floor. At that moment her velocity is zero. We found in part (a) that $v(t) = 0$ when $t = 0, 1, 2$. She starts the test at $t = 0$ and moves right. Since she's moving left for $1 < t < 2$, the momentary pause at $t = 1$ must be when Emilia reached point B. And since $s(1) = 20$, we conclude that point B is 20 feet away.

(d) Since $v(t)$ is continuous, from (5.13):

$$\int_0^2 v(t)\,dt = s(2) - s(0) = 0,$$

since $s(2) = s(0) = 0$ (using the formula for $s(t)$ from part (b)). We conclude that Emilia's displacement during the test is zero feet. ■

Part (d) of this example is a manifestation of a more general interpretation of the Evaluation Theorem in the case when the integrand is a rate (like $v(t)$): *Integrating a rate yields the* ***net change*** *in the underlying function* ($s(t)$ in the case of the example). This is the reason why in the instances when the integrand in (5.13) is a derivative, that equation is sometimes called the **net change theorem**. This physical interpretation of the definite integral in these cases complements its geometric interpretation as the "net signed area" under the curve (recall (5.18)). The suggested exercises below further explore this new interpretation.

Related Exercises 4–5, 16, 32, and 34–35.

APPLIED EXAMPLE 5.29 In many countries income is distributed unevenly among the country's wage earners; for example, in 2013 the bottom 99% of wage earners in the United States received only about 80% of the nation's pre-tax income. Economists quantify this income distribution disparity using a **Lorenz curve** $L(x)$, defined to be the percentage of the nation's income earned by the bottom x% of households (here both x and $L(x)$ are in decimal form).[4] Given a country's Lorenz

[4]For example, the 2013 United States data just mentioned would correspond to $L(0.99) = 0.8$.

cure, its **Gini coefficient** G, defined as

$$G = \int_0^1 [2x - 2L(x)]\,dx, \tag{5.26}$$

can be used to measure the degree of income inequality in that country; the range for G is $0 \le G \le 1$, with higher values indicating greater income inequality. Calculate the Gini coefficient of a country with Lorenz curve $L(x) = x^2$.

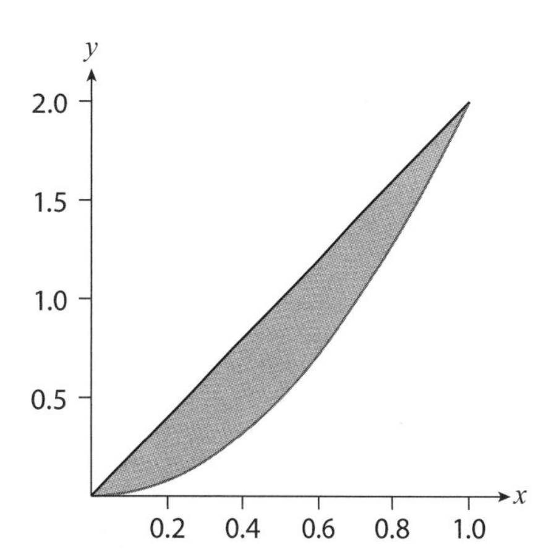

Figure 5.9: The graphs of $y = 2x^2$ (gray) and $y = 2x$ (black) for $0 \le x \le 1$, and the region (blue) between these graphs.

Solution From (5.26):

$$G = \int_0^1 [2x - 2x^2]\,dx. \tag{5.27}$$

This quantity is the area of the shaded region in Figure 5.9. That area is the difference between the area under the graph of $f(x) = 2x$ (the black line in the figure) and that under the graph of $f(x) = x^2$ (the gray curve in the figure). Using what we've learned in this chapter:

$$\int_0^1 [2x - 2x^2]\,dx = 2\int_0^1 x\,dx - 2\int_0^1 x^2\,dx \quad \text{Difference and Constant Multiple Rules, Theorem 5.5}$$

$$= 2\left[\frac{x^2}{2}\right]_0^1 - 2\left[\frac{x^3}{3}\right]_0^1 \quad \text{Using (5.16) and Theorem 5.2}$$

$$= 1 - \frac{2}{3} = \frac{1}{3}. \quad \text{Simplifying}$$

Thus, the country's Gini coefficient is $G = \frac{1}{3}$. ■

Related Exercises 15

For my last act, let me take you back to where it all started: the falling apple problem illustrated in Figure 1.4. Let's generalize things a bit, and consider the problem of determining the position function of a sufficiently heavy object (i.e., not a feather) thrown in the air, neglecting air resistance.

APPLIED EXAMPLE 5.30 Suppose an object is thrown straight up from a height of h meters and with initial vertical velocity v_y (in ft/s). Assuming the object's acceleration function is $a(t) = -g$, where $g \approx 32$ ft/s^2 is the acceleration due to gravity, find the vertical position function $y(t)$ of the object.

Solution

Since $a(t) = v'(t)$, the equivalence (5.15) gives us a formula for the object's velocity:

$$v(t) = \int a(t)\,dt = \int -g\,dt = -g\int 1\,dt = -gt + C,$$

where I've used the properties of integrals and also (5.17). Using $v(0) = v_y$ yields $C = v_y$, so that

$$v(t) = v_y - gt.$$

Now, since $v(t) = y'(t)$, using again the equivalence (5.15) yields

$$y(t) = \int v(t)\,dt = \int (v_y - gt)\,dt = v_y\int 1\,dt - g\int t\,dt = v_y t - \frac{gt^2}{2} + D.$$

Using $y(0) = h$ yields $D = h$. Thus,

$$y(t) = h + v_y t - \frac{1}{2}gt^2. \tag{5.28}$$

(Note: This equation is valid only until the object hits the ground.) ■

Equation (5.28) is quite the accomplishment: It lists the general (vertical) position function of an airborne object assuming that gravity accelerates objects at a constant rate (which was known to Galileo). In the special case that $v_y = 0$ (i.e., the object is dropped from rest from a height h), (5.28) becomes $y(t) = h - d(t)$, where $d(t)$ is the distance function (3.2) from Chapter 3 that helped spur Newton and his contemporaries to invent calculus. (Exercise 33 uses (5.28) to explain why sufficiently heavy objects [e.g., a football] thrown in the air follow parabolic trajectories.)

5.11 Parting Thoughts

We have now come to the end of the chapter. In fact, we've come to the end of the book. The past five chapters have developed the core concepts in calculus: the limit, the derivative, and the integral. So, if you've made it this far, I'm proud of you. There are more topics in Calculus 1 than what's in this book, but those follow-ups ultimately rely on either limits, derivatives, or integrals, which you've now gotten a good deal of training on. So as far as I'm concerned, if these five chapters have made sense, *I would say you have learned calculus.* I stand by this statement even if you have yet to work through the optional content on exponential, logarithmic, and trigonometric functions, because ultimately that content is also just the application of the limit, derivative, and integral concepts we learned to different families of functions. If you *do* have the time, however, I encourage you to work through those optional sections. Transcendental functions are widely applicable, and if you continue studying mathematics (or science) you'll keep running into them (and their calculus).

We have also explored a variety of applications of the calculus concepts we've learned. Let me encourage you to look over the Index of Applications at the end of the book once more; it summarizes all of the applications contained in this book and can guide you to discovering other applications you may not have read about while working through these five chapters.

I hope you have enjoyed this calculus adventure. I have more parting thoughts for you in the Epilogue, but again, congratulations on learning calculus! Best of luck with your future studies in mathematics.

CHAPTER 5 EXERCISES

1–3: Calculate the area functions $A(t)$ for the objects whose speed functions $s(x)$ are given below.

1. $s(x) = 10$

2. $s(x) = 1 - x$, considering only $0 \leq x \leq 1$

3. $s(x)$ given by the graph below.

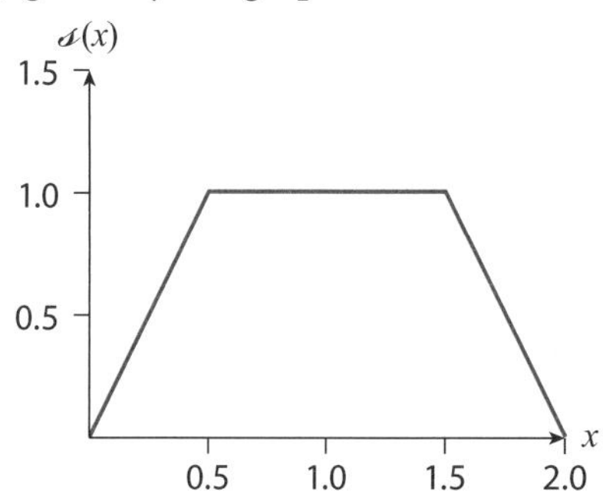

4. Pretend that the above graph is the graph of an object's instantaneous speed function $s(x)$. Use it to calculate the object's change in distance over the following intervals: (a) $[0, 0.5]$, (b) $[0, 1]$, and (c) $[0.5, 2]$.

5. The graph of the velocity of an object is shown below.

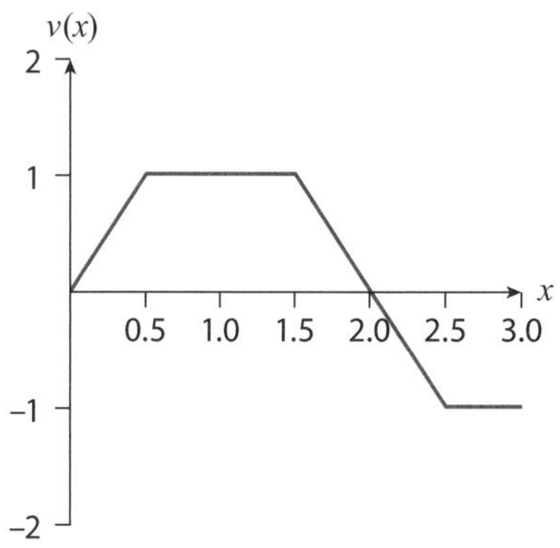

(a) Over what interval(s) is the object moving left? Right?

(b) Calculate $\int_0^1 v(x)\,dx$ and $\int_0^3 v(x)\,dx$, and interpret your results.

6. Consider the area function $A(t) = \int_0^t \sqrt{1+x^2}\,dx$, where $0 \leq t \leq 1$.

(a) Graph $A(t)$ as an area.

(b) Calculate $A'(t)$, and determine on what subinterval(s) of $[0, 1]$ $A(t)$ is increasing.

(c) Calculate $A''(t)$. On what subinterval(s) of $[0, 1]$ is $A(t)$ concave up?

7. Let $A(t) = \int_0^t x\,dx$.

(a) Calculate $A'(t)$.

(b) Let $g(t) = A(t^2)$. Calculate $g'(t)$.

8. Evaluate $\int_{-1}^1 \sqrt{1-x^2}\,dx$ using areas. (*Hint:* Graph the integrand first.)

9. Suppose a differentiable function f satisfies

$$\int_0^t f(x)\,dx = [f(t)]^2$$

for all t. Find possible formulas for f.

10–12: Verify that $F'(x) = f(x)$, and then use Theorem 5.2 to calculate $\int_a^b f(x)\,dx$ for the given a and b values.

10. $F(x) = (x+1)^2, f(x) = 2(x+1), a=0, b=1$

11. $F(x) = -\frac{1}{x}, f(x) = \frac{1}{x^2}, a=1, b=2$

12. $F(x) = \sqrt{x}, f(x) = \dfrac{1}{2\sqrt{x}}, a=1, b=9$

13. This exercise guides you through the proof of the Sum Rule from Theorem 5.5.

(a) Define

$$A_{f+g}(t) = \int_a^t [f(x)+g(x)]\,dx$$
$$A_f(t) = \int_a^t f(x)\,dx$$
$$A_g(t) = \int_a^t g(x)\,dx. \qquad (5.29)$$

What theorem allows us to conclude that $[A_{f+g}(t)]' = f(t)+g(t)$, $A_f'(t) = f(t)$, and $A_g'(t) = g(t)$?

(b) It follows from (a) that

$$\begin{aligned}[A_{f+g}(t)]' &= A_f'(t) + A_g'(t) \\ &= [A_f(t) + A_g(t)]'.\end{aligned}$$

What theorem was used to obtain that last equality?

(c) Following the reasoning in Section 5.1 allows us to conclude from $[A_{f+g}(t)]' = [A_f(t)+A_g(t)]'$ that $A_{f+g}(t) = A_f(t) + A_g(t) + C$. Why does setting $t=a$ finally yield the Sum Rule in Theorem 5.5?

14. Let's return to (5.1), which is equivalent to $A'(t) = d'(t)$.

(a) Explain what $A'(t) = d'(t)$ tells you about the graphs of $d(t)$ and $A(t)$.
(b) Consider now the function $g(t) = A(t) - d(t)$. What can you say about $g'(t)$, and why?
(c) Finally, explain why $g'(t) = 0$ implies that $A(t) = d(t) + C$.

15. Income Inequality Let's return to Applied Example 5.29.

(a) Explain why $L(0) = 0$, $L(1) = 1$, and why both x and $L(x)$ are numbers between 0 and 1.

(b) Explain why the Lorenz curve of a country in which every household has the same income is $L(x) = x$. Show that in this case $G = 0$ (no income inequality).

(c) In reality, every country's Lorenz curve satisfies $L(x) < x$. Explain what this means.

(d) Show that $L(x) < x$ implies $G > 0$ (income inequality is present).

16. Cardiac Output The **cardiac output** F of a person's heart is the volume (measured in liters) of blood the heart pumps per second. Cardiologists measure F by injecting a certain amount A (measured in milligrams) of dye into the right atrium of the heart and monitoring the concentration $c(t)$ (measured in mg/L) of dye in the aorta as the heart pumps. After some time T, all of the injected dye has flowed through the monitoring probe. Assuming F is constant, one can then show that

$$F = \frac{A}{\displaystyle\int_0^T c(t)\,dt}.$$

Suppose $c(t)$ is given in the figure below. Estimate F by estimating the area of the shaded region.

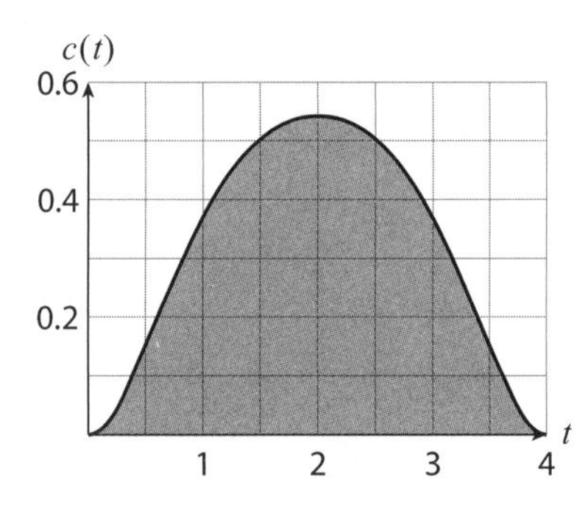

17–26: Evaluate the integral.

17. $\int_{-1}^{1} 4\,dx$

18. $\int_{0}^{2} (x^2 - 1)\,dx$

19. $\int_{0}^{2} \sqrt[3]{x}\,dx$

20. $\int_{0}^{1} x(1 + x^3)\,dx$

21. $\int \frac{x-2}{\sqrt{x}}\,dx$

22. $\int (y-1)(y-2)\,dy$

23. $\int_{0}^{1} (1 + 3z)^2\,dz$

24. $\int_{1}^{2} x\sqrt{x-1}\,dx$

25. $\int \frac{t^2}{\sqrt{1-t}}\,dt$

26. $\int_{0}^{a} x\sqrt{x^2 + a^2}\,dx \ (a > 0)$

27. What's wrong with this calculation:

$$\int_{-1}^{1} x^{-2}\,dx = \left.\frac{x^{-1}}{-1}\right|_{-1}^{1} = -2?$$

28–31: Use the facts that $\int_0^1 f(x)\,dx = 1$ and $\int_0^1 g(x)\,dx = 2$ to calculate the integrals.

28. $\int_{0}^{1} 7f(x)\,dx$

29. $\int_{0}^{1} (2f(x) + 3g(x))\,dx$

30. $\int_{-1}^{0} g(-x)\,dx$

31. $\int_{0}^{1} xf(x^2)\,dx$

32. World Oil Consumption Suppose $r(t)$ is the instantaneous rate of change of the world's oil consumption, measured in barrels of oil per year since 2017. Explain what $\int_0^{10} r(t)\,dt$ represents. Do you expect that quantity to be positive, negative, or zero? Briefly explain.

33. Parabolic Trajectories Sufficiently heavy objects thrown in the air and not straight up follow parabolic trajectories (neglecting air resistance). Here's why.

(a) Denote by $x(t)$ the object's horizontal position (say, distance from you). Ignoring air resistance, explain why $x(t) = v_x t$, where v_x is the object's initial horizontal velocity.

(b) Let $y(t)$ denote the object's vertical position (say, from the ground). Using part (a) and (5.28), show that $y(x) = Ax^2 + Bx + C$, where $A < 0$. (The graph of $y(x)$ is therefore a downward opening parabola.) Identify the constants A, B, and C, and the physical meanings of B and C.

34. Water Clocks This problem will show you how integrals can help you construct a "water clock," an ancient timekeeping device that uses draining water to measure the passage of time. Consider a cylindrical tank of height H feet and cross-sectional area A ft^2. Suppose the tank is filled with water to a depth of d feet ($d \leq H$), and then a small circular opening of area A_h ft^2 is cut out of the bottom. Denoting by h the water level height (in feet) in the tank, a result known as **Torricelli's Law** can be used to show that

$$h'(t) = \frac{8A_h}{A}\left(\frac{4A_h}{A}t - \sqrt{d}\right),$$

where t is the seconds since the water started pouring out of the bottom hole in the tank.

(a) In the case of $A = \pi$, $d = 2$, and where the bottom hole has diameter 1/16 of an inch (make sure to convert this to feet before calculating A_h), integrate the resulting $h'(t)$ equation to show that

$$h(t) = \left(\sqrt{2} - \frac{t}{4(96)^2}\right)^2.$$

(b) How far up from the tank's bottom should a mark be made on the side of the tank to indicate the passage of 1 hour (3600 seconds)? 2 hours (7200 seconds)?

35. The Experience Curve Companies generally get more efficient at producing their products over time, and hence their production costs decrease. Research has quantified this effect via the **experience curve**. Denoting by $P(n)$ the cost (in \$) of producing the n-th unit, a mathematical model for $P'(n)$ is

$$P'(n) = -aP(1)n^{-a-1},$$

where $n \geq 1$ and $a > 0$. In this problem we'll relate this to a popular model for the experience curve. For simplicity, let's assume that $a = 0.23$ and $P(1) = 100$.

(a) Calculate and interpret $P'(100)$.

(b) Calculate $P(n)$. (The function obtained is a particular form of **Henderson's Law**.)

(c) Show that each doubling in the number of units produced decreases production costs by roughly 15% (i.e., as a company gains more experience producing its products, its production costs decrease, hence the name "experience curve" for $P(n)$).

36. This problem will guide you through a more formal proof of (5.8).

(a) Recall that f in (5.8) is assumed continuous, so that it's also continuous on the subinterval $[t, t + \Delta t]$. What theorem from Chapter 4 guarantees that f has a maximum and minimum for some x-values in $[t, t + \Delta t]$?

(b) Denoting the minimum of f on that interval by $f(m)$ and the maximum by $f(M)$, where m and M are x-values in $[t, t + \Delta t]$, we now know that $f(m) \leq f(x) \leq f(M)$ for all x in $[t, t + \Delta t]$. Explain why it follows that

$$\int_t^{t+\Delta t} f(m)\,dx \leq \int_t^{t+\Delta t} f(x)\,dx \leq \int_t^{t+\Delta t} f(M)\,dx.$$

(c) Explain why the result in (b) implies that

$$f(m) \leq \frac{1}{\Delta t}\int_t^{t+\Delta t} f(x)\,dx \leq f(M).$$

(d) Using (5.5), we can rewrite the result in (c) as

$$f(m) \leq \frac{\Delta A}{\Delta t} \leq f(M).$$

Finally, explain why as $\Delta t \to 0$, these inequalities imply that

$$\lim_{\Delta t \to 0} \frac{\Delta A}{\Delta t} = f(t)$$

(which reproduces (5.8)).

37. If f is continuous everywhere and $c \in \mathbb{R}$, prove that

$$\int_{ca}^{cb} f(x)\,dx = c\int_a^b f(cx)\,dx.$$

38. If f is continuous everywhere and $c \in \mathbb{R}$, prove that

$$\int_a^b f(x + c)\,dx = \int_{a+c}^{b+c} f(x)\,dx.$$

39. Suppose f is continuous on $[0, a]$.

(a) If $f(-x) = f(x)$ (f is then called an **even function**), prove that

$$\int_{-a}^a f(x)\,dx = 2\int_0^a f(x)\,dx.$$

(b) If $f(-x) = -f(x)$ (f is then called an **odd function**), prove that

$$\int_{-a}^a f(x)\,dx = 0.$$

40. Suppose f' is continuous on $[a, b]$. Prove that

$$2\int_a^b f(x)f'(x)\,dx = [f(b)]^2 - [f(a)]^2.$$

41. The **average value** of a continuous function f with domain $[a,b]$ is defined by

$$f_{\text{av}} = \frac{1}{b-a}\int_a^b f(x)\,dx.$$

Find the average value of $f(x)=\sqrt{x}$ on $[0,2]$.

Exercises Involving Exponential and Logarithmic Functions

42–47: Evaluate the integral.

42. $\displaystyle\int 3e^{3x}\,dx$

43. $\displaystyle\int 5^x\,dx$

44. $\displaystyle\int_0^1 e^t\sqrt{1+e^t}\,dt$

45. $\displaystyle\int_e^{e^2} \frac{1}{z\ln z}\,dz$

46. $\displaystyle\int \frac{e^x}{\pi+e^x}\,dx$

47. $\displaystyle\int_1^e \frac{(\ln\theta)^2}{\theta}\,d\theta$

48. Suppose $x<0$. Use the Chain Rule to establish that $\frac{d}{dx}[\ln(-x)]=\frac{1}{x}$. Together with Theorem 3.9, this yields (5.20).

49. Use Property 4 from Theorem 5.6, along with the fact that $e^x\geq 1$ for $x>0$, to prove that

$$e^x \geq 1+x \quad \text{for } x\geq 0.$$

Then, use this result to prove that

$$e^x \geq 1+x+\frac{x^2}{2} \quad \text{for } x\geq 0.$$

50. Verify (via the equivalence (5.15)) that for $a\neq 0$,

$$\int te^{-at}\,dt = -\frac{e^{-at}(1+at)}{a^2}+C.$$

51. The **hyperbolic sine** and **hyperbolic cosine** functions, denoted by $\sinh(x)$ and $\cosh(x)$, respectively, are defined by

$$\sinh(x)=\frac{e^x-e^{-x}}{2}, \qquad \cosh(x)=\frac{e^x+e^{-x}}{2}.$$

Show that

$$\int \sinh(x)\,dx = \cosh(x)+C,$$

$$\int \cosh(x)\,dx = \sinh(x)+C.$$

52. The **prime number theorem** states that the number of primes less than or equal to a positive real number x—a function we'll denote by $p(x)$—can be approximated by

$$p(x)\approx \int_2^x \frac{1}{\ln t}\,dt$$

when x is large.

(a) Treating $\approx$ as $=$, calculate $p'(x)$. You'll find that $p'(x)>0$; interpret your result.

(b) Treating $\approx$ as $=$ again, calculate $p''(x)$. You'll find that $p''(x)<0$; interpret your result.

53. Population density Let $p(r)$ denote the number of people (in thousands) per square mile living a distance r miles from a city center, i.e., the **population density**. The total population living within x miles of the city's center is given by

$$P(x)=\int_0^x 2\pi r p(r)\,dr.$$

(a) Assuming p is continuous, use the Fundamental Theorem of Calculus and (4.10) (from Chapter 4) to approximate ΔP for a small change Δx in x. Interpret your result for $\Delta x>0$.

(b) Suppose $p(r)=6e^{-\pi r^2/100}$. Calculate $P(x)$ and $\lim_{x\to\infty} P(x)$; interpret the latter result.

Exercises Involving Trigonometric Functions

54–59: Evaluate the integral.

54. $\int (t^3 - \cos t)\, dt$

55. $\int (\csc^2 x - \sin x)\, dx$

56. $\int 2\sqrt{\cot t}\csc^2 t\, dt$

57. $\int_0^{\pi/4} \sin(3z)\, dz$

58. $\int_0^{\pi/4} \frac{\sin x}{\cos^2 x}\, dx$

59. $\int (1 + \tan\theta)\, d\theta$

60. Referring to Exercise 41, find the average value of $f(x) = \sin x$ on $[0, \pi]$.

61. Consider the integral

$$\int_{\pi/3}^{\pi/2} \sin\theta\sqrt{1 - 4\cos^2\theta}\, d\theta.$$

(a) Show that the substitution $u = 2\cos\theta$ converts the integral into

$$\frac{1}{2}\int_0^1 \sqrt{1 - u^2}\, du.$$

(b) Evaluate the integral in part (a) by using an area formula from geometry. *Hint:* Graph the region of interest—the area under the graph of $y = \sqrt{1 - x^2}$ for $0 \leq x \leq 1$.

62. Lung capacity Let v denote the velocity of air flowing into the lungs of a person at rest (we'll measure v in liters per second) during a respiratory cycle. A reasonable model for v is

$$v(t) = a\sin(bt),$$

where a and b are positive constants, and t denotes the time (measured in seconds) since the respiratory cycle began.

(a) What is the maximum air flow velocity? (Your answer will depend on a.)

(b) How long is one respiratory cycle? (Your answer will depend on b.)

(c) Let t^* equal half the number in part (b). Calculate

$$\int_0^{t^*} v(t)\, dt,$$

and explain what it represents.

63. Average temperature Home thermostats cycle on and off in their attempt to maintain a prescribed indoor temperature. Suppose that, on a hot summer day, your home's thermostat is regulating the temperature inside your house according to the function

$$T(t) = a + b\sin(ct),$$

where $T(t)$ is the temperature (in °F) at time t (measured in hours since midnight), and where a, b, and c are positive real numbers.

(a) Suppose you'd like the maximum temperature to be 76° F and the minimum to be 72° F. Find a and b.

(b) Using the a and b values found in part (a) and Exercise 41, show that the average temperature over a 24-hour period T_{av} is

$$T_{av} = 74 + \frac{1}{12c}\left[1 - \cos(24c)\right].$$

(c) Find the smallest nonzero c-value that makes $T_{av} = 74°$ F.

Epilogue

Let me be the first to congratulate you for working through this book. Calculus is a subject that is viewed by many as difficult, abstract, and inaccessible. I sincerely hope this book gave you the opposite experience. The many applied examples and applied exercises included should have given you a broad sense of how calculus can be applied to real-world phenomena. Some of these applications, as I tried hard to point out, drove the development of calculus. It is no surprise, then, that you will find calculus lurking in many of the physical, life, and social sciences.

Regardless of where and how calculus shows up in your future, I hope you'll also remember the main takeaways from this book. First and foremost, my initial answer to the question "what is calculus?" from Chapter 1:

Calculus is a mindset—a dynamics mindset. Contentwise, calculus is the mathematics of infinitesimal change.

You now have a *much* deeper understanding of what I meant—we used the dynamics mindset over and over again to develop the three pillars of calculus (limits, derivatives, and integrals). Each of these embodies the "infinitesimal change" nature of calculus, and I tried to help you remember the particulars via the chapter titles I chose:

- Limits: how to approach indefinitely (and thus never arrive)
- Derivatives: change, quantified
- Integration: adding up change

We covered a lot in this book. Yet there is always more math to learn. Depending on what field of study you ultimately choose to pursue, you might need more than just first-semester calculus (which was the focus of this book) to understand your discipline of interest. That is why I strongly encourage you to continue studying mathematics. It is the only subject whose results remain true *forever*—for example, Euclid's proofs of various relationships in geometry are as true today as they were millennia ago, and will continue to be so millennia from now. No matter where in the world you live or what language you speak, mathematics provides a unifying language with which to understand our world, our Universe, and our lives.

I hope you enjoyed the book, and that you continue studying mathematics.

Acknowledgments

Writing a book is a lot of work. What I always find surprising is just how many people it involves. With every book I write I forget that it is not just the author who writes the book; the reviewers who make suggestions, the family members who support the author through the writing, and the publication team that produces the book are all instrumental in creating the finished product. These are just a few of the people who helped me bring these words to you. Thank you to Vickie Kearn, my editor at Princeton University Press, and her publication and marketing team. Thank you to Zoraida, Emilia, Alicia, Maria, and the rest of my family for your continuing support and encouragement. Thank you to the reviewers, students and faculty alike, who took the time to provide valuable feedback on early drafts of the book. Thank you especially to Gwen Ncube for her careful reading of an early draft of this book, and her many helpful comments on it. And finally, thank *you*. Ultimately, all of these people's time and energy were directed toward one unifying goal: *to help you learn calculus*. Thank you for letting me be your guide on that adventure.

Appendix A Review of Algebra and Geometry

This appendix is a quick review of much of the algebra and geometry concepts you'll need to have mastered to succeed in calculus. We'll begin with a review of numbers and interval notation, then proceed to basic geometry, and end with a review of algebra. Ready? Let's get started.

A.1 A Quick Review of Numbers

The simplest types of numbers are "whole numbers," also called **natural numbers**: 1, 2, 3, etc. Adding zero and the negatives of the natural numbers to the list produces the **integers**. Next, we can envision dividing one integer by another (nonzero) integer. This yields a **rational number**—a *ratio* of one **integer** (called the **numerator**) with another nonzero integer (called the **denominator**): $\frac{a}{b}$.[1] Note that integers are particular types of rational numbers (namely, rational numbers with denominator 1), and natural numbers are particular types of integers. So thus far, rational numbers are the largest set of numbers we've discussed.

Let's now review the four arithmetic operations that combine two rational numbers (and by extension, integers and natural numbers) to produce another rational number: addition, subtraction, multiplication, and division.

Adding and Subtracting Rationals

We can only add (or subtract) rationals with a common denominator. So, the first step in adding (or subtracting) two rationals, say $\frac{a}{b}$ and $\frac{c}{d}$, is to find a common denominator. One choice that will always work is bd.[2] We then add (or subtract) the rationals by making each denominator bd (in this case that means multiplying the numerator and denominator of $\frac{a}{b}$ by d, and the numerator and denominator of $\frac{c}{d}$ by b) and then adding (or subtracting) the numerators together while keeping the denominator the same:

$$\frac{a}{b}+\frac{c}{d}=\frac{ad}{bd}+\frac{bc}{bd}=\frac{ad+bc}{bd}, \qquad \frac{a}{b}-\frac{c}{d}=\frac{ad-bc}{bd}.$$

EXAMPLE A.1 Add and subtract the rationals $\frac{1}{2}$ and $\frac{3}{7}$.

[1] Sometimes we also refer to a rational number as a **fraction**.

[2] In math, two things written next to each other implies multiplication. For example, bd means b times d.

Solution The common denominator is 14, so

$$\frac{1}{2}+\frac{3}{7}=\frac{7}{14}+\frac{6}{14}=\frac{13}{14}, \qquad \frac{1}{2}-\frac{3}{7}=\frac{7}{14}-\frac{6}{14}=\frac{1}{14}.$$ ■

Multiplying and Dividing Rationals

Multiplying rationals is easy. If $\frac{a}{b}$ and $\frac{c}{d}$ are two rationals, then[3]

$$\frac{a}{b}\cdot\frac{c}{d}=\frac{ac}{bd}.$$

So, you multiply the numerators together, and also the denominators.

Dividing rationals is almost as simple. The first thing to know is that dividing by a rational $\frac{c}{d}$ is the same as multiplying by its **reciprocal** $\frac{d}{c}$. Here's the proof:

$$\frac{1}{\frac{c}{d}}=\frac{\frac{d}{c}}{\frac{c}{d}\cdot\frac{d}{c}}=\frac{\frac{d}{c}}{1}=\frac{d}{c}.$$

We can then use this fact to divide two rationals:

$$\frac{\frac{a}{b}}{\frac{c}{d}}=\frac{a}{b}\div\frac{c}{d}=\frac{a}{b}\cdot\frac{d}{c}=\frac{ad}{bc}. \tag{A.1}$$

EXAMPLE A.2 Multiply and divide the rationals $\frac{1}{2}$ and $\frac{3}{7}$.

Solution

$$\frac{1}{2}\cdot\frac{3}{7}=\frac{1\cdot 3}{2\cdot 7}=\frac{3}{14}, \qquad \frac{\frac{1}{2}}{\frac{3}{7}}=\frac{1}{2}\div\frac{3}{7}=\frac{1}{2}\cdot\frac{7}{3}=\frac{7}{6}.$$ ■

Now that you've gotten some practice with rationals, let's move on to the "real number system."

A.2 The Real Number System and Interval Notation

In simplest terms, the **real number system** is the set of numbers ("real numbers") that can be written as decimals. One of three things can happen:

1. *The decimal terminates after a finite number of decimal places.* Examples: 2.7, 142 (which is 142.0).

2. *The decimal has an infinite number of decimal places, but there's a pattern that repeats.* Examples: 0.333 . . ., 37.146146146 . . .

[3]Here the "·" denotes multiplication. We'll see why we prefer this to the "×" notation when we discuss algebra in a few pages.

3. *The decimal has an infinite number of decimal places, but those numbers don't follow any pattern*. Example: π ("Pi").

Real numbers fitting into the first two of these cases can always be converted to rational numbers. Real numbers fitting into the last case are called **irrational numbers** (since they cannot be expressed as rational numbers). Perhaps the most famous irrational number is $\pi \approx 3.14159\ldots$, the irrational number that relates a circle's circumference C to its diameter d: $C = \pi d$.

Interval Notation

From the decimal expansion for π we know that

$$3 < \pi < 4.$$

The "$<$" sign here is the "less than" sign. Its sibling is the "$>$" sign, the "greater than" sign.[4] There are two similar symbols that are often encountered: $\leq$ ("less than or equal to") and $\geq$ ("greater than or equal to"). These symbols show up frequently when describing the possible values that variables can have. For example, squares with negative side lengths don't exist. Thus, in the area formula $A = x^2$ for a square of side length x, the real number x should satisfy $x \geq 0$. This can also be expressed in **interval notation**:

$$\{x \geq 0\} = [0, \infty).$$

By convention, a bracket next to a number in an interval indicates we include that number in the interval, while a parenthesis means we exclude it. Here are the other intervals that frequrently show up in algebra:[5]

$$(a, b) = \{a < x < b\} \qquad [a, b) = \{a \leq x < b\}$$
$$[a, b] = \{a \leq x \leq b\} \qquad (a, b] = \{a < x \leq b\}$$

Note that in this notation, the set of all real numbers is the interval $(-\infty, \infty)$, the entire real line (also denoted by $\mathbb{R}$).

Related Exercises 1.

Now that I've mentioned area, let's move on to a brief review of the basic geometric formulas.

A.3 A Quick Review of Some Formulas from Geometry

Figure A.1 contains a rectangle, a circle, and a triangle. Let's now review each shape's **perimeter**—the distance all the way around—and **area** formulas.

[4]The two are intimately related, since as our example illustrates, we could interpret $3 < \pi$ as "3 is less than π," or, viewing it as $\pi > 3$, it reads "π is greater than 3."

[5]Unfortunately, the interval notation (a, b) may also refer to a point in the Cartesian plane, but it should be clear from the context what (a, b) means.

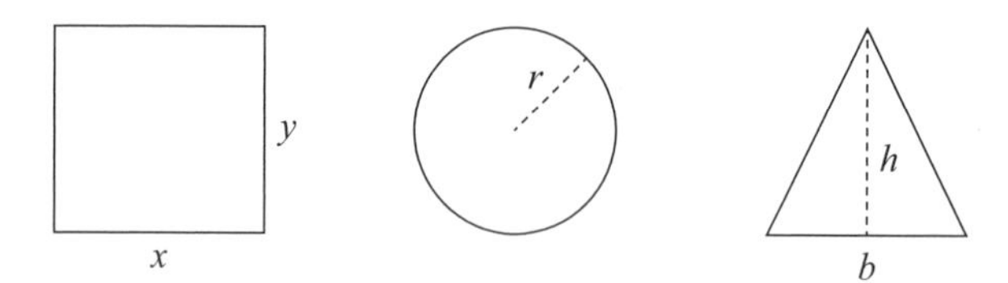

Figure A.1: A rectangle of width x and length y (left), a circle of radius r (middle), and a triangle of base length b and height h (right).

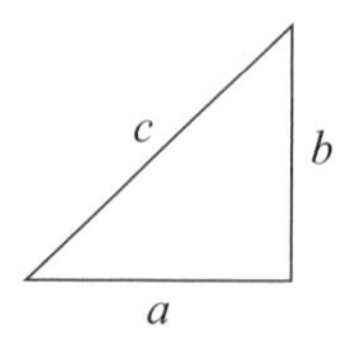

Figure A.2: A right triangle with side lengths a and b, and hypotenuse c.

- The perimeter P of the rectangle in Figure A.1 is

  $$P = 2x + 2y.$$

  This is because as we walk along the edges of the rectangle we walk its width twice (which contributes $2x$ to the perimeter) and its length twice (which contributes $2y$ to the perimeter). The area A of the rectangle is $A = xy$.

- The perimeter of the circle in Figure A.1 is called the **circumference** C, and $C = 2\pi r$, where $\pi \approx 3.14$ is the irrational number mentioned in the previous section. The area A of the circle is $A = \pi r^2$.

- The area of the triangle in Figure A.1 is $A = \frac{1}{2}bh$. The general formula for the perimeter of a triangle is rather ugly, and we won't encounter it in this book. However, let's discuss the perimeter formula for a *right* triangle, a triangle with one right angle (a 90 degree angle). This is really a way for me to remind you of the **Pythagorean Theorem**: For a right triangle with hypotenuse c and side lengths a and b (Figure A.2),

  $$a^2 + b^2 = c^2. \tag{A.2}$$

  The perimeter P of such a triangle is therefore

  $$P = a + b + \sqrt{a^2 + b^2}.$$

These formulas have many, many applications. Exercise 10 at the end of this appendix, for example, uses the Pythagorean Theorem to derive the formula for the distance between two points in the plane. Let me end this section with a simple example that illustrates a real-world application of these geometric formulas.

EXAMPLE A.3 Suppose your friend Bob calls you and says "I just got a dog, I have 20 feet of fence left over from a previous project, and I'd like to build a rectangular play area for my dog that's 6 feet long. Can you help me figure out how wide his play pen will be?" Solve Bob's problem.

Solution The amount of fencing is the perimeter of the rectangle: $P = 20$. We're also given the length: $y = 6$. Using the perimeter formula,

$$20 = 2x + 12 \quad \Longrightarrow \quad x = 4.$$

So, the play pen should be 4 feet wide. ■

Related Exercises 7–9.

Notice that this perimeter question led quite naturally to an algebraic equation. More complicated situations may lead to more complicated algebraic equations (for example, some area problems lead to quadratic equations). Therefore, let's now move on to the algebra review.

A.4 Solving Simple Algebraic Equations

Calculus is full of algebra. There will be many instances where you'll need to solve an equation for the unknown variable, usually denoted by x.[6] For example:

$$x + 2 = 10. \tag{A.3}$$

To solve this equation we undo everything that's being done to x. So, to solve for x we subtract 2 from both sides of the equation:

$$x + 2 - 2 = 10 - 2.$$

That leaves us with $x = 8$. The variable x, which at the start of the problem had an unknown value, has now taken on the particular value 8.

Let's now study a slightly harder example. Say we want to solve

$$2x + 4 = 14. \tag{A.4}$$

To do so, we isolate the term involving x (the $2x$ term) first by subtracting 4 from both sides,

$$2x = 10.$$

We now divide both sides by 2. This yields $x = 5$. Note that in this example, and the previous one too, we could have checked our answer by substituting the x-values we

[6]See Section B.1 for a discussion of what a variable is and the different types of variables you'll encounter in single-variable calculus courses.

found into the original equations and verifying that those equations are satisfied.[7] This is a general characteristic of algebra (and math): you can always check your work.

The highest power of x in equations in (A.3) and (A.4) was 1. Let's now review how to solve equations involving at most squared powers of x.

A.5 Quadratic Equations

First things first: x^2 is *defined* to mean x multiplied by x:

$$x^2 = x \cdot x.$$

Here the "·" denotes multiplication.[8] Equations involving squared powers of x, but not higher powers, are called **quadratic equations**. Here are a few examples:

$$x^2 = 4, \quad (x-1)^2 + 1 = 2, \quad x^2 + 5x + 6 = 0, \quad 3x^2 + 14x + 15 = 0. \qquad \text{(A.5)}$$

Let's review how we'd solve each equation.

EXAMPLE A.4 Solve the quadratic equation $x^2 = 4$.

Solution Here x is being squared and the outcome set equal to 4. Following our "undo what's being done to x" approach thus far, we want to undo the square on x. We do so by taking the square root of both sides. What we get is

$$\sqrt{x^2} = \sqrt{4}.$$

Many students (and even some math teachers) would make the following two claims at this point:

$$\sqrt{4} = \pm 2, \qquad \sqrt{x^2} = x.$$

(The "±" symbol here is "plus or minus.") *But neither of these is true.*[9] Here's what *is* true:

$$\sqrt{4} = 2, \qquad \sqrt{x^2} = |x|.$$

Here $|x|$ is the absolute value of x; $|x|$ equals x when $x \geq 0$, and $-x$ when $x < 0$.[10] Using these facts, we have that

$$\sqrt{x^2} = \sqrt{4} \quad \Longrightarrow \quad |x| = 2.$$

[7] For example, for equation (A.3), it is indeed true that $8 + 2 = 10$, and that 8 is the only such number which, when added to 2, yields 10.

[8] Were we to use the more familiar "×" we'd get the awkward looking $x^2 = x \times x$.

[9] See item 1 of the online supplement to this appendix for a detailed discussion of these incorrect claims, as well as a discussion of the correct ones.

[10] For example, $|5| = 5$ and $|-3| = 3$.

Using now the absolute value definition, this yields the two equations:

$$x=2, \qquad -x=2.$$

These equations yield the two solutions $x=\pm 2$. ■

EXAMPLE A.5 Solve the quadratic equation $(x-1)^2+1=2$.

Solution Subtracting 1 from both sides, and then taking the square root of both sides yields

$$\sqrt{(x-1)^2}=\sqrt{1}.$$

Following our earlier revelations about the square root operation, what we get is

$$|x-1|=1.$$

Using again the absolute value definition, this yields the two equations

$$x-1=1, \qquad -(x-1)=1.$$

These equations have the solutions $x=2$ and $x=0$, respectively. ■

Related Exercises 3 and 4(a).

Let's now return to the third equation in (A.5), the equation $x^2+5x+6=0$. This equation *can* be solved in the same manner as the previous two examples, but doing so requires a technique called **completing the square**. Let me show you how to solve it another way, using **factoring**.

Let me illustrate **factoring by grouping** first, using x^2+5x+6 as an example. To start, let's break 5 into $2+3$. Then,

$$5x=(2+3)x=2x+3x.$$

Here I've used the **distributive property**:

$$a(b+c)=ab+ac, \qquad (a+b)c=ac+bc.$$

Returning to x^2+5x+6:

$$x^2+5x+6=x^2+(2+3)x+6=x^2+2x+3x+6.$$

The blue terms have an x in common, while the black terms have a 3 in common. We can now factor out the x from the blue terms and the 3 from the black terms:[11]

$$x(x+2)+3(x+2).$$

We now have groups of terms, each of which has something in common: $x+2$. The next step is to factor out the $x+2$ out of both groups (that's why this process is called

[11]In this case, factoring involves using the distributive property in reverse, i.e., $x^2+2x=x(x+2)$.

factoring by grouping):

$$x(x+2)+3(x+2)=(x+2)(x+3).$$

We conclude that

$$x^2+5x+6=(x+2)(x+3).$$

Returning now to the third equation in (A.5), we can say that[12]

$$(x+2)(x+3)=0 \quad \Longrightarrow \quad x+2=0 \quad \text{or} \quad x+3=0.$$

Thus, the two solutions are $x=-2$ and $x=-3$.

I started the factoring by grouping process in the above example by breaking 5 into $2+3$. But 5 is also $1+4$. Had I used this instead we would've been left with

$$x^2+(1+4)x+6=x^2+x+4x+6=x(x+1)+2(2x+3).$$

But this time around there is no common "factor" of the form $ax+b$. So how did I know earlier to use $5=2+3$ and not $5=1+4$ to accomplish our factoring by grouping? The short answer is that for factoring by grouping to work for a quadratic expression of the form

$$x^2+bx+c,$$

b must be written as a sum of two numbers that multiply to c. (The numbers that multiply to c are called the **factors** of c.) In practice, it's easier to use this guiding principle in reverse: start with the factors of c, and see which ones sum to b. Let me illustrate that with a couple more examples.

EXAMPLE A.6 Factor the quadratic expressions:

(a) x^2+2x+1 (b) x^2+3x-4

Solution

(a) Here $b=2$ and $c=1$. The only factors of c are 1 and 1. It so happens that $2=1+1$, so

$$\begin{aligned} x^2+2x+1 &= x^2+(1+1)x+1 \\ &= x^2+x+x+1 \\ &= x(x+1)+(x+1) \\ &= (x+1)(x+1)=(x+1)^2. \end{aligned}$$

[12] When the product of two real numbers is zero, either one or both must be zero.

(b) Here $b = 3$ and $c = -4$. So, we're looking for factors of -4 that add to 3. The only such combination is 4 and -1. So:

$$\begin{aligned} x^2 + 3x - 4 &= x^2 + (4-1)x - 4 \\ &= x^2 + 4x - x - 4 \\ &= x(x+4) - (x+4) = (x-1)(x-4) \end{aligned}$$ ■

Related Exercises 2(a).

Thus far we've reviewed how to factor quadratics whose coefficient of x^2 is 1. For quadratics that don't fit that mold, like the last equation in (A.5), factoring them is tougher. A similar approach to what we just discussed works, but an even faster approach involves using the *quadratic formula*.

The Quadratic Formula

Let's say we're trying to solve the quadratic equation

$$ax^2 + bx + c = 0. \tag{A.6}$$

The "completing the square" technique can be used to simplify the left-hand side of (A.6). A few rearrangments later result in the following **quadratic formula.**[13]

The Quadratic Formula

The two solutions to the quadratic equation $ax^2 + bx + c = 0$ are

$$x = \frac{-b + \sqrt{b^2 - 4ac}}{2a} \quad \text{and} \quad x = \frac{-b - \sqrt{b^2 - 4ac}}{2a}.$$

These two solutions can be written in one equation by using the "±" ("plus or minus") symbol:

$$x = \frac{-b \pm \sqrt{b^2 - 4ac}}{2a}. \tag{A.7}$$

Let's now illustrate how to use the quadratic formula by solving the last equation in (A.5).

EXAMPLE A.7 Solve the quadratic equation $3x^2 + 14x + 15 = 0$.

[13] The quadratic formula is derived in item 2 of the online supplement to this appendix.

Solution Comparing $3x^2+14x+15$ to ax^2+bx+c, we see that $a=3$, $b=14$, and $c=15$. Therefore, (A.7) gives

$$x=\frac{-14\pm\sqrt{(14)^2-4(3)(15)}}{2(3)}=\frac{-14\pm\sqrt{16}}{6}=\frac{-14\pm 4}{6}.$$

So, our two solutions are

$$x=\frac{-14+4}{6}=-\frac{5}{3}, \qquad x=\frac{-14-4}{6}=-3. \qquad \blacksquare$$

Related Exercises 4.

Let me end by commenting on three aspects of the quadratic formula.

- If b^2-4ac is negative, then you'll have a negative number under the square root in (A.7). The square root of a negative number doesn't exist, so in these cases we say that the quadratic equation has no solutions.[14] Here's an example: $x^2+1=0$. The quadratic formula gives $x=\pm\sqrt{-1}$. But $\sqrt{-1}$ doesn't exist. Thus, there are no solutions to $x^2+1=0$.[15]
- If you cannot find the square root of b^2-4ac without a calculator, then you may have to use the rules of exponents (see the next section) to simplify your answer.
- Finally, if you get the two answers $x=A$ and $x=B$ from solving $ax^2+bx+c=0$, then that means $ax^2+bx+c=a(x-A)(x-B)$. This is a very useful way to factor ax^2+bx+c. (Try it on Exercises 2(b)–(c).) For example, in Example A.7,

$$3\left(x+\frac{5}{3}\right)(x+3)=3x^2+14x+15.$$

A.6 The Rules of Exponents

We've defined x^2 to mean $x\cdot x$. Similarly, we define $x^3=x\cdot x\cdot x$, and in general for any natural number n,

$$x^n=\underbrace{x\cdot x\cdot x\cdots x}_{n \text{ factors}}.$$

So, if we ever have to multiply x^n by something like x^m (here m is another natural number), we should expect that

$$x^n\cdot x^m=x^{m+n}, \tag{A.8}$$

[14]In a branch of mathemtics called *complex analysis*, $\sqrt{-1}$ is defined to be the new "number" i. Accordingly, any number containing i is called a *complex number*.

[15]We could also just realize that since $x^2\geq 0$, then $x^2+1\geq 1$ and can thus never equal zero.

because x^n is the product of n factors of x, x^m is the product of m factors of x, and so $x^n \cdot x^m$ is the product of $m+n$ factors of x, which is x^{m+n}.

Equation (A.8) is our first rule of exponents. By using similar reasoning we can derive a few other rules:

The Rules of Exponents

For m and n any two real numbers, the following rules hold:

$$\begin{aligned} x^m x^n &= x^{m+n} \\ (x^m)^n &= x^{mn} \\ (xy)^n &= x^n y^n \\ \left(\frac{x}{y}\right)^n &= \frac{x^n}{y^n} \\ x^{-n} &= \frac{1}{x^n} \\ \frac{x^m}{x^n} &= x^{m-n} \end{aligned}$$

These rules explain why we define $x^0 = 1$ for all $x \neq 0$.[16] Finally, we define

$$x^{\frac{a}{b}} = \sqrt[b]{x^a}. \qquad \left(\text{Example: } 3^{\frac{2}{3}} = \sqrt[3]{3^2}.\right)$$

These rules and definitions help us simplify expressions involving **radicals** ($\sqrt{\ }$ symbols) that sometimes show up in the quadratic equation.[17]

EXAMPLE A.8 Simplify the expression: (a) $2^3 2^5$ (b) $(x^2\sqrt{y})^3$ (c) $\left(\frac{x^2}{\sqrt[3]{y^2}}\right)^3$ (d) $\frac{x^{-1}+x}{x+y}$

Solution

(a) $2^{3+5} = 2^8$.

(b) $(x^2\sqrt{y})^3 = (x^2)^3(y^{1/2})^3 = x^6 y^{3/2} = x^6\sqrt{y^3} = x^6 y\sqrt{y}$.

(c)

$$\left(\frac{x^2}{\sqrt[3]{y^2}}\right)^3 = \frac{(x^2)^3}{(y^{2/3})^3} = \frac{x^6}{y^2}.$$

[16] Note that $x^0 = x^{1-1}$, and since $x \neq 0$, the Rules of Exponents yield $x^0 = x^1 x^{-1} = \frac{x}{x} = 1$.

[17] For example, $\sqrt{8} = 8^{1/2} = (4\cdot 2)^{1/2}$, and using the Third Rule, $(4\cdot 2)^{1/2} = 4^{1/2}\cdot 2^{1/2} = \sqrt{4}\sqrt{2} = 2\sqrt{2}$.

(d)

$$\frac{x^{-1}+x}{x+y} = \frac{\frac{1}{x}+x}{x+y} = \frac{\frac{1+x^2}{x}}{x+y} = \frac{1+x^2}{x(x+y)}.$$ ■

Related Exercises 5.

The Rules of Exponents can also be used to simplify expressions of the form $(x+a)^n$. For example, we can apply the first Rule to conclude that

$$(x+a)(x+a) = (x+a)^2.$$

Now, if we multiply out the left-hand side we see that

$$(x+a)^2 = x^2 + 2ax + a^2. \tag{A.9}$$

Similarly,

$$\begin{aligned}(x+a)^3 &= (x+a)^2(x+a)\\ &= (x^2+2ax+a^2)(x+a)\\ &= x(x^2+2ax+a^2) + a(x^2+2ax+a^2)\\ &= x^3+3x^2a+3xa^2+a^3. \end{aligned} \tag{A.10}$$

If we compare the formulas (A.9) and (A.10), we see a pattern: the expansion of $(x+a)^n$ starts with x^n and ends with a^n, and in each term in between the powers of a and x add to n.[18] It turns out that the **coefficients** in these expansions (the numbers in front of the terms involving x and a) also follow a pattern:

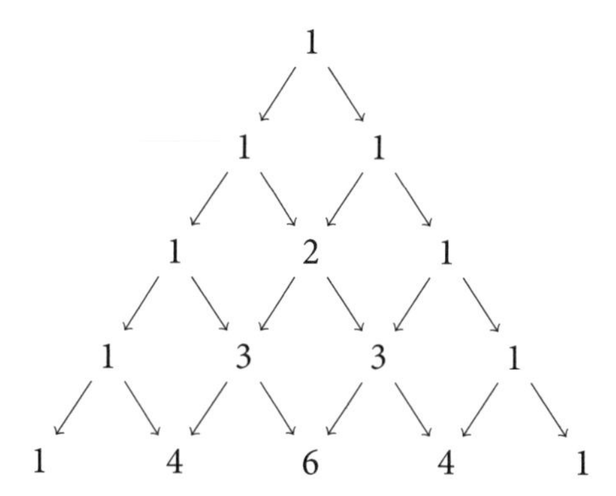

This triangle is called **Pascal's triangle**. The *second* row gives the coefficients in the "expansion" $(x+a)^1 = x+a$. Notice that the 2 in the *third* row is the sum of the two 1s in the second row, and that the third row gives the coefficients of x^2, ax, and a^2 in the expansion (A.9). In general, any number in a row between the two 1s of that row

[18]For example, in the terms $3x^2a$ and $3xa^2$ in (A.10), the powers of a and x add to 3, which is the power of $(x+a)^3$.

is the sum of the two numbers directly above it in the preceding row. For example, using the last row (and our observation about the powers of a and x adding to the "n" in $(x+a)^n$), we can say that

$$(x+a)^4 = 1x^4 + 4x^3a + 6x^2a^2 + 4xa^3 + 1a^4, \tag{A.11}$$

where I've put the coefficients of each term in blue to help you spot the last row of Pascal's triangle.

Related Exercises 6.

That concludes my quick review of the pre-calculus content you should definitely understand before starting Chapter 1.[19] I encourage you to try out a few exercises before starting Chapter 1, just to ensure what we've discussed in this appendix sunk in.

APPENDIX A EXERCISES

1. Write the following inequalities in interval notation (in each case x represents a real number):

(a) $-1 < x < 2$ (b) $x > 3$

(c) $x \leq -7$ (d) $0 \leq x \leq 1$

2. Factor the quadratics below.

(a) $x^2 + 4x + 3$ (b) $6x^2 + 5x + 1$

(c) $3x^2 + 10x + 8$

3. Solve the quadratic equations below.

(a) $x^2 + 2 = 18$ (b) $3x^2 - 5 = 22$

(c) $(x-2)^2 + 2 = 6$

4. Solve the quadratic equations below.

(a) $2x^2 - 5 = 11$ (b) $x^2 + 4x + 4 = 16$

(c) $6x^2 + 5x + 1 = 0$ (d) $x^2 = 7x - 3$

5. Simplify the expressions below.

(a) $(x+1)^{-1}$ (b) $(x+2)(x-2)^{-2}$

(c) $x^{-1} + (x+1)^{-1}$ (d) $(x^2(2x+7))^3$

(e) $\sqrt{16a^4b^5}$ (f) $\dfrac{\sqrt[3]{x^4}}{\sqrt[5]{x^6}}$

6. Write down the next row in Pascal's triangle (Section A.6). Then use it to help you expand $(x+a)^5$.

7. A triangle's height is twice its base length. If its area is 4 ft^2, find the base length.

8. If you double the radius of a circle, by what factor does the circle's area increase?

9. This problem proves the Pythagorean Theorem using the figure below.

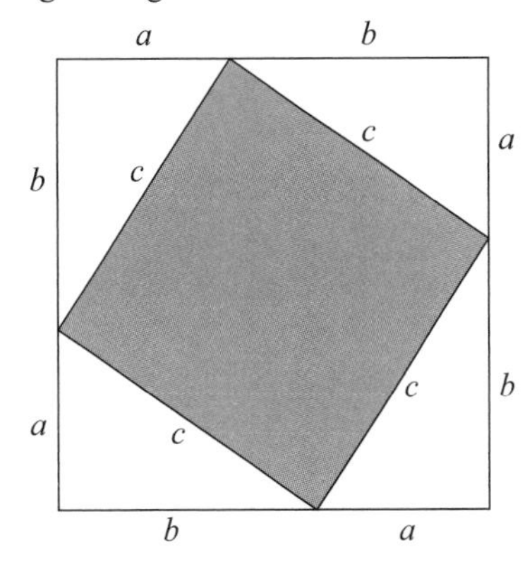

[19]Note: The formula sheet at the beginning of the book contains additional useful algebraic and geometric formulas.

(a) Show that the area of the outer (larger) square is $A = a^2 + 2ab + b^2$.

(b) We can also express A as the sum of the areas of the four white right triangles of base a and height b, and the area of the inscribed square of side length c. Show that doing so yields $A = c^2 + 2ab$.

(c) Compare the area formulas obtained to deduce that $c^2 = a^2 + b^2$.

10. Consider two points (a, b) and (x, y) in the plane, and let d be the straight-line distance between them (see diagram below). Use the Pythagorean Theorem to show that

$$d = \sqrt{(x-a)^2 + (y-b)^2}.$$

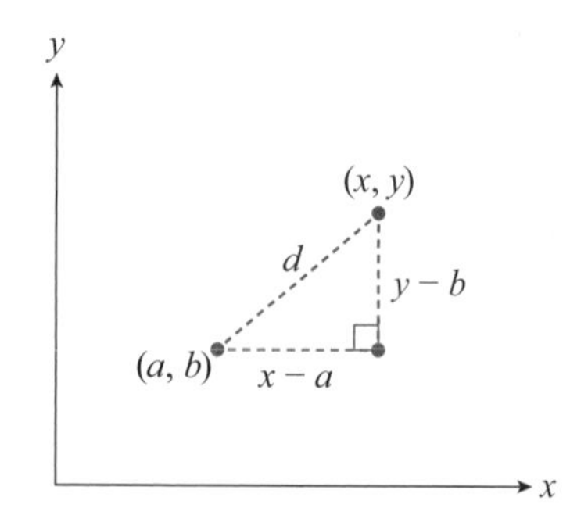

Appendix B Review of Functions

Nearly everything done in calculus is done to a function—we take limits of functions (Chapter 2), we differentiate functions (Chapter 3), and we integrate functions (Chapter 5). But why? What makes functions so central to calculus? And what is a function, anyway? Is it just mathematical jargon or do functions have any interesting real-world applications? This appendix discusses the answers to these questions.

B.1 Let's Talk About Variables (and Pancakes)

A big part of calculus is studying change. In math we quantify change by introducing **variables** (quantities that vary). For example:

- The temperature T inside your house is a variable.
- The amount of money M in your bank account is a variable.
- The area A of a pancake is a variable.

(In math we denote variables by italicized letters, and we often choose letters that remind us of the meaning of the quantity.) Often the variables we're interested in are related to other variables by an equation. Returning to the pancake example, suppose A is the area of a circular pancake of radius r. Then we know

$$A = \pi r^2.$$

Here π is the famous irrational number: $\pi \approx 3.14$. (See Appendix A for the definition of an irrational number.) This equation tells us that A is calculated by squaring the r-value and multiplying the result by π. Since the value of A *depends on* the value of r, we refer to A as the **dependent variable** and r as the **independent variable**. Moreover, we call a particular value of the independent variable an **input** and the resulting particular value of the dependent variable an **output**. (For example, for the input $r = 2$—a pancake of radius 2—the output is $A = \pi(2^2) = 4\pi$, a pancake of area about 12.6 square units.)

There's something special about the $A = \pi r^2$ formula you may not have noticed: *For each input there is exactly one output*. This means you'll never get *two* (or more) A-values for the same r-value. For example, the question "What's the area of a perfectly circular pancake of radius 2 inches?" has only one answer: 4π inches squared. Not all equations involving inputs and outputs have this property, however. For example, in the equation $x^2 + y^2 = 1$ (the equation of a unit circle with center at the origin), when $x = 0$ we get $y^2 = 1$, whose solutions are $y = -1$ and $y = 1$. Thus, the input $x=0$ yields *two different* outputs. Two-variable equations that obey the "for each input there is exactly one output" rule avoid such issues. This is motivation enough for the concept that underlies most of calculus: function.

B.2 What Is a Function Anyway?

Here's the general definition we'll be working with.

> **Definition B.1** Suppose the value of a variable quantity y depends solely on the value of another variable quantity x. This relationship is called a **(single-variable) function** if for each input x there is exactly one output y. We then write
>
> $$y=f(x)$$
>
> and say that "y is a function of x." We call f the function.

The "$f(x)$" notation is read "f of x." It serves two purposes. First, it reminds us that we're dealing with a function (so that we don't have to worry about ambiguities in the outputs). Second, it helps us keep track of the output that results from a particular input: if x is the input, then $f(x)$ is the output (recall $y=f(x)$). Inputting an x-value into a function is called **evaluating a function**. For example, "evaluate $f(x)=x^2$ at $x=2$" means "substitute 2 in for x," which yields $f(2)=4$.

Related Exercises 1–2(a).

B.3 The Domain of a Function

Definition B.1 includes the phrase "for each input." But how do you know which inputs can be substituted into a function? That depends on the **domain** of the function: the collection (set) of allowable inputs.

EXAMPLE B.1 Find the domain:

(a) $g(x)=\dfrac{1}{x}$

(b) $f(x)=x+4$

Solution

(a) The domain of g is all real numbers except for $x=0$.[1]

(b) All real numbers, denoted by the set $\mathbb{R}$, or equivalently the interval $(-\infty,\infty)$. (Appendix A reviews interval notation.) ■

[1] Division by zero is undefined; here's one reason why. Suppose $1/0=a$, some number. Multiplying both sides by zero yields $1=0$, which is not true. So, $1/0$ cannot equal any real number.

Sometimes we also need contextual knowledge to help us determine the domain. To see what I mean, let's briefly return to to our "pancake function" $A(r) = \pi r^2$. *Mathematically* we can substitute in any r-value we'd like. But in the real world pancakes with radii $r \leq 0$ don't exist. So the proper domain in this context is $(0, \infty)$.

B.4 Graphing Functions and Finding Their Range

Analogous to the concept of domain, we call the set of *outputs* that result from using the full set of domain values the **range** of the function. Finding a function's range is often done with the aid of its **graph**: the collection of input–output points $(x, f(x))$ for all values of x in the domain of f.

Notice that graphing a function requires knowing its domain (that's why we talked about domain in the previous section). Once we know the domain, we substitute each allowable input x into the function to get the associated output $f(x)$, and then create the points $(x, f(x))$. Plotting all such points in the xy-plane then generates the function's graph.

EXAMPLE B.2 Graph $f(x) = x^2$ and determine its range.

Solution The domain of f is $\mathbb{R}$, so we can substitute in any real number for x. I've chosen a few such numbers, calculated their $f(x)$-values, and assembled the corresponding points in the table in Figure B.1(a); these points are plotted in Figure B.1(b). Plotting more points would eventually generate the blue parabola in the figure. The figure suggests that the range is the set of y-values satisfying $y \geq 0$. Since $y = x^2$ is never negative and $f(0) = 0$, we conclude that the range is indeed $[0, \infty)$. ■

x	$f(x)$	$(x, f(x))$
0	0	(0, 0)
$\frac{1}{2}$	$\frac{1}{4}$	$(\frac{1}{2}, \frac{1}{4})$
1	1	(1, 1)
–1	1	(–1, 1)
2	4	(2, 4)

(a)

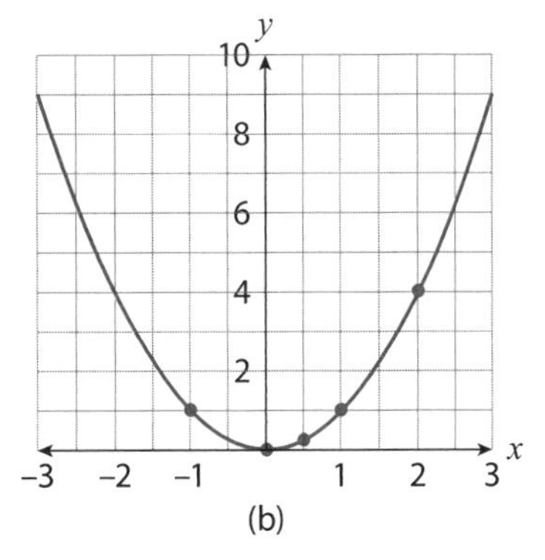

(b)

Figure B.1: (a) A table of values and points for $f(x) = x^2$. (b) The graph of $f(x) = x^2$ for $-3 \leq x \leq 3$.

As Figure B.1(b) illustrates, every point on the graph of a function tells us two things: the x-value and the value of f at that x-value, $f(x)$. Notice too that the $f(x)$-value tells us how far away from the x-axis the point is. For example, point $(2, 4)$ in Figure B.1(b) is 4 units above the x-axis.

EXAMPLE B.3 Consider the function graphed in Figure B.2.

(a) What are $f(0), f(2)$, and $f(5)$?

(b) What is the domain of f?

(c) What is the range of f?

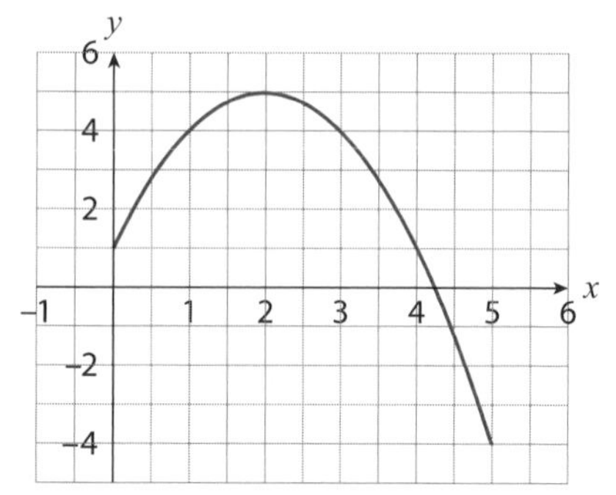

Figure B.2

Solution

(a) $f(0) = 1, f(2) = 5$, and $f(5) = -4$.

(b) The domain of f is $[0, 5]$ $(0 \leq x \leq 5)$.

(c) Every y-value between -4 and 5 (including both of those numbers) is attained for some x-value. Thus, the range of f is $[-4, 5]$. ∎

Related Exercises 2(b)–(c), and 3–6.

The functions graphed in the previous two figures are special cases of a family of functions known as *polynomials*. We'll soon study polynomials in general. But first, let's study the simplest polynomial: the linear function.

B.5 Linear Functions and Their Applications

Definition B.2 A function f is called **linear** if its equation can be written as

$$f(x) = mx + b, \tag{B.1}$$

where m is called the **slope** and $(0, b)$ the y**-intercept**.

Every linear function has domain $\mathbb{R}$ (we can input any real number in for x in (B.1)) and range $\mathbb{R}$. To understand that last claim we need to discuss the interpretations of the slope and y-intercept.

The y-intercept has a simple interpretation: it's the point where the graph of the linear function crosses (intercepts) the y-axis. The slope, on the other hand, measures how steep the line's graph is. (The graphs of linear functions are lines.) To understand that slope–steepness connection let's first discuss how to calculate the slope.

How to Calculate the Slope of a Line

Given two distinct points on a line, (x_1, y_1) and (x_2, y_2), the slope m of the line is

$$m = \frac{y_2 - y_1}{x_2 - x_1}. \tag{B.2}$$

Often (B.2) is expressed in terms of $\Delta y = y_2 - y_1$ and $\Delta x = x_2 - x_1$, the changes in the y-values and x-values, respectively, between the two points used to calculate m in (B.2). (The symbol "Δ" is the uppercase Greek letter delta; in math it typically signifies a change in a quantity.) Making those substitutions in (B.2) yields

$$m = \frac{\Delta y}{\Delta x}. \tag{B.3}$$

Due to this "change in y divided by change in x" definition, the slope is sometimes referred to as the "rise over run." If we multiply both sides of (B.3) by Δx we get

$$\Delta y = m\Delta x. \tag{B.4}$$

To appreciate how the slope–steepness connection arises from this equation, consider the familiar task of helping a friend load a moving truck (illustrated in Figure B.3).

The horizontal distance $\Delta x = 6$ is the same in both scenarios pictured. So, from (B.4) we have $\Delta y = 6m$. And as this equation implies, the steeper ramp (Figure (a)) has the largest slope.

Equation (B.4) also contains many other useful insights about lines and their slopes. For starters, suppose we're at point P on the graph of a line and move to the right 1 unit ($\Delta x = 1$). Equation (B.4) then tells us that the change in y-values, Δy, is $\Delta y = m$. If $m > 0$, this means we move *up* to reach the next point Q on the graph, so that the line tilts *up*. If $m = 0$ we travel horizontally (no tilt). And if $m < 0$ we move *down*, so that the line tilts *down*. Figure B.4 illustrates the $m > 0$ case.

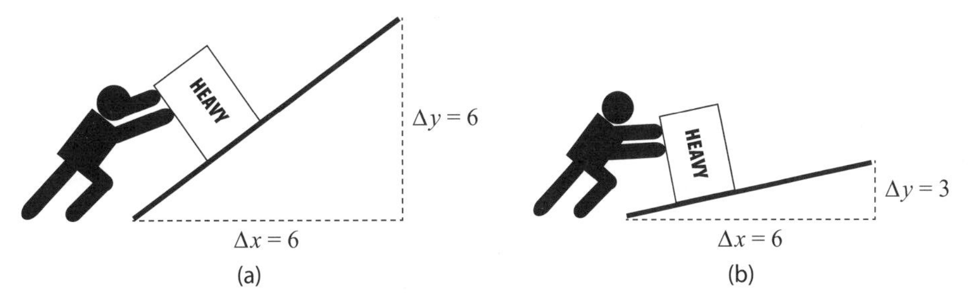

Figure B.3: Moving a heavy box up ramps of slope (a) 1 and (b) 0.5.

Figure B.4 also illustrates how one can graph a linear function. First, plot the y-intercept $(0, b)$. Then, move one unit to the right and either straight across (if $m = 0$), up by m (if $m > 0$), or down by m (if $m < 0$). Connect those two points and you've drawn the graph of the line.

Finally, equation (B.4) is useful for determining a convenient form for the equation of a line. If we return to Figure B.4 and suppose P has coordinates (x_1, y_1) and Q coordinates (x, y), then $\Delta y = y - y_1$ and $\Delta x = x - x_1$. Substituting these into (B.4) yields the familiar *point–slope* equation of a line.

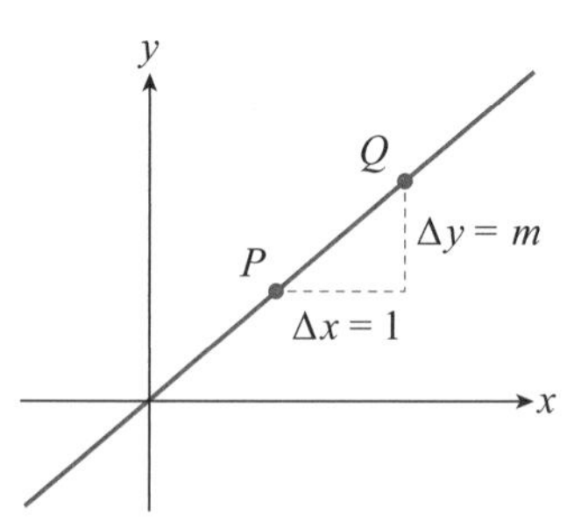

Figure B.4: The graph of a certain linear function with slope $m > 0$. Moving right 1 unit from point P (i.e., $\Delta x = 1$) results in a change in y of $\Delta y = m$, and so a move up to point Q.

The Point-Slope Equation of a Line

The equation of the line with slope m and passing through (x_1, y_1) is

$$y - y_1 = m(x - x_1). \tag{B.5}$$

EXAMPLE B.4 Find the equation of:

(a) The line passing through the two points $(-1, 1)$ and $(1, 3)$

(b) The line that has slope -3 and passes through $(1, 6)$

Solution

(a) We calculate the slope via (B.2):

$$m = \frac{3-1}{1-(-1)} = 1.$$

Then, using this and the point $(1, 3)$, the point–slope equation yields

$$y - 3 = (1)(x - 1) \quad \Longrightarrow \quad y = x + 2.$$

(b) We're given a point and a slope, so let's use (B.5):

$$y - 6 = -3(x - 1) \Longrightarrow y = -3x + 9.$$

■

Related Exercises 7–11.

In addition to applications to moving boxes, linear functions have many applications to the social and physical sciences (and beyond). Let's briefly explore these now (the exercises at the end of this appendix explore more applications of linear functions).

Applications of Linear Functions

Often linear functions are used to describe mathematically the relationship between two variables in a real-world problem. In general, "mathematizing" a real-world problem is referred to as **mathematical modeling**. (Mathematical Modeling is discussed in greater detail in Chapter 4.) In the contexts linear functions show up in, the slope and y-intercept often have useful real-world interpretations. The next example illustrates this.

APPLIED EXAMPLE B.5 Suppose you're traveling to Europe from the United States. Temperature in Europe is measured in Celsius (let's denote that by C). Luckily, the conversion back to Fahrenheit (denoted by F) is given by the linear function

$$F(C) = \frac{9}{5}C + 32. \tag{B.6}$$

(a) Identify the y-intercept and the slope. Then graph the function.

(b) Interpret the y-intercept and the slope.

Solution

(a) Comparing (B.6) to (B.1) we see that the slope is $\frac{9}{5} = 1.8$ and the y-intercept is $(0, 32)$. To plot (B.6) we first plot the y-intercept $(0, 32)$. Then, we move one unit right and 1.8 units up (the slope) and plot the resulting point, $(1, 33.8)$. Connecting these two points yields the graph of the line (Figure B.5).

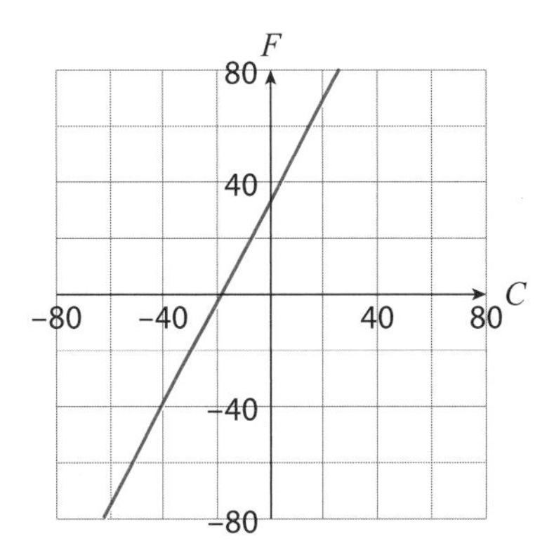

Figure B.5: The graph of the Celsius to Fahrenheit conversion equation $F(C) = \frac{9}{5}C + 32$.

(b) The y-intercept is easy to interpret: 0° Celsius converts to 32° Fahrenheit. To interpret the slope, note that (B.4) in this case reads

$$\Delta F = 1.8\Delta C.$$

When $\Delta C = 1$ this yields $\Delta F = 1.8$ (the slope). This says that a 1° Celsius increase in temperature equates to a 1.8° Fahrenheit increase. ■

The last useful thing to know about slopes in real-world contexts is their units. From (B.3) it follows that the units of the slope (m) are the units of the output y divided by the units of the input x. In the case of the previous example, for instance, the slope of 1.8 has units of degrees Fahrenheit divided by degrees Celsius. Due to the slope's units being a *ratio* of the units of the output and input, the slope is an example of a **rate of change**. (This is an important fact that underlies one of the main concepts in calculus—the derivative; see Chapters 3 and 4.) Here's another illustration of that idea.

APPLIED EXAMPLE B.6 Your friend is running in a 100-meter dash today. You're at the event, standing 200 meters away cheering her on. At noon the runners take off. The distance d (measured in meters) your friend is away from you turns out to be the function of time given by

$$d(t) = 200 - 3.9t,$$

where t is measured in seconds since noon. (Your friend is running toward you.)

(a) Identify the slope and y-intercept.

(b) Graph $d(t)$.

(c) Interpret your answers to (a) in the context of this problem.

(d) When does your friend finish the race?

Solution

(a) The slope is -3.9 and the y-intercept $(0, 200)$.

(b) To plot $d(t)$ we first plot the y-intercept $(0, 200)$. If we then move one unit right, since the slope is -3.9 we must move 3.9 units down. By connecting those two points we obtain the graph in Figure B.6.

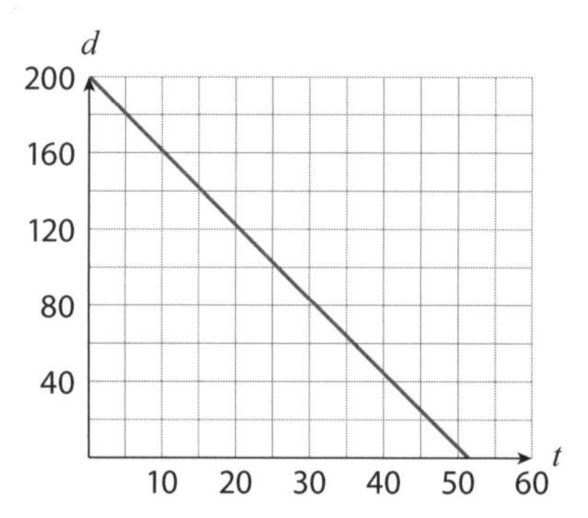

Figure B.6: The graph of $d(t) = 200 - 3.9t$.

(c) The y-intercept tells us that at noon your friend was 200 meters away from you. To interpret the slope, let's use (B.4), which now reads $\Delta d = -3.9\Delta t$. Therefore, for every second that passes ($\Delta t = 1$) the distance between you and your friend decreases 3.9 meters (since the slope is negative). Moreover, since the output (d) has units of meters and the input (t) units of seconds, the slope of -3.9 has units of meters/second, the units of velocity. These two conclusions tell us that the slope is the velocity with which your friend is running toward you.

(d) Your friend finishes the race when she reaches the 100-meter mark. At that time, she is 100 meters away from you, so $d(t) = 100$. Thus

$$100 = 200 - 3.9t \quad \Longrightarrow \quad t = \frac{100}{3.9} \approx 25.6 \text{ seconds}.$$

(The symbol $\approx$ used above means "is approximately.") ■

Related Exercises 18 and 24.

B.6 Other Algebraic Functions

Linear functions are examples of **algebraic functions**: functions that consist of a finite number of sums, differences, multiples, quotients, and radicals involving x^n (powers of x). Algebraic and nonalgebraic functions make up the usual cast of characters in the calculus story. Let's discuss algebraic functions now (nonalgebraic functions are discussed in the last two (optional) sections of this appendix).

Polynomials

The simplest algebraic functions are *polynomials*. These functions are finite sums of terms of the form ax^n, where a is a real number and n a nonnegative integer. Here's the general definition.

Definition B.3 A function f is called a **polynomial** if

$$f(x) = a_n x^n + a_{n-1}x^{n-1} + \cdots + a_2 x^2 + a_1 x + a_0,$$

where $n \geq 0$ is an integer and $a_0, a_1, \ldots, a_n$ are real numbers called the **coefficients**. The highest power of x present is called the **degree** of the polynomial.

The graphs of the constant polynomial ($n = 0, f(x) = a_0$) and the linear polynomial ($n = 1$, $f(x) = a_1 x + a_0$) are lines. All other polynomials' graphs are curves. Figure B.7(a) shows a few representative examples. As those graphs hint to, all polynomials have domain $\mathbb{R}$.

Power Functions

Polynomials are sums of terms of the form x^n. If we now allow n to be any real number we get the following new family of functions.

Definition B.4 A function f is called a **power function** if

$$f(x) = ax^b,$$

where a and b are real numbers.

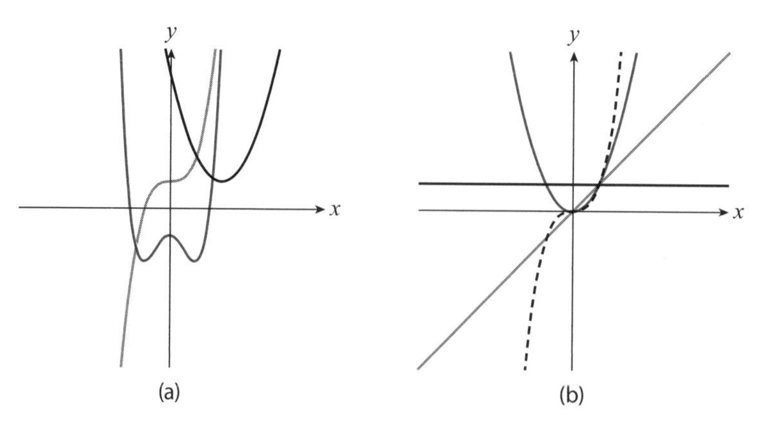

Figure B.7: (a) Portions of the graphs of $f(x) = x^2 - 4x + 5$, $g(x) = x^3 + 1$ (gray), and $h(x) = x^4 - 2x^2 - 1$. (b) Portions of the graphs of $f(x) = 1$, $f(x) = x$ (gray), $f(x) = x^2$, and $f(x) = x^3$ (dashed).

Power functions' graphs are heavily influenced by the type of number b is. Let's briefly discuss the three most interesting cases (we'll set $a = 1$ to make things easier).

Case 1: b is a nonnegative integer

In this case the power functions become 1, x, x^2, x^3, etc. Figure B.7(b) shows graphs of these first four; all power functions in this case have domain $\mathbb{R}$.

Case 2: b is a positive rational number

Here $b = \frac{m}{n}$, where m and n are integers ($n \neq 0$), and $m > 0$. The power function then takes the form

$$f(x) = x^{\frac{m}{n}} = \sqrt[m]{x^n}.$$

(One special case of this is $x^{\frac{1}{2}} = \sqrt[2]{x^1} = \sqrt{x}$; Appendix A reviews the Rules of Exponents used in these simplifications.) Figure B.8(a) shows the graphs of a few members of this family of functions. As those graphs suggest, the domain of this subfamily depends on m and n.

Case 3: b is a negative integer

In this case we can express b as $b = -n$, where n is a positive integer. The power function then takes the form

$$f(x) = x^{-n} = \frac{1}{x^n}.$$

The graphs of these functions are variants of the $n = 1$ (i.e., $b = -1$) case, the power function $f(x) = \frac{1}{x}$; Figure B.8(b) shows a portion of its graph. Since we

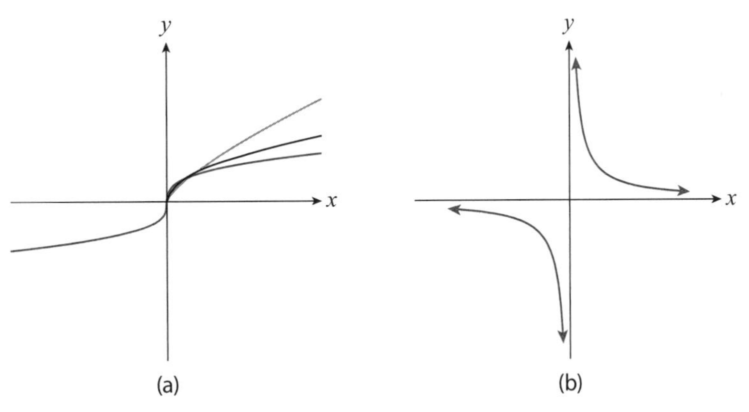

Figure B.8: (a) Portions of the graphs of $f(x)=\sqrt{x}$, $f(x)=\sqrt[4]{x^3}$ (gray), and $f(x)=\sqrt[3]{x}$. (b) A portion of the graph of $f(x)=\frac{1}{x}$.

can't divide by zero there is no y-value associated with $x=0$ for this subfamily of functions (we say that $f(0)$ is **undefined**). And because of this, the graph of the functions in this subfamily never cross the vertical line $x=0$ (the y-axis). (The functions *are* defined at every other x-value though, so their domain is all nonzero real numbers.) The vertical line $x=0$ is actually a **vertical asymptote** of all the functions in this subfamily.[2] These "do not cross" lines are common in the next family of functions we'll discuss (rational functions). But before leaving the power functions family, let's discuss one interesting application of them.

APPLIED EXAMPLE B.7 Many biological charateristics of organisms roughly follow power law functions of the organism's mass x, measured in grams (these are called **allometric scaling laws**). For example, mammals' lifespans L (in years) and heart rate H (in beats per minute) roughly follow the power laws

$$L(x)=2.33x^{0.21}, \quad H(x)=1,180.32x^{-0.25},$$

with humans seeming to be the only exception.[3] Use these functions to show that a mammal's heart beats about 1.5 billion times in their lifetime. (Humans achieve closer to 2.5 billion heartbeats in a lifetime.)

Solution Since 1 year contains 525,600 minutes, $525,600H(x)$ has units of heart beats per year. The total number of heart beats in a mammal's lifetime is then

$$525,600H(x)L(x)=(1.45\times 10^9)x^{-0.04}.$$

[2] We will discuss vertical asymptotes in detail in Chapter 2.

[3] These equations were derived in [4, 5], respectively. See also [3] for discussions of other allometric scaling laws.

This is another power law function. But since the exponent $b = -0.04$ is close to zero and since we define $x^0 = 1$,

$$525,600H(x)L(x) \approx (1.45 \times 10^9)x^0 \approx 1.5 \text{ billion heart beats.}$$

■

Related Exercises 12–15, 19–22, and 28.

Rational Functions

Definition B.5 A function f is called a **rational function** if

$$f(x) = \frac{p(x)}{q(x)},$$

where p and q are polynomials, and $q(x) \neq 0$.

The domain of a rational function must exclude all x-values for which $q(x) = 0$. Sometimes there are no such values (as in Figure B.9(a)), while other times there may be multiple such values (as in Figure B.9(b)). The graph in Figure B.9(b) has vertical asymptotes at $x = \pm 1$, the same x-values that make the denominator of the function, $x^2 - 1$, zero. The same phenomenon happened with $f(x) = \frac{1}{x}$. You may be tempted to conclude from these examples that rational functions have vertical asymptotes at the x-values that make their denominators zero. *But that's not always true*. For example, the rational function $f(x) = \frac{x}{x}$ doesn't have a vertical asymptote at $x = 0$. In fact, the definition of a vertical asymptote requires the calculus concept of limits; in Chapter 2 we'll discuss that definition of a vertical asymptote.

EXAMPLE B.8 Find the domain:

(a) $f(x) = \dfrac{x}{x-1}$

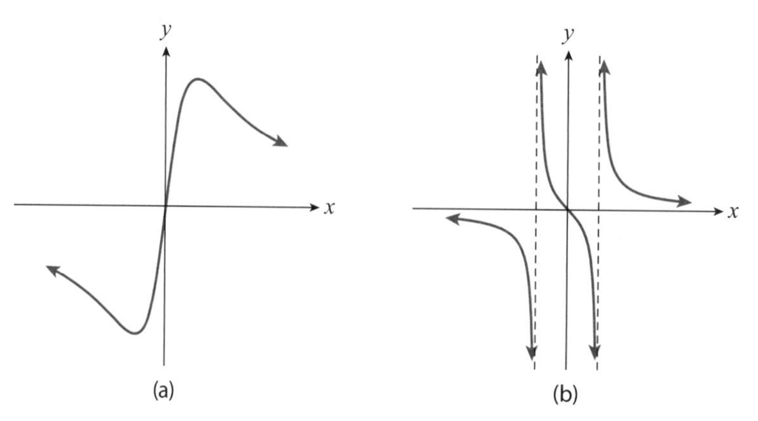

Figure B.9: Portions of the graphs of (a) $f(x) = \frac{x}{x^2+1}$ and (b) $g(x) = \frac{x}{x^2-1}$; the dashed lines in graph (b) are the vertical asymptotes at $x = \pm 1$.

(b) $g(x) = \dfrac{x^3+2}{x^4+x^3-4x^2-4x}$

Solution

(a) The domain is all real numbers except for $x = 1$.

(b) Let's first factor the denominator:

$$\begin{aligned} x^4+x^3-4x^2-4x &= x(x^3+x^2-4x-4) \\ &= x[x^2(x+1)-4(x+1)] \\ &= x(x+1)(x^2-4). \end{aligned}$$

The domain must exclude all x-values that make that last expression zero. Thus, the domain is all real numbers except $x = -2, -1, 0, 2$. ■

Related Exercises 16–17, 26, and 29.

B.7 Combinations of Algebraic Functions

We can combine members of the families of functions we've discussed to produce other functions. Some examples:

$$f(x) = x^2 + \sqrt{x}, \qquad g(x) = \frac{\sqrt{x}}{x^2+1}, \qquad h(x) = \sqrt{x^2+1}.$$

These functions are sums, quotients, roots, and compositions of polynomials and power functions. We've yet to discuss compositions of functions—what's happening in $h(x)$—so let's do that now.

Definition B.6 Suppose f and g are functions. The function $f \circ g$ defined by

$$(f \circ g)(x) = f(g(x))$$

is called the **composite** of f with g.

This definition says that the composite of f with g is obtained by replacing x in $f(x)$ with $g(x)$.[4] For example, if we replace x by $g(x) = x^2+1$ in the function $f(x) = \sqrt{x}$ we obtain the composite function

$$f(g(x)) = \sqrt{g(x)} = \sqrt{x^2+1},$$

the $h(x)$ function at the start of this section. (I've colored $g(x)$ to help you see what's going on.) Because finding composite functions is so important (as we'll see in Chapter 2), let's work through a few examples.

[4]Sometimes we call g the "inner function" and f the "outer function."

EXAMPLE B.9 Let $f(x)=x-1$ and $g(x)=\frac{1}{1+x}$. Find: (a) $f\circ g$, (b) $g\circ f$, (c) $f\circ f$

Solution

(a) We replace x in $f(x)$ with $g(x)$:

$$f(g(x))=g(x)-1=\left(\frac{1}{1+x}\right)-1=\frac{1-(1+x)}{1+x}=-\frac{x}{1+x}.$$

(b) We replace x in $g(x)$ with $f(x)$:

$$g(f(x))=\frac{1}{1+f(x)}=\frac{1}{1+(x-1)}=\frac{1}{x}.$$

(c) We replace x in $f(x)$ with $f(x)$:

$$f(f(x))=f(x)-1=(x-1)-1=x-2.$$

■

The previous example illustrates two facts about composite functions: (1) $f\circ g$ and $g\circ f$ are generally not the same, and (2) the domain of $f\circ g$ may differ from the domains of f and g.

Related Exercises 23 and 27.

The final way we will discuss to combine functions is to "combine them in pieces." Here's what I mean.

Definition B.7 A function f defined by two or more other functions, each with their own domain, is called a **piecewise function**.

As this definition suggests, a piecewise function is made up of other pieces of functions. Here's an example:

$$f(x)=\begin{cases} 2x, & 0\le x\le 1 \\ 3x-1, & 1<x\le 3 \end{cases}$$

(See Figure B.10 for its graph.) The only slight complication that arises in working with piecewise functions is keeping track of which function to use when. For example, to evaluate $f(1)$ for the piecewise (linear) function above we'd use the $2x$ "piece" of f, since 1 is in the first domain, $0\le x\le 1$. (We'd get $f(1)=2$.) But to find $f(2)$ we'd use the $3x-1$ piece, since 2 is in the second domain, $1<x\le 3$. (We'd get $f(2)=5$.)

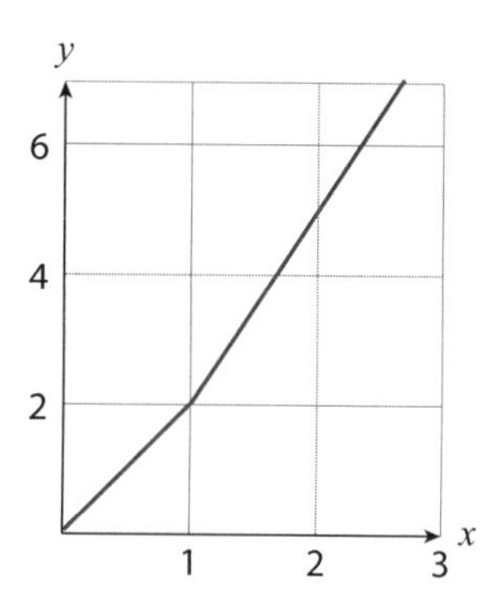

Figure B.10

B.8 Exponential and Logarithmic Functions

Historically speaking, logarithms appeared in the mathematical literature before exponential functions did.[5] But let's follow the modern approach to these topics and discuss exponential functions first.

Exponential Functions

Here's an intriguing question. Suppose a genie appears and gives you two choices:

(A) Receive \$10 million right now.

(B) After 30 days, receive the result of doubling each previous day's balance of an account starting with 1 dollar today.

Which option would you choose? If you've heard the saying "good things come to those who wait," you can predict the better choice. To explain it mathematically, let's denote by M the dollar amount resulting from option (B), and by x the number of days from today. Then,

$$M(0)=1, \quad M(1)=2, \quad M(2)=4, \quad M(3)=8, \quad \ldots$$

You may have already spotted the pattern in these numbers:

$$M(x)=2^x. \tag{B.7}$$

And as it turns out,

$$M(30)=2^{30}=\$1,000,737,418.23,$$

one *billion* dollars! So, tell that genie to come back in 30 days!

The rapid growth of $M(x)$ stems from it being an *exponential function.*

Definition B.8 Exponential Functions. Suppose a and b are real numbers, with $a \neq 0$, $b > 0$, and $b \neq 1$. The function

$$f(x)=ab^x \tag{B.8}$$

is called the **exponential function** with **base** b and **initial value** a. Moreover, x is called the **exponent**.

[5]The mathematician John Napier (also known as Jhone Neper) introduced logarithms in 1614; exponential functions seem to have first been discussed between 1661 and 1691 by the mathematicians Huygens and Leibniz (the co-inventor of calculus).

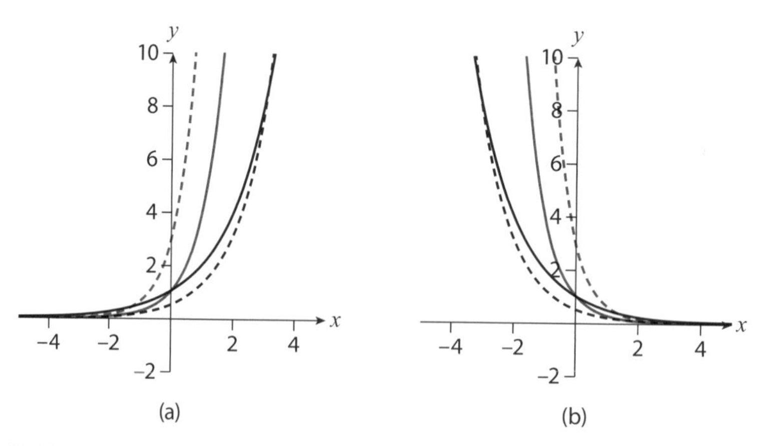

Figure B.11: (a) Portions of the graphs of $f(x) = 2^x$, $f(x) = 0.5(2.5)^x$ (dashed), $f(x) = 4^x$, and $f(x) = 3(5)^x$ (dashed). (b) Portions of the graphs of $f(x) = \left(\frac{1}{2}\right)^x$, $f(x) = 0.5\left(\frac{2}{5}\right)^x$ (dashed), $f(x) = \left(\frac{1}{4}\right)^x$, and $f(x) = 3\left(\frac{1}{5}\right)^x$ (dashed).

A few quick comments on this definition.

- Since $f(0) = ab^0$ and $b^0 = 1$ (see Appendix A.6), we get $f(0) = a$. Thus, the "first" y-value of $f(x) = ab^x$ is a, which explains why that's called the initial value of the exponential function.
- Notice that $f(x+1) = ab^{x+1} = ab^x b^1 = bf(x)$, where we've used the Rules of Exponents (see Appendix A.6). Thus, with each one-unit increase in x the y-value gets *multiplied* by b. If $b > 1$ we get larger y-values, whereas if $0 < b < 1$ we get smaller ones. We conclude that exponential functions' graphs *increase* if $b > 1$ and *decrease* if $b < 1$.[6] Accordingly, we call the $b > 1$ and $a > 0$ cases of (B.10) **exponential growth** and the $0 < b < 1$ and $a > 0$ cases **exponential decay**; Figures B.11(a) and (b) show a few examples of each case.
- Exponential functions are indeed functions (in that they satisfy Definition B.1). Moreover, the domain of every exponential function is all real numbers; the range is $(0, \infty)$ if $a > 0$ and $(-\infty, 0)$ if $a < 0$. Importantly, $f(x) = ab^x$ is *never equal to zero*. (In fact, $y = 0$ is a *horizontal asymptote* of all exponential functions; in Chapter 2 we will discuss horizontal asymptotes via the limit concept in calculus.)

Exponential functions show up naturally in money matters, as the $M(x)$ function from (B.7) suggests. Here's another example.

APPLIED EXAMPLE B.10 Suppose you open a savings account that pays r% interest each year and deposit \$100 into it. Denote by $M(t)$ the amount of money you have t years later.

[6] This agrees with what we found in (B.7), an exponential function with $b = 2$.

(a) Show that $M(t) = 100\,(1+r)^t$. (Here r is expressed as a decimal.)

(b) What is the account balance after 10 years, assuming $r = 5\%$?

Solution

(a) At the end of the first year you'll have

$$M(1) = 100 + 100r = 100\,(1+r)$$

dollars in the account. (Note that $M(1)$ is your starting deposit of \$100 plus the interest earned on that \$100.) At the end of the second year you'll have

$$M(2) = M(1) + rM(1) = 100\,(1+r)^2.$$

Therefore, at the end of year t,

$$M(t) = 100\,(1+r)^t.$$

(b) We have

$$M(10) = 100\,(1+0.05)^{10} \approx \$162.89.$$

That's about a 63% gain over 10 years. This is *larger* than the 50% gain one would naively assume (5% gains every year for 10 years). The **compounding effect** is responsible for the larger gain (interest earned on the \$100 deposit itself earns interest, i.e., compounds). ■

Related Exercises 30–32, 41–44.

Exponential functions also show up in economics. In particular, they help us understand **inflation**, the "general increase in the overall price level of the goods and services in the economy" [6]; see Section 3.2.1 of [7] for more information.

It turns out that one particular base of the exponential function shows up all throughout mathematics and science: e, **Euler's number**. This irrational number is approximately equal to 2.718, and is defined in terms of the calculus concept of limit (this is discussed in Sections A2.3 and A2.6 of the online appendix to Chapter 2). Using e, we can express any exponential function with base b as an exponential function with base e:

$$ab^x \iff ae^{rx}. \tag{B.9}$$

The number r needed to make this equivalence work is $r = \ln b$, the **natural logarithm** of b. This is a special case of the more general notion of a logarithmic function, so let's discuss this next.

Logarithmic Functions

Let me introduce logarithms by returning to (B.7) and the genie's two choices. Here's a question: How many days would it take for the "doubling the past day's account balance" choice to match the \$10 million choice? To find the answer we'd need to solve

$$2^x = 10,000,000. \tag{B.10}$$

To do so, we have to "unexponentiate" to get an equation of the form $x =$ answer. This is precisely what *logarithms* do:

$$2^x = 10,000,000 \iff x = \log_2(10,000,000) \approx 23.3. \tag{B.11}$$

So, by the 24th day the genie's second option already yields a sum of money greater than \$10 million.

Thus far, the notation "$\log_2$" is just that—notation for the number x for which $2^x = 10,000,000$. We could run through a similar analysis and calculate, for example, the number x for which $2^x = 15$. (That would yield $\log_2(15) \approx 3.9$.) If we now considered the outputs of these calculations as y-values and the inputs as x-values, we could generate the plot of $\log_2 x$. This plot would pass the **Vertical Line Test** (see Exercise 25), making $\log_2 x$ into a function. And since there was nothing special about the base $b = 2$, we could similarly generate the functions $\log_b x$ for any base b (recall that $b > 0$ and $b \neq 1$). This leads to the following definition.

Definition B.9 Logarithmic Functions. Suppose b is a real number, with $b > 0$ and $b \neq 1$. The function

$$f(x) = \log_b x \tag{B.12}$$

is called the **logarithm base** b of x. If $b = 10$ we write $\log_{10} x$ as $\log x$ and call this the **common logarithm**. If $b = e$ (Euler's number) we write $\log_e x$ as $\ln x$ and call this the **natural logarithm**.

Just to make sure, we say the right-hand side of (B.12) out loud as "log base b of x." Armed with logarithms, we can now "unexponentiate" the exponential equation $b^x = c$:

$$b^x = c \iff x = \log_b c. \tag{B.13}$$

Thus: *Logarithms are the exponents in an equation involving exponential functions.* That is literally what (B.13) says, since $\log_b c = x$ and x is the exponent in $b^x = c$. In particular, (B.13) yields $b^r = e \iff r = \ln b$, explaining the equivalence (B.9).

The equivalence (B.13) contains even more insights. First, let's insert the two equations in (B.13) into each other:

$$b^{\log_b c} = c, \qquad \log_b(b^x) = x. \tag{B.14}$$

The first equation says that if we start with b and raise it to the log base b of c, we get back c. The second says that if we take the base b logarithm of b-raised-to-the-x-power, we get back x. So, b^x and $\log_b x$ undo each other's operations (exponentiation and unexponentiation, respectively); we call pairs of such functions **inverse functions**. This relationship gives us information about the domain and range of logarithmic functions. Here are the important takeaways.

- As we previously discussed, the range of b^x is $(0,\infty)$. Therefore, in the leftmost equation in (B.14), the inputs c of $\log_b c$ are positive numbers. We conclude that *the domain of* $\log_b x$ *is* $(0,\infty)$.
- As we previously discussed, the domain of b^x is all real numbers. Therefore, in the rightmost equation in (B.14), the outputs of $\log_b x$ are real numbers. We conclude that *the range of* $\log_b x$ *is all real numbers.*
- We now see that *the domain and range swap* between the functions b^x and $\log_b x$. Therefore, the points (x,y) on their graphs swap their coordinates. Indeed, since (B.13) tells us that

$$y=b^x \iff \log_b y=x,$$

the point (x,y) on the graph of b^x becomes the point (y,x) on the graph of $\log_b x$. And since (y,x) is the point (x,y) reflected about the line $y=x$, we conclude that *the graphs of* b^x *and* $\log_b x$ *are reflections of each other about the line* $y=x$. (This also teaches us that $x=0$ is a *vertical asymptote* of all logarithmic functions; vertical asymptotes are discussed in detail in Chapter 2.) Figure B.12 illustrates this for the $b=2$ case.

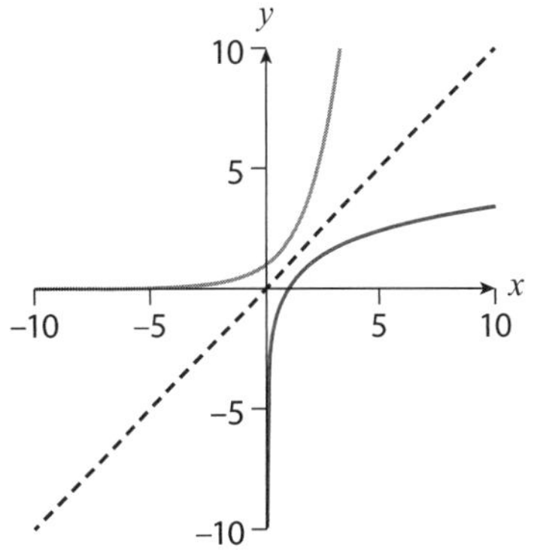

Figure B.12: Portions of the graphs of $f(x)=2^x$ (in gray) and $f(x)=\log_2 x$; the dashed line is the graph of $y=x$.

Let's now work through a few examples of how logs are used in practice.

EXAMPLE B.11 Solve each equation for x.

(a) $3\log x=1$ (b) $5^{2x+3}=7$

(c) $\log_3(2x+1)=1$ (d) $\log x+\log(x+3)=1$

Solution

(a) Let's use (B.13). Using that $\log x=\log_{10} x$,

$$\log_{10} x=\frac{1}{3} \implies x=10^{\frac{1}{3}}=\sqrt[3]{10}\approx 2.15.$$

(b) Using (B.13):

$$5^{2x+3}=7 \implies 2x+3=\log_5 7.$$

Solving for x yields

$$x = \frac{1}{2}\left[(\log_5 7) - 3\right] \approx -0.89.$$

(c) Using (B.13):

$$\log_3(2x+1) = 1 \quad \Longrightarrow \quad 2x+1 = 3^1 \quad \Longrightarrow \quad x = 1.$$

(d) Since the equation involves $\log = \log_{10}$, let's exponentiate each side of the equation:

$$10^{\log x + \log(x+3)} = 10^1 \quad \Longrightarrow \quad 10^{\log x} 10^{\log(x+3)} = 10^1,$$

where I've used the exponent property $10^{a+b} = 10^a 10^b$ (see the Rules of Exponents in Appendix A.) The leftmost equation in (B.14) tells us that $10^{\log x} = x$ and $10^{\log(x+3)} = x+3$. Thus, our equation becomes $x(x+3) = 10$, or $x^2 + 3x - 10 = 0$. This equation factors into $(x+5)(x-2) = 0$, and thus $x = -5$ and $x = 2$. *However*, we must reject the $x = -5$ answer, since $\log(-5)$ (what would be the first term in the original equation we were asked to solve) is not defined (recall that the domain of $\log_b x$ is $(0, \infty)$). Our final answer, then, is $x = 2$. ■

Part (d) of this example illustrates the need to keep the domain of the functions involved in mind. The example also illustrates how the Rules of Exponents can help simplify log calculations. In fact, the following Rules of Logarithms can be derived from the Rules of Exponents.

Theorem B.1 Rules of Logarithms. Let x and y be positive real numbers, and r be any real number. Then,

1. $\log_b(xy) = \log_b x + \log_b y$ 2. $\log_b\left(\frac{x}{y}\right) = \log_b x - \log_b y$
3. $\log_b(x^r) = r \log_b x$

Rule 3 can be used to derive the **change of base formula** (see Exercise 47):

$$\log_b c = \frac{\log_a c}{\log_a b}. \tag{B.15}$$

(This formula converts a base b logarithm into a base a logarithm.) Here's an applied example illustrating these new log properties.

APPLIED EXAMPLE B.12 Any vibrating object (say, a radio speaker) results in compressions and rarefactions of air molecules, creating a "pressure wave" that our ears detect as sound. The loudness L of a sound wave of pressure p can be measured by the function

$$L(p) = \ln(50,000p),$$

where the units of p are Pascals (Pa) and the units of L Nepers (Np).[7]

(a) Solve $L(p) = 0$. Then, using the fact that 2×10^{-5} Pa is about the threshold of human hearing,[8] interpret your answer.

(b) We're much more familiar with the **decibel** scale (dB). Given that 1 dB = $0.05 \ln 10$ Np, show that

$$L(p) = 20 \log(50,000p) \text{ dB}.$$

(Note the change from ln to log.)

(c) If the answer to part (b) is rewritten in the form $L(p) = A \log p + B$, what are A and B, and what does B represent physically?

Solution

(a) First, recall from Definition B.9 that $\ln x = \log_e x$. So, $L(p) = \log_e(50,000p)$. To solve $L(p) = 0$ we employ (B.13):

$$\log_e(50,000p) = 0 \quad \Longleftrightarrow \quad 50,000p = e^0.$$

Therefore, $p = \frac{1}{50,000} = 2 \times 10^{-5}$ Pa (since $e^0 = 1$). We conclude that $L(2 \times 10^{-5}) = 0$, which tells us that $L(p)$ has been calibrated so that the threshold of human hearing corresponds to a loudness of 0 Np.

(b) To convert to dB we multiply $L(p)$ by $1/(0.05 \ln 10)$:

$$\begin{aligned} \log_e(50,000p) \text{ Np} \cdot \frac{1 \text{ dB}}{0.05 \log_e 10 \text{ Np}} &= \frac{\log_e(50,000p)}{(0.05) \log_e 10} \text{ dB} \\ &= 20 \frac{\log_e(50,000p)}{\log_e 10} \text{ dB}. \end{aligned}$$

Using now (B.15):

$$\frac{\log_e(50,000p)}{\log_e 10} = \log_{10}(50,000p) = \log(50,000p).$$

Thus,

$$L(p) = 20 \log(50,000p) \text{ dB}.$$

(c) Using Theorem B.1:

$$20 \log(50,000p) = 20[\log(50,000) + \log p] = 20 \log p + 20 \log(50,000).$$

[7]The unit is named after Napier, the mathematician who introduced logarithms.

[8]This roughly corresponds to the sound of a mosquito flapping its wings about 10 feet away from you.

Therefore, $L(p) = A \log p + B$, where $A = 20$ and $B = 20\log(50,000)$. Note that $L(1) = B$. Thus, B is the loudness (in dB) corresponding to a sound wave of pressure $p = 1$ Pa. ■

Related Exercises 33–40, and 45–46.

Logarithms have many other applications. One very real—and useful—such application is to estimating how much longer you'll need to work before you can live off your accumulated savings. This **financial independence number** turns out to depend on the logarithm of the ratio of your yearly expenses to your yearly savings; spend less each year (and save more) and you'll decrease your number (see Section 3.3 in [7] for more details).

B.9 Trigonometric Functions

"Trigonometry" has its origins in the Greek words *trias* ("three"), *gonia* ("angle"), and *metron* ("measure"). To the ancient Greeks, therefore, trigonometry was the study of the relationships between sides and angles of triangles. And since the basic trigonometric functions—sine, cosine, and tangent—are defined using triangles, let's talk about angles and triangles.

Angles and Triangles

Figure B.13 shows a circle of radius r. Imagine now rotating the radius OA through an angle θ (called a **central angle**), measured counterclockwise from OA. (By convention, angles measured counterclockwise are positive, and those measured clockwise are negative.) The tip of the radius OA would trace out an arc on the circle of some length s (called **arc length**) between OA and the new radius OB. If θ is large enough, eventually we'll have $s = r$. The special central angle for which that is true is what defines a "radian," the preferred unit of angle measure in calculus.

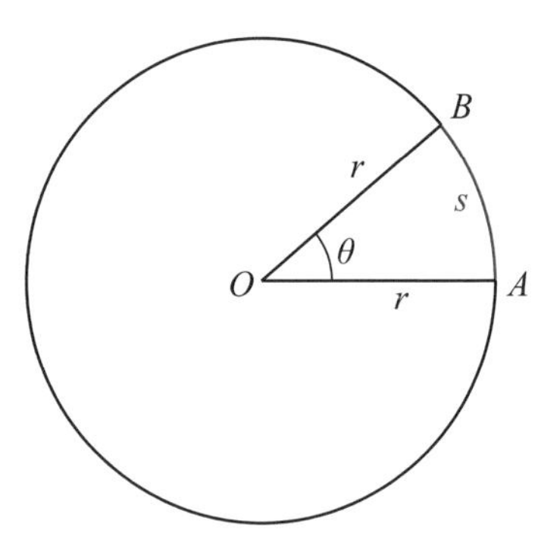

Figure B.13: The arc length s generated by a central angle θ on a circle of radius r.

> **Definition B.10 Radian Angle Measure.** We define an angle of 1 radian ("1 rad") to be the central angle of the arc whose length is equal to the radius of the circle.

Put another way, 1 rad is the central angle of the arc of length r on the circle of radius r.

Let's now return to Figure B.13. Imagine continuing to rotate OB counterclockwise until we go all the way around the circle and reach OA. We then say that we've traversed the **circumference** C of the circle. Ancient civilizations had long noticed that the ratio of C to d—the **diameter** of the circle, defined by $d=2r$—is constant, regardless of the size of the circle. Today we denote this constant by the Greek letter "π":

$$\frac{C}{d}=\pi \iff C=\pi d \iff C=2\pi r. \tag{B.16}$$

Returning to the terminology of Definition B.10, if 1 rad is the central angle of the arc of length r on the circle of radius r, then 2π rad is the central angle of the arc of length $2\pi r$ on the circle of radius r. But since $2\pi r$ is the circumference of that circle, we conclude that the central angle subtended by a circle of radius r is 2π radians. You are no doubt familiar with the alternative unit for measuring angles: **degrees**. In those units, one full revolution around the circle measures 360°. (This dates back to at least ancient Babylonian astronomers (circa 1900–1500 BC), who observed that the Sun takes about 30 days to move through each of the 12 constellations that make up the Zodiac. Since $30(12)=360$, each degree therefore roughly corresponds to one day's motion of the sun through the Zodiac.) Since we now know that one full revolution corresponds to an angle of 2π radians, we get the nifty conversion

$$\pi \text{ rad}=180^\circ \implies 1 \text{ rad}=\left(\frac{180^\circ}{\pi}\right)\approx 57.3^\circ. \tag{B.17}$$

Returning now to Figure B.13, we can set up the following proportion:

$$\frac{s}{2\pi r}=\frac{\theta}{2\pi} \iff s=r\theta. \tag{B.18}$$

This equation gives us the arc length s swept out by a central angle θ on a circle of radius r. Note: This equation ($s=r\theta$) should be used only when θ is measured in radians.

EXAMPLE B.13 Consider a circle of radius 4. Find the arc length subtended by the central angle $\theta=45^\circ$. What fraction of the circle's circumference is your answer?

Solution To use (B.18) we first need to convert 45° into radians. Using (B.17):

$$45^\circ \cdot \left(\frac{\pi \text{ rad}}{180^\circ}\right)=\frac{\pi}{4} \text{ rad}.$$

Then (B.18) yields

$$s=4\left(\frac{\pi}{4}\right)=\pi.$$

The circle's circumference is $C=2\pi(4)=8\pi$ (using (B.16)), so the arc is $\frac{\pi}{8\pi}=0.125$ (12.5%) of the circle's circumference. ■

Equation (B.18) also has many real-world applications, as the following example illustrates.

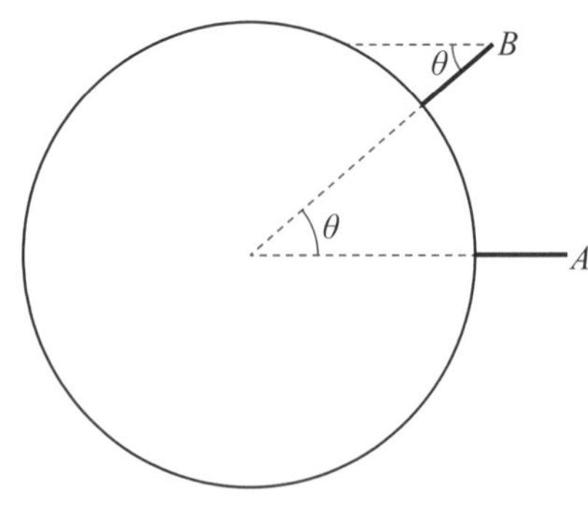

Figure B.14

APPLIED EXAMPLE B.14 Erastosthenes—a Greek mathematician who lived circa 250 BC—was perhaps the first person to accurately estimate the Earth's radius and circumference. (He assumed the Earth was a sphere.) Here's how he did it. Erastosthenes knew that on the summer solstice, the Sun shone directly overhead at noon in Syene (what is today Aswan, Egypt); this is location A in Figure B.14. On the same day, Erastosthenes was in Alexandria (location B in the figure) at noon and measured the angle θ that the Sun's rays made with the tip of the shadow of a pole. He measured $\theta \approx 7.12°$.

(a) Assuming the Sun's rays are parallel, explain why the two angles in Figure B.14 are equal.

(b) Convert 7.12° to radians.

(c) Erastosthenes knew that the distance from Alexandria to Syene was 5,000 stadia. A "stadion" was the length of an athletic stadium. A Greek stadion converted to about 607 feet, but an Egyptian stadium converted to about 517 feet. Given that 1 mile converts to 5,280 feet, convert 5,000 stadia to miles in each case.

(d) Interpreting the 5,000 stadia distance as arc length, estimate the radius of the Earth using (B.18) and your results from parts (b) and (c). (A modern-day estimate is 3, 959 miles.)

Solution

(a) Euclid proved that when two parallel lines are cut by a transversal, alternate angles are equal. Assuming the Sun's rays are parallel, the line in Figure B.14 connecting the center of the Earth and location B is a transversal; thus the two angles in the Figure are equal.

(b) The radian measure is

$$\frac{7.12\pi}{180} \approx 0.124.$$

(c) Using the Greek units, 5, 000 stadia converts to:

$$5,000 \text{ stadia} \cdot \frac{607 \text{ ft}}{1 \text{ stadion}} \cdot \frac{1 \text{ mile}}{5,280 \text{ ft}} \approx 574.81 \text{ miles.}$$

Using the Egyptian units, 5, 000 stadia converts to:

$$5,000 \text{ stadia} \cdot \frac{517 \text{ ft}}{1 \text{ stadion}} \cdot \frac{1 \text{ mile}}{5,280 \text{ ft}} \approx 489.58 \text{ miles.}$$

(d) From (B.18) we have $r = \frac{s}{\theta}$. Using the Greek units:

$$r \approx \frac{574.81}{0.124} \approx 4,635 \text{ miles.}$$

Using the Egyptian units:

$$r \approx \frac{489.58}{0.124} \approx 3,948 \text{ miles.}$$

The Greek units overestimate Earth's radius by about 17%. But the Egyptian units underestimate by about 0.28%! Either way, these are impressive estimates given the times (circa 200 BC!). ■

Related Exercises 48–52.

Okay, let's now move on to discussing triangles—shapes containing three ("tri") angles.

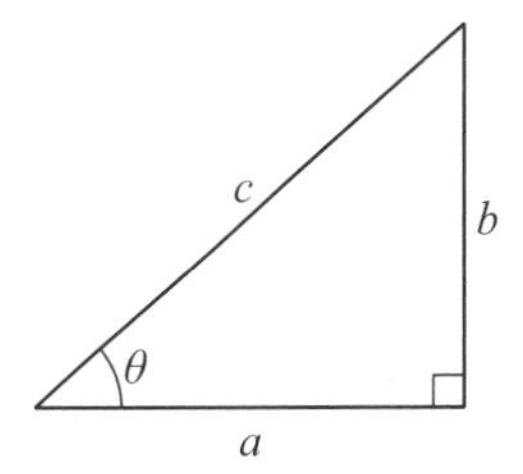

Figure B.15: A right triangle with side lengths a and b, hypotenuse c, and acute angle θ.

The ancient Greek mathematician Euclid—considered by many to be the "father of geometry"—studied and proved many results about triangles in his important treatise on geometry titled *Elements* (circa 300 BC). Among these are properties of **right triangles**—triangles including an angle measuring 90° (a "right angle"); Figure B.15 shows one such triangle. As Euclid proved, the sum of the angles in a triangle is 180°, so we know that the angle $\theta < 90°$ in the Figure. (Such angles are called **acute** angles.) The three sides of the triangle in the Figure have special names depending on their location relative to the right angle and the angle θ: the side of length c is called the **hypotenuse** (since it's the side opposite the right angle), the side of length a is the angle's **adjacent** side, and the side of length b is the angle's **opposite** side. Euclid proved in *Elements* that

$$a^2 + b^2 = c^2. \tag{B.19}$$

This is now known as the **Pythagorean Theorem**. (Greek writings attribute this theorem to Pythagoras (circa 500 BC).) The Pythagorean Theorem only relates the squares of the side lengths of a right triangle. Later on, the Greeks also studied the relationships between a, b, and c themselves, eventually leading to what we today recognize as the three basic trigonometric ratios:

$$\sin\theta = \frac{\text{opp.}}{\text{hyp.}} = \frac{b}{c}, \quad \cos\theta = \frac{\text{adj.}}{\text{hyp.}} = \frac{a}{c}, \quad \tan\theta = \frac{\text{opp.}}{\text{adj.}} = \frac{b}{a}. \tag{B.20}$$

A few quick comments:

- The abbreviations "adj," "opp," and "hyp" refer to the adjacent, opposite, and hypotenuse side lengths in Figure B.15 relative to the angle θ in the figure.
- Be careful with the notation: "$\sin\theta$" is read "sine of theta." It is not "sin times theta." Indeed, "sin " by itself is meaningless in math.
- Note that $\tan\theta = \frac{\sin\theta}{\cos\theta}$.

EXAMPLE B.15 Refer to Figure B.15 to answer the following questions.

(a) Calculate $\sin\theta$ and $\cos\theta$ if $a = b = 1$. What is θ (give both radian and degree measures)?

(b) Given that $\sin 30° = \frac{1}{2}$, calculate $\cos 30°$ and $\tan 30°$.

Solution

(a) From (B.19) we get that $c^2 = 1^2 + 1^2$, so that $c = \sqrt{2}$. (We ignore the $c = -\sqrt{2}$ solution to $c^2 = 2$ because c is a distance.) It follows from (B.20) that

$$\sin\theta = \frac{1}{\sqrt{2}} = \frac{\sqrt{2}}{2}, \qquad \cos\theta = \frac{\sqrt{2}}{2}, \qquad \tan\theta = 1.$$

Since $a = b$ (when this happens the triangle is said to be **isosceles**), then the non-right interior angles of the triangle are equal. (Euclid proved that the base angles of isosceles triangles are equal.) Since these two angles must add to 90°, we conclude that $\theta = 45° = \frac{\pi}{4}$.

(b) We can use the given sine value to construct a right triangle in which $b = 1$ and $c = 2$. This won't be the only triangle for which $\sin\theta = \frac{1}{2}$, but every other triangle will be "similar" to that one. (We say that two triangles are **similar** if their angles are pairwise equal and their corresponding sides are in constant proportion.) We can then find a from (B.19):

$$a^2 = c^2 - b^2 = 3,$$

which yields $a = \sqrt{3}$. It then follows from (B.20) that

$$\cos 30° = \frac{\sqrt{3}}{2}, \qquad \tan 30° = \frac{\sqrt{3}}{3}.$$

■

Related Exercises 53, and 58–60.

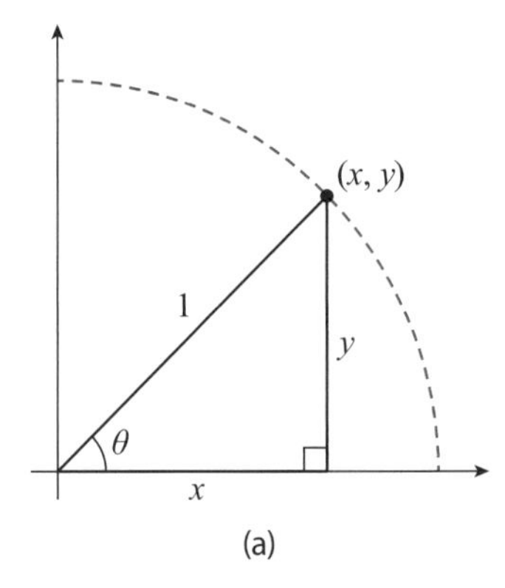

(a)

(x, y)	θ	$\cos\theta$	$\sin\theta$	$\tan\theta$
$(1, 0)$	$0\ (0^\circ)$	1	0	0
$\left(\frac{\sqrt{3}}{2}, \frac{1}{2}\right)$	$\frac{\pi}{6}\ (30^\circ)$	$\frac{\sqrt{3}}{2}$	$\frac{1}{2}$	$\frac{\sqrt{3}}{3}$
$\left(\frac{\sqrt{2}}{2}, \frac{\sqrt{2}}{2}\right)$	$\frac{\pi}{4}\ (45^\circ)$	$\frac{\sqrt{2}}{2}$	$\frac{\sqrt{2}}{2}$	1
$\left(\frac{1}{2}, \frac{\sqrt{3}}{2}\right)$	$\frac{\pi}{3}\ (60^\circ)$	$\frac{1}{2}$	$\frac{\sqrt{3}}{2}$	$\sqrt{3}$
$(0, 1)$	$\frac{\pi}{2}\ (90^\circ)$	0	1	und.

(b)

Figure B.16: (a) A circle of radius 1 (in blue) with an inscribed right triangle. (b) Points (x, y) on the unit circle and their associated $x = \sin\theta$ and $y = \cos\theta$ values for $0 \leq \theta \leq \frac{\pi}{2}$; note that $\tan 90^\circ$ is undefined (since $\cos 90^\circ = 0$).

As the previous example shows, we can use Euclidean geometry to extract lots of values of sine, cosine, and tangent. Moreover, if we embed the triangle in Figure B.15 in a circle we can extract *all* their values. Here's how.

First, let's insert the right triangle in Figure B.15 into the Cartesian plane and set $a = x$, $b = y$, and $c = 1$ (see Figure B.16(a)). Then (B.20) becomes

$$\sin\theta = y, \qquad \cos\theta = x, \qquad \tan\theta = \frac{y}{x}. \tag{B.21}$$

If we now imagine θ varying, the hypotenuse in Figure B.16(a) would produce a circle of radius 1 (the blue dashed curve in Figure B.16(a)). Any point (x, y) on this circle, therefore, would generate particular $\sin\theta$, $\cos\theta$, and $\tan\theta$ values via (B.21). Importantly, this teaches us that $\cos\theta$ *is the x-coordinate of a point on the unit circle, while* $\sin\theta$ *is that same point's y-coordinate*. Therefore, we can associate points on the unit circle with values of $\sin\theta$ and $\cos\theta$! The table in Figure B.16(b) illustrates this fact.

Let's now discuss how to derive the values of the trio of trigonometric ratios for θ-values in Quadrants II–IV (see Figure B.17(a) for a reminder of the angle ranges corresponding to each quadrant of the plane). To start, note that each point (x, y) on the unit circle is naturally associated with three other points:

(a) $(-x, y)$, the point horizontally opposite from (x, y)

(b) $(-x, -y)$, the point (x, y) reflected about the origin

(c) $(x, -y)$, the point vertically opposite from (x, y).

Figure B.17(b) illustrates these sister points. As the figure's caption indicates, the angle measures associated with each of the three latter points are the angle measures associated with (x, y) plus an integer multiple of $\pi/2$. Therefore, as θ varies from 0°

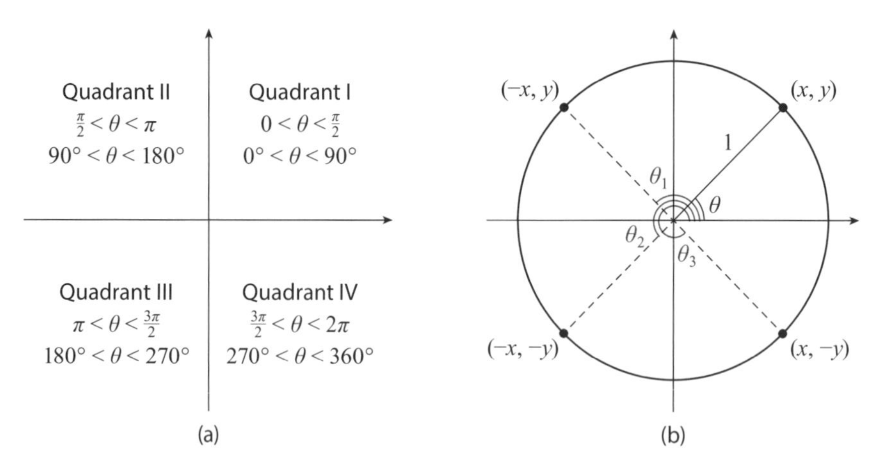

Figure B.17: (a) Angle ranges for each quadrant. (b) A point (x, y) on the unit circle at an angle θ, and the points: $(-x, y)$ at $\theta_1 = \theta + \frac{\pi}{2}$, $(-x, -y)$ at $\theta_2 = \theta + \pi$, and $(x, -y)$ at $\theta_3 = \theta + \frac{3\pi}{2}$.

to 90° in Quadrant I, the values of $\sin\theta$, $\cos\theta$, and $\tan\theta$ generated also generate values in Quadrants II–IV. Moreover, the symmetry of Figure B.17(b) hints at relationships between the trigonometric ratios corresponding to the four points in the figure. For example,

$$\sin\theta = \sin\theta_1,$$

since (x, y) and $(-x, y)$ have the same y-value, and since $\sin\theta$ is the y-coordinate of a point on the unit circle. Similarly, $\cos\theta = \cos\theta_3$. We'll soon use these insights to help us understand the functional nature of $\sin\theta$, $\cos\theta$, and $\tan\theta$, and also to help us graph them.

Before getting there, let me briefly discuss one more insight our unit circle embedding of a right triangle yields. Applying the Pythagorean Theorem to the triangle in Figure B.16 yields $x^2 + y^2 = 1$. Using (B.21), this becomes

$$(\cos\theta)^2 + (\sin\theta)^2 = 1.$$

By convention, we write $(\sin\theta)^2 = \sin^2\theta$ and $(\cos\theta)^2 = \cos^2\theta$, which yields the trigonometric identity

$$\sin^2\theta + \cos^2\theta = 1. \tag{B.22}$$

Many other trigonometric identities can be derived from the unit circle, including the following two (which we'll use in Chapter 3):

$$\sin(a + b) = \sin(a)\cos(b) + \sin(b)\cos(a) \tag{B.23}$$

$$\cos(a + b) = \cos(a)\cos(b) - \sin(a)\sin(b). \tag{B.24}$$

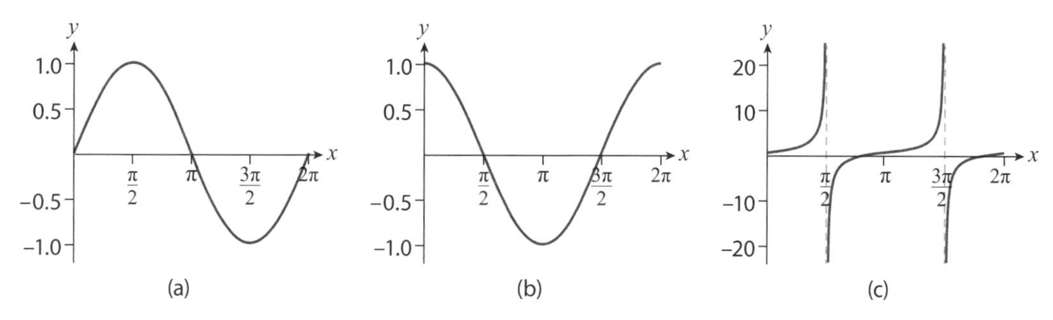

Figure B.18: The graphs of (a) $f(\theta)=\sin\theta$, (b) $f(\theta)=\cos\theta$, and (c) $f(\theta)=\tan\theta$ for $0\leq\theta\leq 2\pi$.

Trigonometric Functions

Returning to Figures B.16 and B.17, it's clear that to each angle θ there corresponds a unique point (x, y) on the unit circle. We conclude that for each θ there is a unique $\cos\theta$ value and a unique $\sin\theta$ value. Thus, $\cos\theta$ and $\sin\theta$ satisfy Definition B.1 and define functions. And since $\tan\theta=\frac{\sin\theta}{\cos\theta}$ (recall (B.21)), we conclude that $\tan\theta$ is also a function. Figures B.18(a)–(c) show the resulting graphs of $\sin\theta$, $\cos\theta$, and $\tan\theta$ for $0\leq\theta\leq 2\pi$. A few of their features deserve comment.

- The graphs consider only $0\leq\theta\leq 2\pi$. That's because once $\theta>2\pi$ we've already gone around the circle once. That means we'd start repeating values of sine, cosine, and tangent (recall the correspondence between points on the unit circle and values of those trigonometric functions). We conclude that

  $$\sin\theta=\sin(\theta+2\pi),\quad \cos\theta=\cos(\theta+2\pi),\quad \tan\theta=\tan(\theta+2\pi). \tag{B.25}$$

  Any function f for which $f(x)=f(x+c)$ for all x and some constant c is called **periodic**, and the smallest value of c is called the **period**. In this terminology, the first two equations in (B.25) tell us that sine and cosine are periodic functions with period 2π. The last equation also tells us that tangent is a periodic function, though with period π. (This follows from the fact that $\tan\theta=\tan(\theta+\pi)$ for every θ; recall that the period is the *smallest* c such that $f(x)=f(x+c)$ for all x.) Therefore, *the graphs of sine, cosine, and tangent go on forever*, and look the same on any interval of length 2π (in the case of sine and cosine) or π (in the case of tangent).
- The graphs of sine and cosine indicate that $-1\leq\sin\theta\leq 1$ and $-1\leq\cos\theta\leq 1$. These inequalities follow from the functions' interpretations as x- and y-coordinates of points on the unit circle (and also from the identity (B.22)).
- The function $\tan\theta$ is undefined when $\theta=\frac{\pi}{2}=90°$ and $\theta=\frac{3\pi}{2}=270°$, since its denominator, $\cos\theta$, is zero at those θ-values. Moreover, unlike sine and cosine, the outputs of $\tan\theta$ are unbounded.

Thus far we've only discussed measuring angles counterclockwise from the x-axis. Let's now discuss what happens when we measure angles *clockwise*.

According to our sign convention, this corresponds to negative angle values. Figure B.19 shows a representative triangle on the unit circle along with its positive-angle counterpart (in blue). The symmetry of the unit circle implies that the endpoints of the hypotenuse of both triangles have the same x-coordinate, and that their y-coordinates are negatives of each other. Recalling again that $\cos\theta$ is the x-coordinate on the unit circle and $\sin\theta$ the y-coordinate, we conclude that

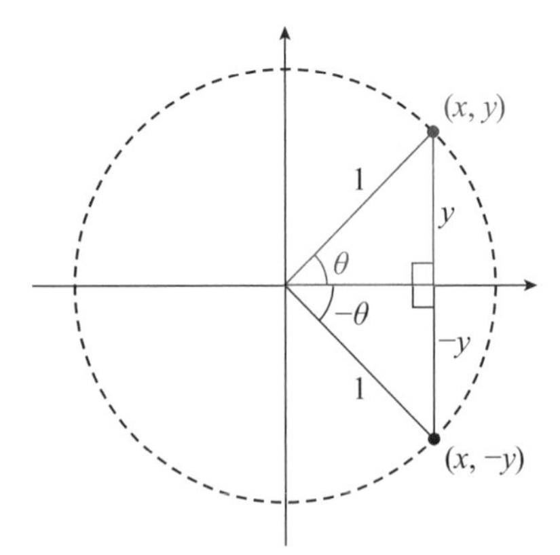

Figure B.19: Two mirror image right triangles on the unit circle with interior angle θ.

$$\cos(-\theta) = \cos\theta, \quad \sin(-\theta) = -\sin\theta. \tag{B.26}$$

Thus, the graphs of $\sin\theta$, $\cos\theta$, and $\tan\theta$ go on forever *in both directions*! See Figure B.20.

As the first two graphs illustrate, sine and cosine functions have maximum and minimum values. Let's end this section with a discussion of how to spot those values from the equations of the trigonometric functions. To wit, consider the functions

$$f(\theta) = A\sin(B\theta) + C, \quad g(\theta) = A\cos(B\theta) + C. \tag{B.27}$$

The constants A, B, and C here have the following interpretations.

- C is called the **midline**; it's found by taking the average of the maximum and minimum y-values. The line $y = C$ is the horizontal line exactly half the distance to the maximum and minimum values of the function.

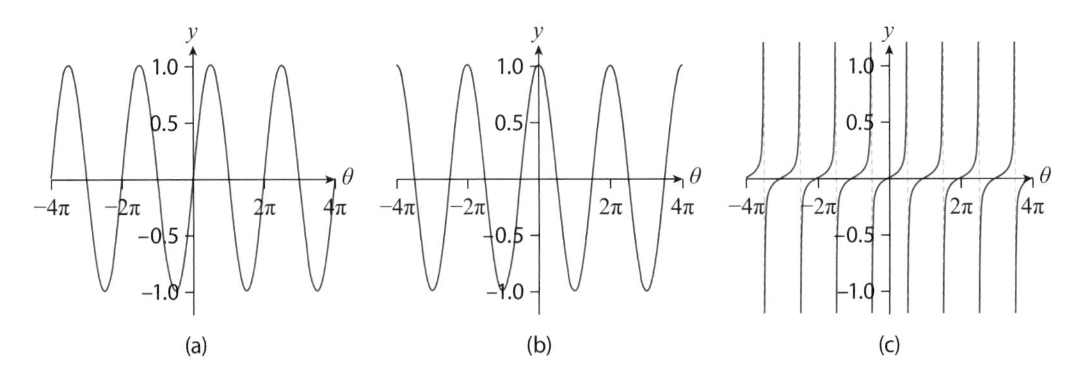

Figure B.20: The graphs of (a) $f(\theta) = \sin\theta$, (b) $f(\theta) = \cos\theta$, and (c) $f(\theta) = \tan\theta$ for $-4\pi \leq \theta \leq 4\pi$.

- $|A|$ is called the **amplitude**; it equals the difference between the maximum y-value and the midline (or equivalently, the difference between the midline and the minimum y-value). Therefore, $C+|A|$ and $C-|A|$ are the maximum and minimum y-values attained, respectively.
- B is called the **angular frequency**; it gives the number of complete oscillations in an interval of length 2π. Two related concepts are the **period** $T=\frac{2\pi}{B}$, the distance (on the θ-axis) it takes to complete one full oscillation, and the **frequency** $f=1/T$. From $T=\frac{2\pi}{B}$ it follows that $f=\frac{B}{2\pi}$. Therefore, f gives the number of complete oscillations in a unit interval (an interval of length 1).

For example, referring back to Figure B.18(a), we see that since the line $y=0$ is half the distance to extrema of the function, the midline is $C=0$. Moreover, since the maximum y-value is 1, and $1-0=0$, then the amplitude is $A=1$. In addition, since one complete oscillation takes 2π units of θ, the period is $T=2\pi$. Therefore, the angular frequency is $B=\frac{2\pi}{2\pi}=1$. Inserting these data into (B.27), we conclude that the function graphed in Figure B.18(a) is either $f(\theta)=\sin\theta$ or $g(\theta)=\cos\theta$. Since $(0,0)$ is a point on the graph in Figure B.18(a), we conclude that *that* is the graph of $f(\theta)=\sin\theta$.

The terminology surrounding the constants in (B.27) is especially useful for developing mathematical models of physical phenomena. The following example illustrates this.

APPLIED EXAMPLE B.16 The average low temperatures over the past few decades for New York City closely follow the curve in Figure B.21(a) (data source: weather.com), oscillating between a minimum of 23° F and a maximum of 68° F. Let t denote the number of months since January 1 and L denote the corresponding average low temperature in New York City. The data considered thus far suggest that

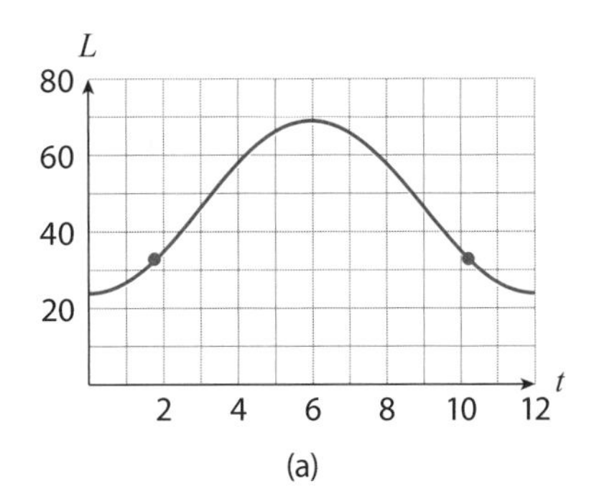

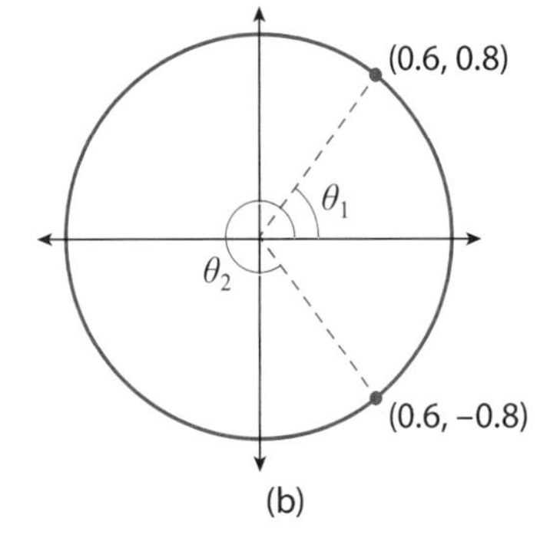

Figure B.21: (a) Average low temperatures L (in Fahrenheit) over the past few decades for New York City as a function of months t since January 1. (b) Two angles on the unit circle with the same x-coordinate.

a reasonable model for L is

$$L(t) = A\cos(Bt) + C.$$

(a) Find A, B, and C.

(b) Estimate the times of year for which the average low in New York City is above freezing (32° F).

Solution

(a) We find C by averaging 23 and 68, which yields $C = 45.5$. Then, since the maximum value is 68 we know that $|A| = 68 - 45.5 = 22.5$. (We'll discuss whether $A = 22.5$ or $A = -22.5$ shortly.) Finally, since we presume the weather pattern repeats every 12 months and t is measured in months, the period $T = 12$. Therefore, the angular frequency is $B = \frac{2\pi}{12} = \frac{\pi}{6}$ and we have the following options for $L(t)$:

$$22.5\cos\left(\frac{\pi t}{6}\right) + 45.5, \qquad -22.5\cos\left(\frac{\pi t}{6}\right) + 45.5.$$

We must reject the first option, since at $t = 0$ it yields 68°, which is certainly not the average low on January 1st in New York City. We conclude that $L(t) = -22.5\cos(\pi t/6) + 45.5$.

(b) Setting $L(t) = 32$ yields

$$-22.5\cos\left(\frac{\pi t}{6}\right) + 45.5 = 32, \quad \text{so that} \quad \cos\left(\frac{\pi t}{6}\right) = 0.6.$$

As Figure B.21(b) shows, $\cos\theta = 0.6$ for two angles between 0° and 360°: $\theta_1 \approx 53.1° \approx 0.92$ radian in Quadrant I, and also $\theta_2 \approx 306.9° \approx 5.36$ radian in Quadrant IV. Thus, we need to solve

$$\frac{\pi t}{6} = 0.92 \quad \text{and} \quad \frac{\pi t}{6} = 5.36.$$

We get $t \approx 1.75$ and $t \approx 10.25$ (the left- and right-most blue dots in Figure B.21(a)). The first solution is about two-thirds into the month of February; the second is about one-quarter into November. We conclude that from about late February to early November, the average low in New York City is above freezing. ■

Related Exercises 54–57.

Virtually any real-world phenomenon that repeats (or oscillates) is amenable to modeling with trigonometric functions; this includes sound, light (an electromagnetic wave), radio waves, and even human sleep cycles (see Chapter 1 of [1]). The applied exercises in the next few pages explore some of these applications.

APPENDIX B EXERCISES

1. True or False: $y = \pm\sqrt{1-x^2}$ defines a function.

2. Two functions f and g are graphed below.

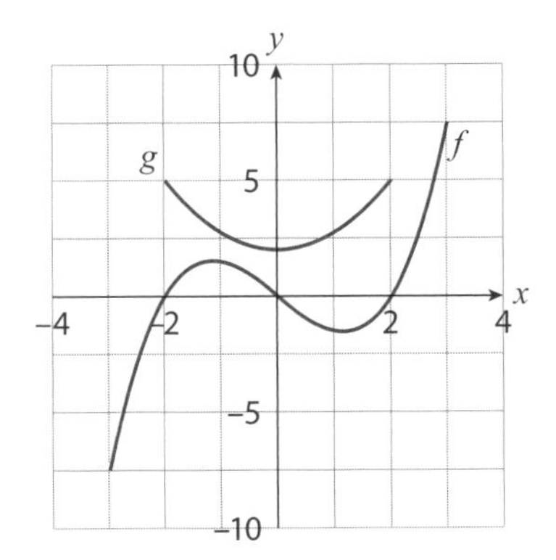

(a) Evaluate $f(0), f(2)$, and $g(2)$.

(b) What is the domain of g?

(c) What is the range of f?

3–6: Find the domain of the function.

3. $f(x) = \sqrt{1-x^2}$ **4.** $g(x) = 1 + \sqrt[3]{x^2}$

5. $h(t) = t^2 - 5$ **6.** $m(s) = \sqrt{s} + \sqrt{2-s}$

7–10: Identify the slope and y-intercept of each line, and graph each line.

7. $y = 2x - 3$ **8.** $y = -5x + 4$

9. $3y = 6x + 9$ **10.** $2x + 4y = 0$

11. Two lines are **perpendicular** if their slopes are negative reciprocals of each other. (For example, the lines $y = 2x$ and $y = -\frac{1}{2}x$ are perpendicular.) Use this fact to find the equation(s):

(a) Of all lines perpendicular to $y = 5x + 4$.

(b) Of a line perpendicular to $y = 5x + 4$ and passing through $(1, 1)$.

(c) Of a line perpendicular to $y = x + 2$ and with y-intercept 3.

12–15: Classify each function as a polynomial function (and, if applicable, a linear one) or power function.

12. $f(x) = 4$ **13.** $g(x) = (1+x)(1-x)$

14. $f(x) = 5\sqrt[3]{x^2}$ **15.** $h(t) = t^3 + 2t^2 + 1$

16–17: Find the domain.

16. $f(x) = \dfrac{1}{x^2 - x}$ **17.** $f(x) = \dfrac{x^2}{x-1}$

18. Monthly Cell Phone Bill Zoraida got a new cell phone for the holidays. It costs \$20 to activate the phone and \$50 per month for service.

(a) Find the total cost of service C as a function of the number of months m.

(b) After how many months will Zoraida have spent \$500 paying her cell phone bill?

19. Football Suppose a six foot tall quarterback throws a football with an initial vertical velocity of 50 ft/s. Neglecting air resistance, the ball's height y above the ground (measured in feet) is then

$$y(t) = 6 + 50t - 16t^2,$$

where t is the time (in seconds) since the ball was thrown. When does the ball hit the ground?

20. Maximum Heart Rate Loosely speaking, an individual's **maximum heart rate** M is the highest heart rate that can be sustained during prolonged exercise. One popular formula is $M_1(a) = 220 - a$, where a is your age (in years). Another, more accurate formula was developed in [2]:

$$M_2(a) = 192 - 0.007a^2.$$

(a) Calculate $M_1(20)$ and $M_2(20)$ and interpret your results.

(b) Find the age(s) where $M_1(a) = M_2(a)$. For what ages is the linear formula an overestimate of the (more accurate) quadratic formula? For what ages is it an underestimate?

21. Linguistics The American linguist George Zipf discovered that if you make a list of the most common words in a book, the r-th word on the list appears in the book with the frequency f roughly given by

$$f(r) = 0.1r^{-1}.$$

For example, if the word "the" is the most common word in the book you choose, it will have r-value 1, and since $f(1) = 0.1$, Zipf's law predicts that roughly 10% of the book consists of the word "the."

(a) Evaluate $f(2)/f(1)$ and interpret your result.

(b) Find a formula for $f(r+1)/f(r)$ and use your prior work to interpret the result.

22. Heart Health The volume of blood flowing per unit time inside an artery is called the **volumetric flow rate** (let's denote this by Q). If the artery is straight enough to resemble a cylindrical pipe, then we can approximate Q using **Poiseuille's Law**:

$$Q(r) = aPr^4,$$

where a is a constant that depends, among other things, on the length of the pipe, P is the change in pressure between the ends of the pipe, and r is the pipe's radius.

(a) Suppose the artery narrows to a new radius r_n. The heart must now must pump harder to maintain the same volumetric flow rate. Show that the new pressure P_n that results is

$$P_n = P\left(\frac{r}{r_n}\right)^4.$$

(b) This equation predicts that even small reductions in artery radius lead to large changes in pressure. For example, show that a 16% reduction in artery radius *doubles* the pressure.

23. Distance to Horizon Suppose you're looking at the horizon on a clear day. The distance d to the horizon (measured in miles) is a function of your height h above sea level (let's measure that in feet). A good approximation for the function is $d(h) = \sqrt{1.5h}$.

(a) This is a composite function: $d(h) = f(g(h))$. Identify the two functions f and g.

(b) Find d for a 5-foot tall person: (1) standing on a beach at sea level, and (2) 1,000 feet up inside a skyscraper.

24. Estimating the Age of the Universe Today physicists have reason to believe that the universe began with an explosion of epic proportions—the **Big Bang Theory**—out of which all matter and energy sprang into existence. One piece of supporting evidence is Edwin Hubble's discovery in 1929 that distant galaxies are moving away from us at velocities v (in km/s) related linearly to their distance d (in "megaparsecs") from us:

$$v(d) = Hd,$$

where $H \approx 67.8$ km/s per megaparsec is **Hubble's constant**; the equation above is now known as **Hubble's Law**. Below is Hubble's original plot of the relationship for his particular dataset [12].

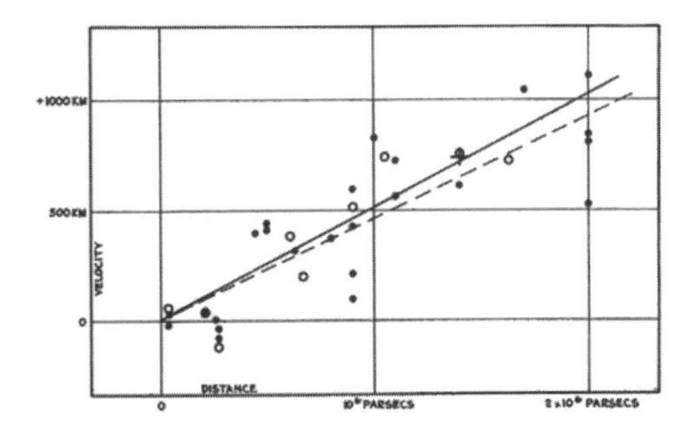

(a) Identify the slope and y-intercept of Hubble's Law.

(b) Let's assume that galaxies move with a constant velocity. Then their distance away from us is $d = vt$, where t is the time (in seconds) they've been in motion—*the age of the Universe*. Using this and Hubble's Law, show that $t = \frac{1}{H} \approx 0.0147$ megaparsecs per km/s. Since 1 megaparsec is about 3.08×10^{19}, $t \approx 4.55 \times 10^{17}$ seconds $= 14.4$ billion years. (The current estimate of the age of Universe is about 13.8 billion years.)

25. Suppose you're given a curve in the xy-plane. Prove that if no vertical line intersects the curve at two or more points, the curve is the graph of a function. (This is called the **Vertical Line Test**.)

26. Let $f(x) = \frac{x^2-1}{x+1}$ and $g(x) = x - 1$. Is $f(x) = g(x)$?

27. Suppose $f(x) = ax + c$ and $g(x) = dx + e$. Show that $f(g(x))$ is a linear function, and find its slope.

28. Let r_1 and r_2 denote the two solutions to the quadratic equation $ax^2 + bx + c = 0$, where $a \neq 0$. Prove that $r_1 + r_2 = -\dfrac{b}{a}$, and that $r_1 r_2 = \dfrac{c}{a}$.

29. Prove that the only rational functions satisfying $f(x) = \dfrac{1}{f(x)}$ are $f(x) = 1$ and $f(x) = -1$.

Exercises Involving Exponential and Logarithmic Functions

30–32: Determine if the exponential function represents exponential growth or decay, and identify the initial value and base.

30. $f(x) = 10^x$ **31.** $h(t) = 4e^t$ **32.** $g(z) = 2^{-z}$

33–36: Find the exact value.

33. $e^{2-\ln 4}$ **34.** $\ln\left(\frac{3}{e}\right)$ **35.** $\log_3\left(\frac{1}{9}\right)$

36. $\log_5 25$

37–38: Combine the expression into one logarithm.

37. $\ln 2 + 3\ln 4$ **38.** $\ln(x-y) - \ln(x+y)$

39–40: Sove for x.

39. $e^{8-4x} = 4$ **40.** $\log x + \log(x-1) = \log 2$

41. Find the equation of the exponential function passing through the points $(1, 6)$ and $(3, 24)$.

42. Population Growth Since 2010, the United States population has been growing by about 0.75% each year. Given that the country's 2010 population was 309.3 million, find the exponential function describing its population P (in millions) as a function of years t since 2010. Assuming the growth rate stays the same, estimate the population in 2025.

43. Terminal Velocity of Raindrop A typical raindrop falls from about 13,000 feet and takes about 3 minutes to reach the ground. As it falls, it fuses with other raindrops and gains both mass and velocity. Its increasing surface area, however, leads to a larger force of air resistance. This air drag force is eventually balanced out by the accelerating force of gravity and the raindrop reaches a **terminal velocity**. A realistic model (see Chapter 3 of [1]) of an average raindrop's velocity v (in ft/s) as a function of time t (in seconds) since it began falling is

$$v(t) = 13.92(1 - e^{-2.3t}).$$

(a) Calculate and interpret $v(0)$.

(b) Rewrite $v(t)$ in the form $c - ab^t$.

(c) Use part (b) to explain why $v(t)$ is increasing.

(d) Plot $v(t)$ and *guess* the terminal velocity from your graph. (See Exercise 48 at the end of Chapter 2 for how to *calculate* the terminal velocity.)

44. Saving for Retirement via the Compounding Effect Suppose that you currently have $B(0)$ dollars in a savings account earning an r% return each year, and that you add an additional s dollars each year to that savings account. It can be shown (see Chapter 4 of [1]) that your account's balance B some number of years t from now is

$$B(t) = \left(B(0) + \frac{s}{r}\right) e^{rt} - \frac{s}{r}.$$

Suppose $B(0) = 0$, $r = 0.07$ (7% is about the average stock market return over 20-year investment horizons), and $s = \$1,000$.

(a) Calculate $B(t)$ and $B(40)$.

(b) The total deposits after t years is $D(t) = 1,000t$. The fraction of the balance those deposits constitutes is $D(t)/B(t)$. Calculate that new function, plot it for $1 \leq t \leq 40$, and interpret the graph.

(c) Show that after 20 years, the yearly deposits constitute about 46% of the account's balance, but that after 40 years, the yearly deposits constitute just 18% of the account's balance. Conclusion: *Start saving for retirement early*; that allows more time for investment gains to compound.

45. Radiocarbon Dating All animals contain **radiocarbon**, a radioactive isotope of carbon that undergoes radioactive decay over time. If N_0 is the initial number of radiocarbon in a sample, then the amount N in the sample t years later is given by

$$N(t) = N_0 e^{-\lambda t},$$

where $\lambda > 0$ is the **decay constant** and is related to the isotope's **half-life** T, the years it takes for half of the initial sample to decay.

(a) Show that $T = \frac{\ln 2}{\lambda}$.

(b) The accepted value for the half-life of radiocarbon is $T = 5,730$ years. Use this and part (a) to calculate the decay constant.

(c) Suppose the radiocarbon in the remains of an animal has decayed to 70% its initial value. Estimate the age of the animal. (This technique is called **radiocarbon dating**.)

46. Calculating a Loan's Payoff Time Suppose you obtain a loan (e.g., a student loan) of $\$L$ at an annual interest rate of r%. If you pay $\$M$ each month, it can be shown (see Chapter 3 in [7]) that it'll take n months to pay off the loan, where

$$n = \frac{\log\left(\frac{M}{M - Lc}\right)}{\log(1 + c)},$$

where $c = \frac{r}{12}$, and r is expressed as a decimal. Suppose the loan is a credit card balance of \$1,000 with a minimum monthly payment of \$20 and an annual interest rate of $r = 12\%$.

(a) How many months will it take to pay off the loan?

(b) If we now leave M as a variable, show that

$$n \approx 231.4 \log\left(\frac{M}{M - 10}\right).$$

(c) Plot n for $20 \leq M \leq 50$ and calculate $n(40)$. How much sooner would \$40 monthly payments pay off the loan compared to \$20 monthly payments?

47. Solve $b^x = c$ using a logarithm with base b. Then resolve the equation using a logarithm with base a. Use your results to prove the change of base formula (B.15).

Exercises Involving Trigonometric Functions

48–51: Convert the angle from degrees to radians, or radians to degrees.

48. $120°$ **49.** $36°$ **50.** $\frac{7\pi}{2}$ **51.** $\frac{3\pi}{8}$

52. What is the length of the arc on a circle of radius 2 inches subtended by the central angle $20°$?

53. Find $\sin\theta$, $\cos\theta$, and $\tan\theta$ from the following triangles. (You'll need the Pythagorean Theorem to find the missing side lengths.)

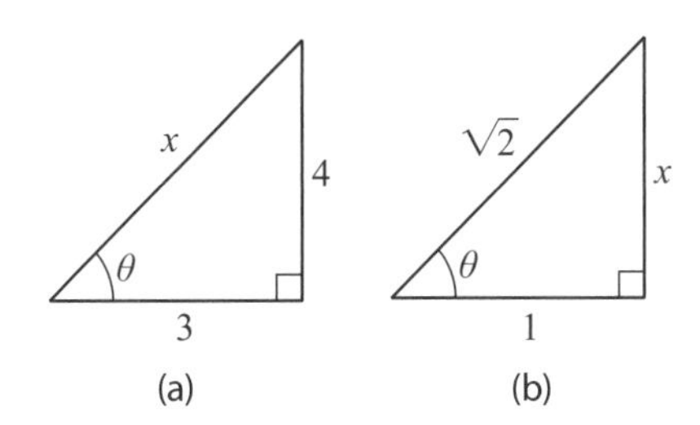

54. The Trigonometry of Colors Light is an **electromagnetic wave**, a wave in which electric and magnetic fields oscillate perpendicular to the direction of propagation (such waves are called **transverse waves**). Therefore, light can be modeled using trigonometric functions. The **wavelengths** (period) λ of *visible* light—which range from 400 nanometers (nm) to 700 nanometers—determine the light's color:

Color	Violet	Blue	Green	Yellow	Orange	Red
Wavelength (nm)	380–450	450–495	495–570	570–590	590–620	620–750

Each color can be represented by the function

$$C(t) = \sin(Bt), \qquad \lambda = \frac{2\pi}{B}.$$

(a) Find a $C(t)$ function for red using 700 nm for red.

(b) Which color is modeled by the equation $C(t) = \sin(\pi t/200)$?

55. The Trigonometry of Music Recall from Applied Example B.12 that sound is a pressure wave. The simplest sound waves can be modeled via the function

$$S(t) = \sin(2\pi f t),$$

where f is the frequency, measured in Hertz (Hz). (Such sounds are called **pure tones.**) For example, the musical note A above middle C has a frequency of 440 Hz. The **chromatic scale**, which forms the foundation for Western music, then determines the frequencies of the subsequent notes on a musical scale according the following *twelfth root of 2* progression:

$$440 \cdot 2^0,\ 440 \cdot 2^{\frac{1}{12}},\ 440 \cdot 2^{\frac{2}{12}},\ \ldots,$$
$$440 \cdot 2^{\frac{12}{12}} = 880.$$

In order from left to right, the musical notes produced are A, A# ("A sharp"), and so on up to A2 (the note with frequency 880 Hz), the note one octave above A. (Visually, each of these notes corresponds to one key on a piano between A and A2.)

(a) Write down the trigonometric function associated with a pure tone C note whose frequency is $440\sqrt[4]{2}$.

(b) Let $f(x) = 440 \cdot 2^{\frac{x}{12}}$ for $0 \leq x \leq 12$. What does $f(0)$ represent? And what can you conclude from the fact that $f(12) = 2f(0)$?

56. The Trigonometry of Electric Current Today electric current is delivered as **alternating current** (AC). (This wasn't always the case, and the story involves Thomas Edison's reluctance to embrace new technology; see Chapter 1 of [1] if you're interested.) AC current is delivered via a voltage V (measured in volts (V)) which oscillates according to the function

$$V(t) = \sqrt{2}A\sin(Bt).$$

(a) A typical wall outlet delivers a peak voltage of $120\sqrt{2}$ V. Use this information to find A.

(b) Standard household AC currents oscillate at a frequency of 60 Hz. Use this to find B.

57. Use (B.23) to prove that $\sin\left(\theta + \frac{\pi}{2}\right) = \cos\theta$.

58. Use the figure below to help show that the area of the outer triangle shown is $A = \frac{1}{2}ab \sin\theta$.

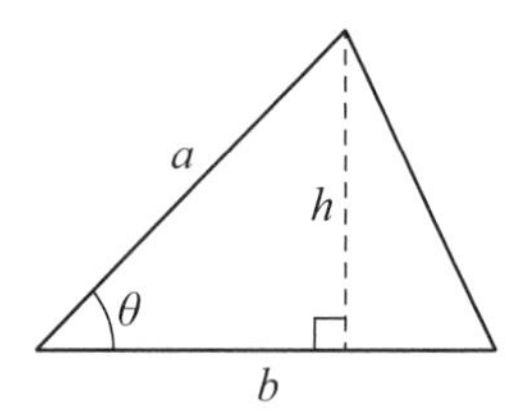

59. Trigonometry provides us with an alternative—somewhat more natural—approach to understanding the slope of a line. To see how, let's return to Figure B.4. Let θ be the angle between the line and the x-axis ($-\frac{\pi}{2} < \theta < \frac{\pi}{2}$). Show that

$$m = \tan\theta.$$

Thus, the slope of the line (m) is directly related to the angle of inclination of the line (θ).

60. Suppose we inscribe n isosceles triangles of equal base r in a circle of radius r. (The figure below illustrates one such triangle.)

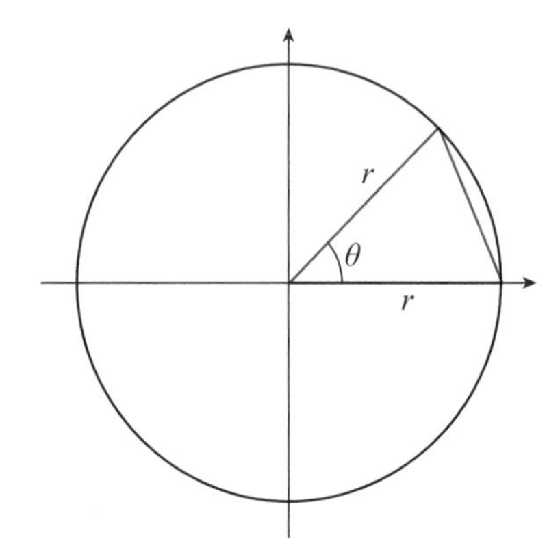

(a) Explain why $\theta = \frac{2\pi}{n}$.

(b) Let $A(n)$ be the sum of the areas of the n inscribed triangles. Show that

$$A(n) = \frac{1}{2}nr^2\sin\left(\frac{2\pi}{n}\right).$$

(c) Calculate $A(4)$, $A(10)$, and $A(100)$. You'll find that $A(n)$ better approximates the area of the circle as n gets larger. (This will be proven using calculus in Exercise 60 in Chapter 2.)

Appendix C Additional Applied Examples

APPLIED EXAMPLE C.1 Einstein's *Special Theory of Relativity* teaches us that certain physical quantities we thought were independent of velocity are in fact not. One example is the length of an object. For instance, imagine a train moving along a straight track. If L_0 denotes the length of the train when it's at rest, Einstein discovered that its length L when it's moving with velocity v is

$$L(v) = L_0\sqrt{1 - \frac{v^2}{c^2}},$$

where $c > 0$ is the speed of light. The takeaway: *Length is relative.*

(a) Describe what happens to L as v increases from zero to a nonzero value. (The phenomenon you will discover is called **Lorentz contraction**.)

(b) What is the largest value v can attain, and why?

(c) Calculate $\lim_{v \to c^-} L(v)$ and interpret your result; why was a left-hand limit necessary?

Solution

(a) First note that $L(0) = L_0$ (i.e., the train's length at rest is L_0). As v increases from zero, v^2/c^2 increases too, $1 - v^2/c^2$ becomes smaller than 1, and so does $\sqrt{1 - v^2/c^2}$. Conclusion: $L(v) < L_0$ for $v > 0$ (i.e., the train's length in motion is less than its length at rest).

(b) The square root of a negative number is not defined, so we must have $1 - v^2/c^2 \geq 0$. This implies that $v^2/c^2 \leq 1$, so that $v^2 \leq c^2$. The largest velocity value satisfying this inequality is $v = c$ (i.e., the speed of light).

(c) Note that $L(v) = f(g(v))$, where $f(x) = L_0\sqrt{x}$ and $g(v) = 1 - v^2/c^2$. The function g is continuous from the left at $v = c$, and $g(c) = 0$. As $v \to c^-, g(c) \to 0^+$, and since f is continuous from the right at $x = 0$, we conclude that

$$\lim_{v \to c^-} L(v) = L_0\sqrt{1 - c^2/c^2} = 0.$$

This implies that as the train's velocity approaches the speed of light, the train's length shrinks to zero! Finally, the left-hand limit was necessary because, as discussed in Part (b), $L(v)$ is undefined for $v > c$. ■

APPLIED EXAMPLE C.2 Another result from Einstein's *Special Theory of Relativity* (introduced in Example C.1) is that if $m_0 > 0$ is the mass of a particle when it's at rest, then its mass m when it's moving with velocity v is

$$m(v) = \frac{m_0}{\sqrt{1 - v^2/c^2}},$$

where c is the speed of light. The takeaway: *mass is relative.*

(a) Calculate $m(v)$ for $v = 0, 0.25c$, and $0.5c$.

(b) What happens when $v = c$?

(c) Find the limit as $v \to c^-$ and interpret your answer, both mathematically and physically.

Solution

(a) We have

$$m(0) = m_0, \qquad m(0.25c) = \frac{m_0}{\sqrt{\frac{15}{16}}} \approx 1.03m_0, \qquad m(0.5c) = \frac{m_0}{\sqrt{\frac{3}{4}}} \approx 1.15m_0.$$

(b) If $v = c$ then $m(c)$ is undefined (since the denominator is zero when $v = c$).

(c) In the course of solving part (c) of Applied Example C.1 we showed that

$$\lim_{v \to c^-} \sqrt{1 - v^2/c^2} = 0.$$

This tells us that the denominator of $m(v)$ is approaching a tiny positive number as $v \to c^-$. And since the numerator of $m(v)$ is the constant m_0, $m(v)$ is thus becoming large and positive as $v \to c^-$. Conclusion: $m(v) \to \infty$ as $v \to c^-$. Mathematically, this means that the graph of $m(v)$ has a vertical asymptote at $v = c$ (recall Definition 2.5). Physically, this result implies that as the particle's velocity approaches the speed of light it *gains* mass without bound! ■

APPLIED EXAMPLE C.3 Suppose you and your partner need to decide on how to divide up a total T of something, say money. How can the $\$T$ be divided in a fair and unbiased way? Mathematician John Nash—known for virtually inventing game theory and profiled in the movie *A Beautiful Mind*—considered this "bargaining problem" and devised a procedure to solve it. Nash's solution requires quantifying one's **utility functions**, measures of the preferences each party has as his or her share of the money changes. In [7], Section 6.2, Nash's approach was adapted to the case when "utility" is replaced by "happiness level" and each party's happiness levels increase linearly as the amount of the money he or she receives increases. Let's work through the solution that results.

Suppose that happiness is measured on a scale from 0 to 10 (with 10 representing "happy" and 0 "unhappy"). Let M denote your happiness level were you to receive all of the money to be divided, and N your partner's. Finally, let Y_d and P_d be the happiness levels you and your partner, respectively, would experience in the event no agreement can be reached.[1]

(a) Let x be the share of the money you get, and z the share your partner gets. Assuming we want to split all of T, explain why

$$x+z=T, \quad x\geq 0, \quad z\geq 0.$$

(b) Let $Y(x)$ denote your happiness level when receiving x of the money and $P(z)$ your partner's when receiving z of the money. Assume that

$$Y(x)=\frac{M}{T}x, \qquad P(z)=\frac{N}{T}z.$$

Interpret what these functions mean.

(c) The **Nash product** H is defined by

$$H=(Y-Y_d)(P-P_d). \tag{C.1}$$

Show that when H is expressed as a function of x we get

$$H(x)=-\left(\frac{M}{T}x-Y_d\right)\left(\frac{N}{T}x-(N-P_d)\right). \tag{C.2}$$

(d) Nash's algorithm is to find maximize H over all possible combinations of Y and P satisfying $Y\geq Y_d$ and $P\geq P_d$. Show that these two constraints yield the constraint

$$\frac{TY_d}{M}\leq x\leq T-\frac{TP_d}{N}. \tag{C.3}$$

Explain why it follows that Nash's algorithm applies only when

$$\frac{Y_d}{M}+\frac{P_d}{N}\leq 1. \tag{C.4}$$

(e) Assuming (C.4) holds, show that the absolute maximum of $H(x)$ on the interval (C.3) occurs at

$$x=\frac{T}{2}\left(1+\frac{Y_d}{M}-\frac{P_d}{N}\right). \tag{C.5}$$

[1] Note that M, N, Y_d, and P_d are all numbers between 0 and 10 (due to the happiness scale adopted).

Then, show that the corresponding share for your partner is

$$z = \frac{T}{2}\left(1 + \frac{P_d}{N} - \frac{Y_d}{M}\right).$$

Solution

(a) We have $x + z = T$ because $x + z$ is the total amount of the money there is, which we've labeled T. Moreover, because no party can receive more than \$$T$ (we are dividing up \$$T$), we have $x \leq T$ and $z \leq T$. Substituting in $x = T - z$ and $z = T - x$ yields $z \geq 0$ and $x \geq 0$, respectively.

(b) Y is a linear function with slope $\frac{M}{T}$. This means that for each unit increase in your share x of \$$T$ your happiness level increases by $\frac{M}{T}$. Also, since the y-intercept of $Y(x)$ is $(0, 0)$, this means that were you to receive none of the money, your happiness level would be zero. Similar interpretations hold for $P(z)$.

(c) Since $z = T - x$, we have

$$P(x) = \frac{N}{T}(T - x) = -\frac{N}{T}x + N.$$

Substituting this, along with $Y = \frac{M}{T}x$, into (C.1) yields

$$H(x) = \left(\frac{M}{T}x - Y_d\right)\left(-\frac{N}{T}x + (N - P_d)\right).$$

Factoring out a negative one from the second parenthetical term yields (C.2).

(d) Since $Y = \frac{M}{T}x$, then $Y \geq Y_d$ becomes

$$\frac{M}{T}x \geq Y_d \quad \Longrightarrow \quad x \geq \frac{TY_d}{M}.$$

Since $P = \frac{N}{T}z = \frac{N}{T}(T - x)$, then $P \geq P_d$ becomes

$$\frac{N}{T}(T - x) \geq P_d \quad \Longrightarrow \quad x \leq \frac{T}{N}(N - P_d).$$

Putting the two bounds for x together yields (C.3). Since we've converted Nash's $Y \geq Y_d$ and $P \geq P_d$ requirements to (C.3), only x-values *between* the bounds in (C.3) are considered by Nash's algorithm. Such x-values exist only if the left-hand number is less than or equal to the right-hand number:

$$\frac{TY_d}{M} \leq \frac{T}{N}(N - P_d).$$

This simplifies to (C.4).

(e) We now have a continuous function (a quadratic polynomial) on a closed interval, so let's apply the procedure from Box 4.4.We start with $H'(x)$. After some simplification,

$$H'(x) = \frac{M}{T}(N - P_d) + \frac{N}{T}Y_d - \frac{2MN}{T^2}x.$$

The only critical number occurs when $H'(x) = 0$, which yields

$$x = \frac{T^2}{2MN}\left[\frac{M}{T}(N - P_d) + \frac{N}{T}Y_d\right] = \frac{T}{2}\left[\frac{N - P_d}{N} + \frac{Y_d}{M}\right],$$

which simplifies to (C.5). The corresponding z value is obtained from $z = T - x$. ■

Reference [7] (Section 6.2) discusses how the Nash algorithm preserves fairness and unbiasedness. I also worked with TIME magazine to convert these equations into a fun interactive demonstration you can use to try out the Nash solution (C.5) for yourself; see [8] if you're interested.

APPLIED EXAMPLE C.4 Consider an asset whose value V (in \$) increases with time t (in years). A popular model for $V(t)$ is

$$V(t) = V_0 e^{\sqrt{t}}, \qquad V_0 > 0.$$

The **present value** P of this asset is the future sum of money $V(t)$ expressed in today's dollars, and can be modeled by

$$P(t) = V(t)e^{-rt} = V_0 e^{\sqrt{t} - rt},$$

where $r > 0$ is the prevailing yearly interest rate, expressed as a decimal.

(a) What is the initial value of the asset?

(b) Show that

$$\lim_{t \to \infty} P(t) = 0$$

and interpret your result.

(c) When is the optimal time to sell the asset? (Economists refer to this as the **optimal holding time**.)

(d) Suppose the asset is a house. Assuming $r = 3\%$, when is the optimal time to sell it?

Solution

(a) $P(0) = V_0$.

(b) We can rewrite $P(t)$ as

$$P(t) = V_0 e^{\sqrt{t}(1-r\sqrt{t})}.$$

For large enough t-values the exponent of $P(t)$ is negative.[2] As $t \to \infty$ then, $P(t)$ decays to zero. This tells that eventually the present value of the asset approaches zero. (This is an illustration of the **time value of money**. Briefly stated, this concept encapsulates the notion that \$100 today is worth more than \$100 a decade from now.)

(c) We are seeking the absolute maximum of $P(t)$ on the interval $[0, \infty)$. But we just showed that for $t > 1/r^2$ the exponent of $P(t)$ is negative: $\sqrt{t}(1 - r\sqrt{t}) < 0$. It follows that for large enough t,

$$P(t) = V_0 e^{\sqrt{t}(1-r\sqrt{t})} < V_0 e^0 = V_0.$$

In other words, after time $t^* = 1/r^2$ the present value is less than the initial present value V_0. So, let's focus our search for the absolute maximum to the interval $[0, t^*]$. We'll need $P'(t)$:

$$P'(t) = V_0 e^{\sqrt{t}-rt}\left(\frac{1}{2\sqrt{t}} - r\right) = \left(\frac{1 - 2r\sqrt{t}}{2\sqrt{t}}\right) V_0 e^{\sqrt{t}-rt}.$$

This yields two critical numbers: (1) $t = 0$ (since $P'(0)$ is undefined), and (2) $t = \frac{1}{4r^2}$ (found by solving $P'(t) = 0$). Following the procedure in Box 4.4 once again:

$$P(0) = V_0, \qquad P\left(\frac{1}{4r^2}\right) = V_0 e^{1/(4r)}, \qquad P(t^*) = V_0 e^{\sqrt{t^*}-rt^*} = V_0.$$

Since $e^{1/(4r)} > 1$ for any r-value,[3] we conclude that $P(t)$ has an absolute maximum at $t = 1/(4r^2)$. Note: this answer is independent of V_0, the initial value of the asset.

(d) According to our calculations in part (c), the optimal selling time is

$$t = \frac{1}{4(0.03)^2} \approx 278 \text{ years from now.}$$

■

APPLIED EXAMPLE C.5 Figure C.1 shows a larger blood vessel of radius r splitting off at an angle θ into a smaller blood vessel of radius r_2. This splitting will impede the flow of blood up the smaller blood vessel. A reasonable model for the resistance R blood flowing from the larger vessel up into the smaller would

[2] Specifically, if $t > 1/r^2$ then $1 - r\sqrt{t} < 0$ and the exponent of $P(t)$ is negative.

[3] This follows from the fact that $e^x > 1$ when $x > 0$, and the fact here $x = 1/(4r) > 0$.

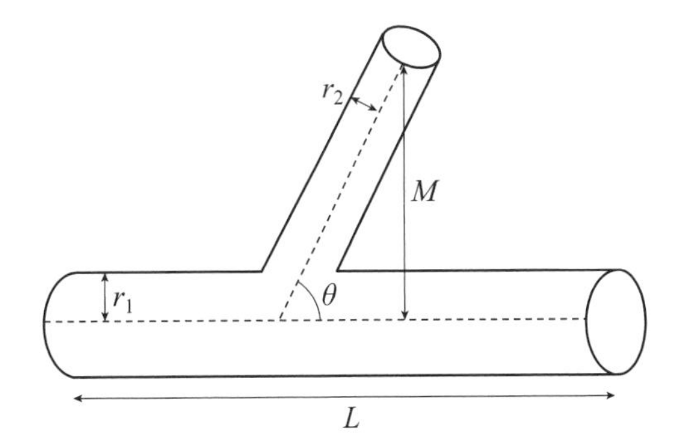

Figure C.1: Schematic of a larger blood vessel branching off into a smaller blood vessel.

experience is

$$R(\theta) = c\left(\frac{L - M\cot\theta}{r_1^4} + \frac{M\csc\theta}{r_2^4}\right), \quad 0 < \theta \leq \frac{\pi}{2}, \tag{C.6}$$

where $c > 0$ is a positive constant. (This equation is derived in [1], equation (44).)

(a) Show that the critical number(s) of R occurs when

$$\cos\theta = \left(\frac{r_2}{r_1}\right)^4. \tag{C.7}$$

(b) Explain why (C.7) yields a unique θ-value, and why R has an absolute minimum at this θ-value.

Solution

(a) The derivative $R'(\theta)$ is

$$\begin{aligned} R'(\theta) &= c\left[\frac{M}{r_1^4}\csc^2\theta - \frac{M\csc\theta\cot\theta}{r_2^4}\right] \\ &= \frac{cM}{r_1^4 r_2^4 \sin^2\theta}\left[r_2^4 - r_1^4\cos\theta\right]. \end{aligned} \tag{C.8}$$

Since $R'(\theta)$ is continuous on the interval of interest, $(0, \pi/2)$, then the only critical number occurs when the expression in the brackets in (C.8) is zero:

$$r_2^4 - r_1^4\cos\theta = 0 \quad \Longrightarrow \quad \cos\theta = \left(\frac{r_2}{r_1}\right)^4.$$

(b) Since we're assuming $r_2 < r_1$, then $0 < \frac{r_2}{r_1} < 1$. Thus, to solve (C.7) we look for where the graph of $\cos\theta$ intersects the horizontal line $y = (r_2/r_1)^4$. A quick glance at Figure B.18(b) shows that this happens only once in the interval

$0 < \theta < \pi/2$ (the interval of interest). We conclude that (C.7) yields a unique solution in the interval $(0, \pi/2)$. Thus, there is only one critical number of R on $(0, \pi/2)$. Let's denote by θ^* that unique θ-value solving (C.7).

Now, returning to (C.8), we see that $R'(\theta) > 0$ when the expression in the brackets in (C.8) is positive, i.e., $r_4^2 - r_1^2 \cos\theta > 0$. This simplifies to $\cos\theta < (r_2/r_1)^4$, which in turn happens when $\theta > \theta^*$. This follows because $\cos\theta$ is a decreasing function on $(0, \pi/2)$ (recall Figure B.18(b)), and so when $\theta > \theta^*$, then $\cos\theta < \cos\theta^* = (r_2/r_1)^4$. Similarly, $R'(\theta) < 0$ if $\cos\theta > (r_2/r_1)^4$, which happens when $\theta < \theta^*$. Part (b) of Theorem 4.2 then tells us that R has a local minimum at θ^*. And since R is continuous on $(0, \pi/2)$, it follows from Theorem 4.4 that R has an absolute minimum at θ^*. ■

APPLIED EXAMPLE C.6 Suppose a U.S. presidential candidate wins p percentage of the popular vote in the presidential election. Political scientists have discovered that the proportion of seats $H(p)$ in the U.S. House of Representatives won by the president's party is well approximated by the "cube rule":

$$H(p) = \frac{p^3}{p^3 + (1-p)^3}, \qquad 0 \le p \le 1.$$

Figure C.2 shows the graph of $H(p)$.

(a) Calculate $H'(p)$ and $H''(p)$.

(b) Determine the interval(s) on which is H concave up/down using (1) Definition 4.5 and Figure C.2, and (2) Theorem 4.7.

(c) Are there any inflection points of H?

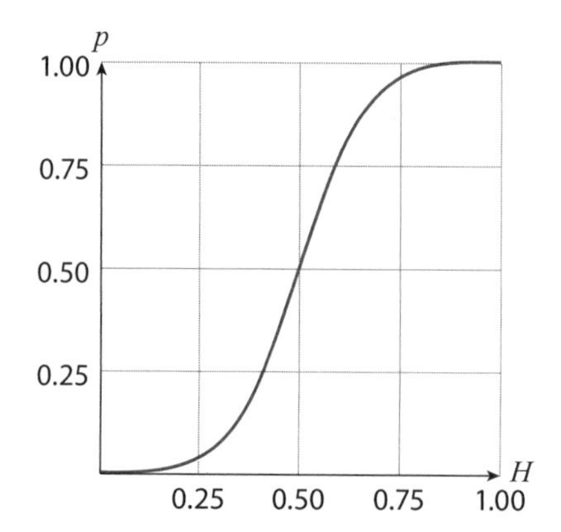

Figure C.2: The graph of $H(p)$.

Solution

(a) Let's first use formula (A.10) to rewrite the function as

$$H(p) = \frac{p^3}{3p^2 - 3p + 1}.$$

Using now the Quotient Rule:

$$H'(p) = \frac{3p^2[3p^2 - 3p + 1] - p^3[6p - 3]}{(3p^2 - 3p + 1)^2} = \frac{3p^2(p-1)^2}{(3p^2 - 3p + 1)^2}.$$

Using the Quotient Rule again and simplifying yields

$$H''(p) = \frac{6p(2p^2 - 3p + 1)}{(3p^2 - 3p + 1)^3}. \tag{C.9}$$

(b) (1) Using Definition 4.5 and referring to Figure C.2, we see that the tangent lines to the graph of H are below the graph when $0 < p < 0.5$ and above the graph when $0.5 < p < 1$. So, H is concave up on the interval $(0, 0.5)$ and concave down on the interval $(0.5, 1)$.

(2) First, we look for p-values for which $H''(p)$ is undefined. This will happen if the denominator in (C.9) is zero. However, since $3p^2 - 3p + 1$ is never zero (setting $3p^2 - 3p + 1 = 0$ yields no real solutions), there are no p-values for which $H''(p)$ does not exist. Next, we look for p-values for which $H''(p) = 0$:

$$H''(p) = 0 \quad \Longrightarrow \quad 6p(2p^2 - 3p + 1) = 0.$$

This yields $p = 0$, $p = 0.5$, and $p = 1$. But since $p = 0$ and $p = 1$ are the endpoints of the interval on which $H(p)$ is defined, we cannot assess if they're points of inflection. (To do so would require information about $H''(p)$ for $p < 0$ and $p > 1$ to determine whether $H''(p)$ changes sign.) That leaves $p = 0.5$. Let's create a sign chart to see if H'' changes sign as we cross $p = 0.5$. Choosing the test points $p = 0.25$ and $p = 0.75$ confirms the sign chart below (i.e., $H''(0.25) > 0$ and $H''(0.75) < 0$):

$H''(p)$: + + + + + | − − − − −
0.5

We conclude from Theorem 4.7 that H is concave up on the interval $(0, 0.5)$ and concave down on the interval $(0.5, 1)$.

(c) Since H changes concavity at $p = 0.5$, it follows from Definition 4.6 that H has an inflection point at $p = 0.5$. ■

APPLIED EXAMPLE C.7 Many real-world phenomena grow nearly exponentially at first, and then experience slower growth due to a variety of reasons. (Two examples are population growth and the spread of infectious diseases.) Such phenomena can be modeled with the the **logistic function**:

$$q(t) = \frac{aq_0}{bq_0 + (a - bq_0)e^{-at}},$$

where a and b are positive constants, $q(t)$ denotes the quantity at time $t \geq 0$ of the phenomenon being modeled (e.g., population), and $q_0 > 0$ denotes the initial quantity present; we'll assume $q_0 < \frac{a}{b}$, since this is typically the case in the real-world phenemona mentioned earlier. This exercise will explore the shape of these logistic curves.

(a) Show that $q(t)$ is increasing for all $t \geq 0$.

(b) The second derivative of q is

$$q''(t) = \frac{a^3 q_0(a - bq_0)e^{at}\left[(a - bq_0) - bq_0 e^{at}\right]}{(bq_0 e^{at} + a - bq_0)^3}.$$

Show that $q(t)$ has an inflection point only if

$$q_0 < \frac{a}{2b},$$

that the inflection point is unique in this case, and that at this inflection point $q(t)$ changes from concave up to concave down.

(c) Calculate

$$\lim_{t \to \infty} q(t).$$

The result is called the **carrying capacity** of the phenomenon being modeled.

(d) Suppose $q(t)$ denotes the number of individuals that catch a cold during a 1-hour calculus lecture (where t is measured in hours). Assuming that: (1) $q_0 = 5$ people in the room already have a cold, (2) $a = 0.4$, and (3) that if everyone stayed in the room forever all 20 people would catch the cold, find an equation for $q(t)$ and plot your result.

(e) Continuing part (d), estimate the number of people at the end of the lecture who have a cold.

Solution

(a) We first find $q'(t)$. Rewriting $q(t)$ as

$$q(t) = aq_0\left(bq_0 + (a - bq_0)e^{-at}\right)^{-1},$$

the Chain Rule then yields

$$q'(t) = -aq_0\left(bq_0 + (a-bq_0)e^{-at}\right)^{-2}\left[-a(a-bq_0)e^{-at}\right].$$

This simplifies to

$$\frac{a^2 q_0(a-bq_0)e^{-at}}{(bq_0 + (a-bq_0)e^{-at})^2}.$$

The denominator is always positive, since we've assumed $q_0 < \frac{a}{b}$. Moreover, that assumption implies $a - bq_0 > 0$. Thus, the numerator is always positive too. We conclude that $q'(t) > 0$ for all $t \geq 0$. It follows from Theorem 4.1 that $q(t)$ is increasing for all $t \geq 0$.

(b) The only possibly zero expression in $q''(t)$ is the expression in brackets. Setting that expression equal to zero and solving yields the unique solution

$$(a-bq_0) - bq_0e^{at} = 0 \quad \Longrightarrow \quad t = \frac{1}{a}\ln\left(\frac{a-bq_0}{bq_0}\right). \tag{C.10}$$

Since $t \geq 0$, we need the parenthetical term to be greater than 1. (Recall $\ln x > 0$ only when $x > 1$.) Thus, we need

$$\frac{a-bq_0}{bq_0} > 1 \quad \Longrightarrow \quad a - bq_0 > bq_0,$$

which yields $q_0 < \frac{a}{2b}$. Working backwards, when this is true the right-most equation in (C.10) produces the unique solution to $q''(t) = 0$; let's call that t-value t^*. Moreover, since $(a-bq_0) - bq_0e^{at}$ is positive for $t < t^*$ and negative for $t > t^*$, we conclude that $q(t)$ is concave up on $(0, t^*)$ and concave down on (t^*, ∞).

(c) Since $a > 0$, $e^{-at} \to 0$ as $t \to \infty$. Thus,

$$\lim_{t\to\infty} q(t) = \frac{a}{b}.$$

(d) The information from (3) is equivalent to the carrying capacity, so that $\frac{a}{b} = 20$. Since we're given that $a = 0.4 = \frac{2}{5}$, we conclude that $b = \frac{a}{20} = \frac{1}{50}$. And since $q_0 = 5$, we get that

$$q(t) = \frac{2}{\frac{1}{10} + \left(\frac{2}{5} - \frac{1}{10}\right)e^{-0.4t}} = \frac{20}{1 + 3e^{-0.4t}}.$$

Figure C.3 shows the graph of $q(t)$.

(e) After 1 hour,

$$q(1) = \frac{20}{1 + 3e^{-0.4}} \approx 6.6,$$

meaning that almost seven people have a cold at the end of the lecture. ■

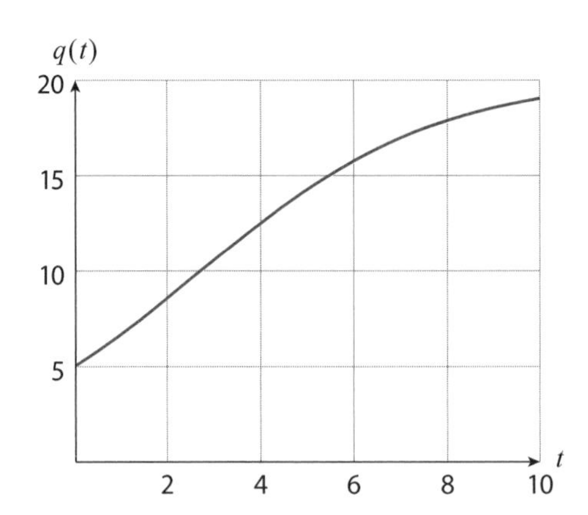

Figure C.3: Graph of the logistic function in Applied Example C.7.

APPLIED EXAMPLE C.8 A jet airplane sits on a runway waiting to take off. Suppose the airplane starts from rest and accelerates at the constant rate of 16,876 mi/hr^2.

(a) Calculate the plane's distance function $d(t)$.

(b) If the plane must reach a speed of about 150 mi/hr before it can safely take off, what is the minimum runway length needed?

Solution

(a) We're given that $s'(t) = 16,876$, so that $s(t) = 16,876t$. Mimicking Applied Example 5.1, it follows that

$$d(t) = \frac{1}{2}(t)(16,876t) = 8,438t^2.$$

(b) Since it takes the airplane $t = \frac{150}{16,876}$ hours to reach its takeoff velocity, by that time it has traveled

$$d\left(\frac{150}{16,876}\right) = 8,438\left(\frac{150}{16,876}\right)^2 \approx 0.67 \text{ miles.}$$

Therefore, the runway should be at least two-thirds of a mile long. (Typical runway lengths range from about 1.1 miles to 1.5 miles.) ■

Answers to Appendix and Chapter Exercises

Appendix A Exercises

1. (a) $(-1, 2)$ (b) $(3, \infty)$ (c) $(-\infty, -7]$ (d) $[0, 1]$

2. (a) $(x+3)(x+1)$ (b) $(2x+1)(3x+1)$ (c) $(x+2)(3x+4)$

3. (a) $x = \pm 4$ (b) $x = \pm 3$ (c) $x = 0, x = 4$

4. (a) $x = \pm 2\sqrt{2}$ (b) $x = -6, x = 2$ (c) $x = -1/2, x = -1/3$ (d) $x = \frac{7 \pm \sqrt{37}}{2}$

5. (a) $\frac{1}{x+1}$ (b) $\frac{x+2}{(x-2)^2}$ (c) $\frac{1}{x} + \frac{1}{x+1} = \frac{2x+1}{x(x+1)}$ (d) $x^6(2x+7)^3$ (e) $4a^2b^2\sqrt{b}$ (f) $x^{2/15}$

6. The next row in Pascal's triangle is $1, 5, 10, 10, 5, 1$. So: $(x+a)^5 = x^5 + 5ax^4 + 10a^2 x^3 + 10a^3x^2 + 5a^4x + a^5$

7. 2 feet **8.** By a factor of 4

Appendix B Exercises

1. False. There are two y-values for many x-values. (For example, for $x = 0$, $y = \pm 1$.) Another way to see this is to plot $y = \pm\sqrt{1 - x^2}$. What you get is the unit circle (see Figure A, from which it is visually clear that there are *many* x-values that are associated with *two* y-values.

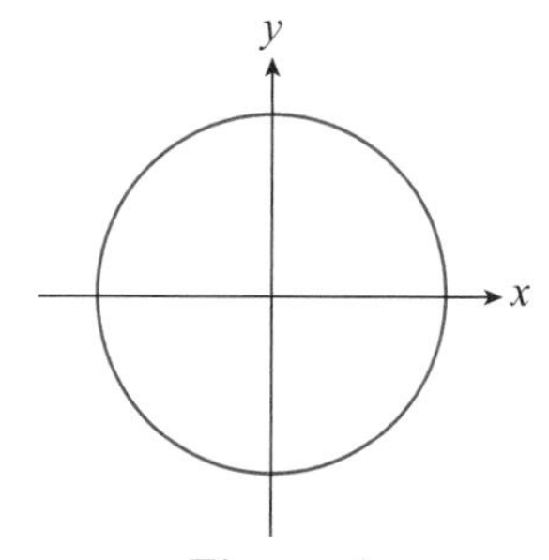

Figure A

2. (a) $f(0) = f(2) = 0$, $g(2) = 5$ (b) $[-2, 2]$ (c) $[-7.5, 7.5]$

3. $[-1, 1]$ **4.** $\mathbb{R}$ **5.** $\mathbb{R}$ **6.** $[0, 2]$

Figure B shows the graphs for exercises 7–10.

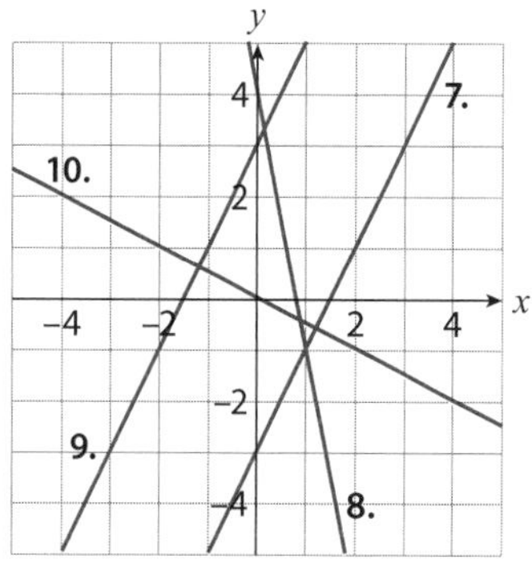

Figure B

7. $m = 2$, $(0, -3)$ **8.** $m = -5$, $(0, 4)$

9. $m = 2$, $(0, 3)$ **10.** $m = -1/2$, $(0, 0)$

11. (a) $y = -\frac{x}{5} + b$ (b) $y = -\frac{x}{5} + \frac{6}{5}$ (c) $y = -x + 3$

12. (a) Constant polynomial (also a linear function with $m = 0$)

13. Polynomial (quadratic) **14.** Power $(5x^{\frac{2}{3}})$

15. Polynomial (cubic)

16. All real numbers except for $x = 0$ and $x = 1$

17. All real numbers except $x = 1$

18. (a) $C(m) = 20 + 50m$ (b) 9.6 months

19. About 3.24 seconds later

20. (a) $M_1(20) = 200$, $M_2(20) = 189.2$. These give the maximum heart rate at age 20, calculated using the linear and quadratic models, respectively, given in the exercise.

(b) The two answers are $a \approx 38.2$ and $a \approx 104.6$. M_1 is an overestimate for $a < 38.2$ (approximately) and $a > 104.6$ (approximately); it is an underestimate for $38.2 < a < 104.6$ (approximately).

21. (a) $\frac{1}{2}$; the second most common word appears half as often as the first most common word.

(b) $\frac{r}{r+1}$; the frequency of appearance of the $r+1$-th most common word is $\frac{r}{r+1}$ times the frequency of appearance of the r-th most common word.

22. (a) Solving $aP_n r_n^4 = aPr^4$ for r_n yields the formula

(b) Substituting $r_n = 0.84r$ into P_n yields $P_n \approx 2P$.

23. (a) $f(x) = \sqrt{x}$, $g(h) = 1.5h$

(b) (1) About 2.74 miles (2) About 38.83 miles

24. (a) Slope: H, y-intercept: $(0, 0)$

(b) This follows from solving $v = Hvt$ for t.

26. No; $f(-1)$ does not exist, yet $g(-1) = -2$

27. The slope is $m = ad$.

30. Growth with base 10 and initial value 1

31. Growth with base e and initial value 4

32. Decay with base 2^{-1} and initial value 1

33. $\frac{e^2}{4}$ **34.** $\ln 3 - 1$ **35.** -2 **36.** 2

37. $\ln 128$ **38.** $\ln \frac{x-y}{x+y}$ **39.** $2 - \frac{1}{2}\ln 2$

40. 2 **41.** $y = 3 \cdot 2^x$

42. $P(t) = 309.3(1.0075)^t$; $P(15) \approx 346$ million

43. (a) $v(0) = 13.92$ is the initial velocity of the raindrop (b) $v(t) = 13.92 - 13.92(e^{-2.3})^t$

(c) $e^{-2.3} < 1$, so $13.92(e^{-2.3})^t$ decays exponentially to zero, making $v(t)$ larger as t gets larger

(d) The plot below suggests the terminal velocity is 13.92 ft/s.

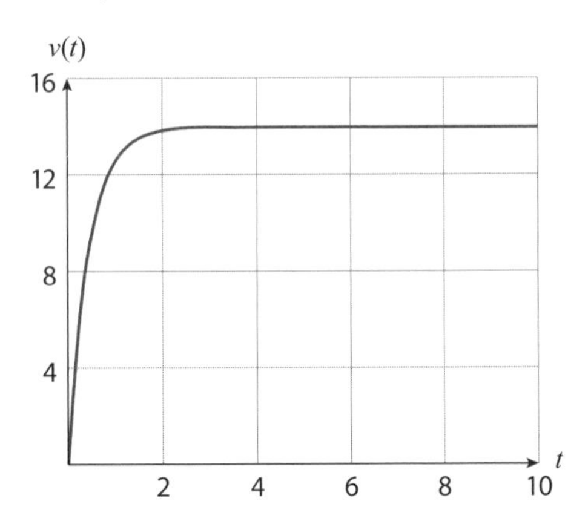

44. (a) $B(t) = \frac{100{,}000}{7}(e^{0.07t} - 1)$; $B(40) \approx \$220{,}638$

(b) $\frac{D(t)}{B(t)} = \frac{7t}{100(e^{0.07t}-1)}$; here's the plot:

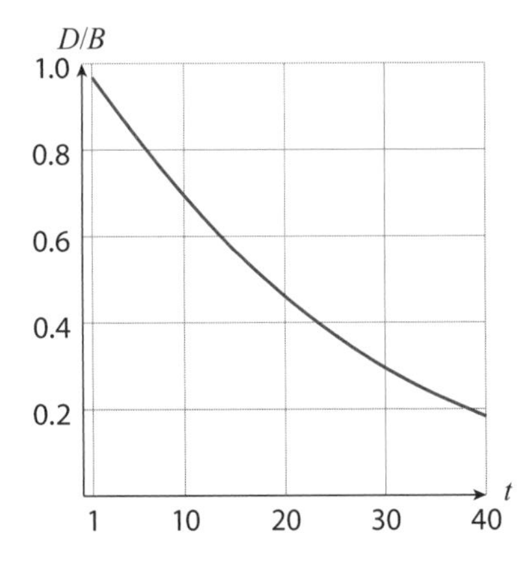

The graph shows that as time increases the deposits make up a smaller fraction of the balance (i.e., most of the increase in balance comes from the return on the invested sum of money).

(c) $\frac{D(20)}{B(20)} \approx 45.8\%$; $\frac{D(40)}{B(40)} \approx 18.1\%$

45. (b) $\lambda \approx 0.00012$ (c) $\approx 2{,}949$ years old

46. (a) About 69.7 months (c) Here's the plot:

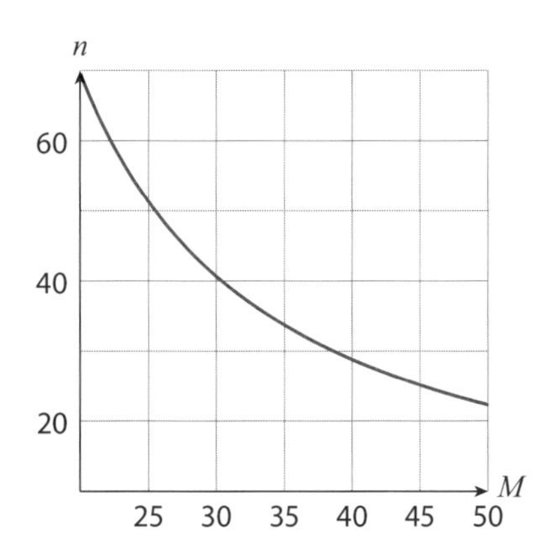

We have $n(40) \approx 28.9$, so the debt is paid off about $69.7 - 28.9 = 40.8$ months earlier.

48. ≈ 2.09 rad **49.** ≈ 0.63 rad **50.** $630°$

51. $67.5°$ **52.** ≈ 0.7 inches

53. (a) $\sin\theta = \frac{4}{5}$, $\cos\theta = \frac{3}{5}$, $\tan\theta = \frac{4}{3}$
(b) $\sin\theta = \cos\theta = \frac{\sqrt{2}}{2}$, $\tan\theta = 1$

54. (a) Using $\lambda = 700$, $C(t) = \sin\left(\frac{\pi t}{350}\right)$
(b) violet ($\lambda = 400$)

55. (a) $S(t) = \sin(880\sqrt[4]{2}\pi t)$
(b) $f(0)$ is the frequency (in Hz) of the A note above middle C; $f(12) = 2f(0)$ says that the frequency of A2 is double the frequency of A.

56. (a) $A = 120$ (b) 120π

60. (a) The n central angles of the n inscribed triangles must add to 2π, so $n\theta = 2\pi$.
(c) $A(4) = 2r^2$, $A(10) \approx 2.9r^2$, $A(100) \approx 3.14r^2$

Chapter 2 Exercises

1. When the limits are equal, the left- and right-hand portions of the graph of $y = f(x)$ near $x = c$ are joined at $x = c$ by either an open circle (as happens at $x = 1$ in Figure 2.3) or a closed circle. When the limits are unequal, there is a jump in y-values as one crosses $x = c$ (as happens at $x = -1$ in Figure 2.3).

2. (a) (i) 1 (ii) -1 (iii) DNE (left- and right-hand limits are not equal) (iv) 2 (v) 2 (vi) 2

(b) False, because $f(2)$ DNE

(c) The function is continuous for all x-values inside the following subintervals: $(-1, 0)$, $(0, 2)$, and $(2, 3)$.

3. (a) (i) K (ii) K (iii) DNE (no portion of graph directly to the left of $x = c$) (iv) DNE (no portion of graph directly to the right of $x = c$) (v) N (vi) M (vii) K (viii) DNE (one-sided limits DNE) (ix) DNE (one-sided limits exist but are not equal)

(b) False, because the one-sided limits are unequal

4. (a) 3 (b) 2 (c) 2 (d) $\frac{1}{2}$ (e) 1 (f) -2

5. 1 **6.** 0 **7.** $\sqrt{2}$ **8.** -1 **9.** DNE **10.** 0

11. 1 **12.** 1 **13.** $a = -1$

14. (a) $x \neq -1$ (b) $(-\infty, -1)$ and $(-1, \infty)$

15. (a) $[0, \infty)$ (b) $[0, \infty)$

16. (a) $[0, \infty)$ (b) $[0, \infty)$

17. (a) $[0, 1]$ (b) $[0, 1]$ **18.** $\mathbb{R}$

19. $(-\infty, 1)$ and $(1, \infty)$

20. $(-\infty, 0)$, $(0, 1)$, $(1, \infty)$

21. $-\infty$ **22.** $-\infty$ **23.** 0 **24.** -3

25. $\frac{\sqrt{3}}{3}$ **26.** 0 **27.** 0

28. Vertical: $x = 3$; horizontal: $y = 3$

29. Vertical: $x = \pm 1$; horizontal: $y = \frac{1}{2}$

31. (a) $\frac{1}{x}$ is never equal to zero, so in particular it's never equal to zero for any $x \neq 0$ in an interval containing 0.

32. (a) $T(0) = t$ (when the train is at rest, the passage of time is measured the same by you and by the stationary observer outside the train), $T(0.5c) = \sqrt{\frac{4}{3}}t$ (when the train is moving at velocity $0.5c$, t seconds measured relative to your watch is measured as roughly $1.15t$ seconds by the stationary observer outside the train (i.e., a 15% *longer* passage of time).

(b) As the train's velocity approaches the speed of light, t seconds measured relative to your watch is measured as larger and larger multiples of t by the stationary observer outside the train.

(c) $T(v)$ is not defined for $v > c$

33. (a) Continuous (b) and (c) not continuous

34. (a) Here's the graph:

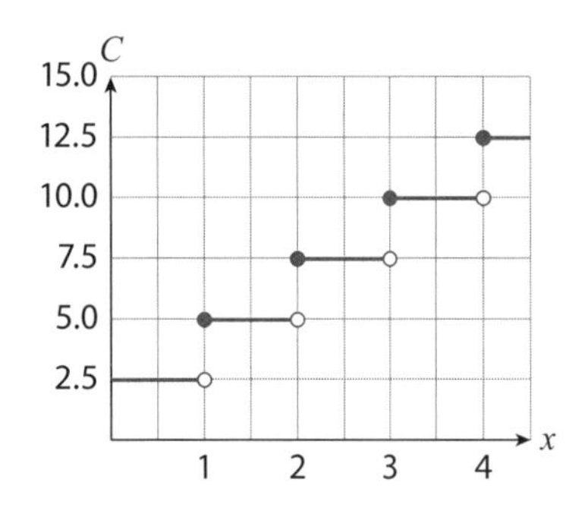

(b) No; the cost of the trip goes up by \$2.50 the instant an additional mile is traveled.

(c) Jump; the graph of $C(x)$ jumps from one y-value to another as we pass through the four discontinuities $x = 1, 2, 3, 4$.

35. (a) As r increases from zero, the gravitational force increases (linearly). When $r = R$ (the radius of the Earth), it reaches its maximum value. As r increases past R, the gravitational force decreases.

(b) Both limits are equal to $\frac{GMm}{R^2}$.

(c) Yes, because the left- and right-hand limits exist, are equal, and they are equal to $F(R)$ (and all of these quantities are real numbers).

(d) $[0, \infty)$.

37. 2 **38.** 1 and 3 **39.** 1 **40.** $\frac{1}{1+\sqrt{2}}$

41. 16 **42.** 0 **43.** 0 **44.** DNE

45. (c) Here's the graph:

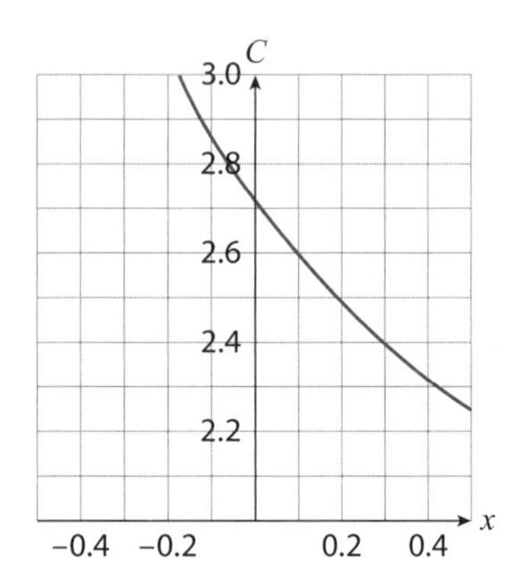

46. (e) Since $n = r/x$, $x \to 0^+$ implies $n \to \infty$. The result in (e) then says that as the number of times in a year the account is compounded approaches infinity (i.e., continuous componding), the balance of the account at the end of year y approaches M_0e^{rt}.

47. (b) Here's the graph:

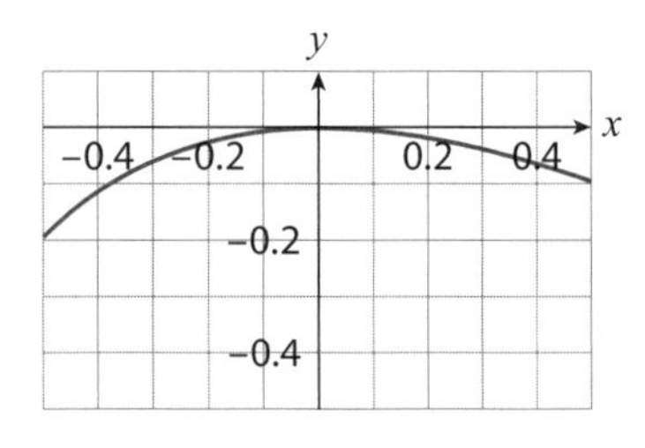

48. 13.92 **51.** 1 **52.** $\sqrt{2}$ **53.** 3 **54.** 0

55. $\frac{1}{3}$ **56.** $\frac{1}{2}$

57. $-1 \leq f(x) \leq 1$ because the amplitude of $\sin\left(\frac{1}{x}\right)$ is 1. So, $|f(x)| \leq 1$. Then $|xf(x)| \leq |x|$. So if $|x| \leq d$ it follows that $|xf(x)| \leq d$.

58. $a = \pm\sqrt{2}$

61. (a) Here's the graph:

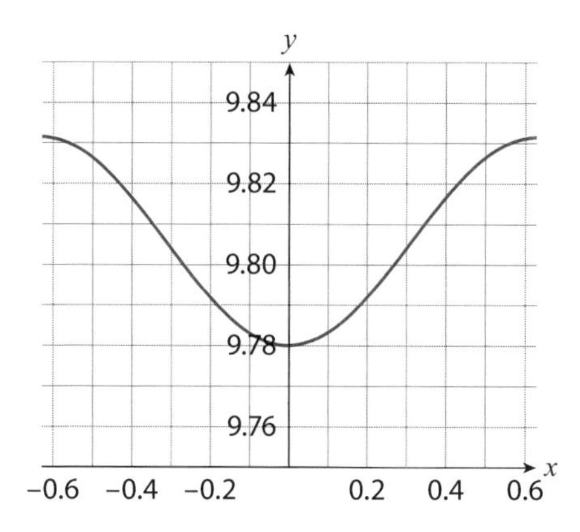

The graph indicates that g is smallest at 0° latitude (the equator) and largest at $\pm\frac{\pi}{2}$° latitude (the North and South Poles).

(b) a (the numeric value given in the exercise); interpretation: as one approaches the equator, the acceleration of gravity approaches a.

Chapter 3 Exercises

1. 0 **2.** 1 **3.** 4 **4.** -2 **5.** -4

6. $\frac{1}{2}$ **7.** $f(x) = \sqrt{x}$, $a = 16$

8. For Exercise 1: $y = 0$; for Exercise 2: $y = x + \frac{9}{2}$

9. $f'(2) = 2, f(2) = 8$ **10.** (a) 16 (b) 16

11. (a) $s(a) = 0$ (the distance function is constant, so the object is not in motion) (b) $s(a) = 2a$ (c) $s(a) = 3a^2$

12. (a) $s(a) = -2$ (the slope of the linear function $d(t)$)

13. (a) $y = 220 - t$ (b) $y = 194.8 - 0.28t$
(c) $H(t)$ predicts a constant decline of 1 bpm every year regardless of age. But a more realistic model would feature a progressively larger decline in MHR as the individual ages. This is what the $M(t)$ model predicts (its graph is a downward opening quadratic function whose tangent line slopes get more negative as t increases).

14. f is differentiable on that domain.

15. $x = 0, x = 2$

16.

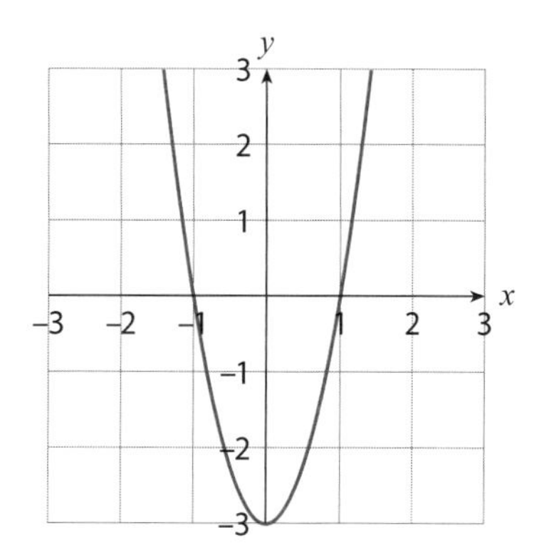

17.

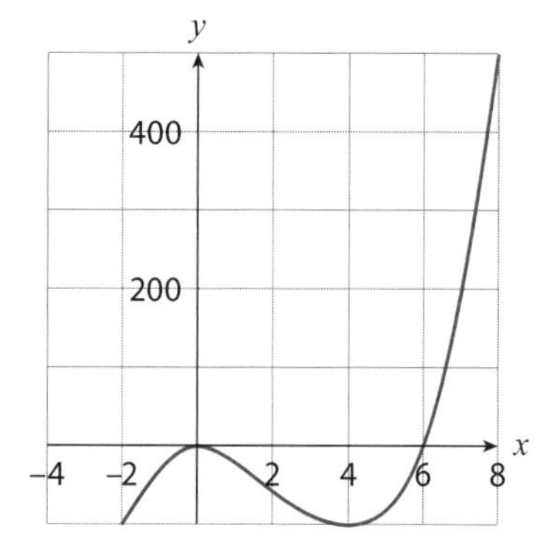

18.

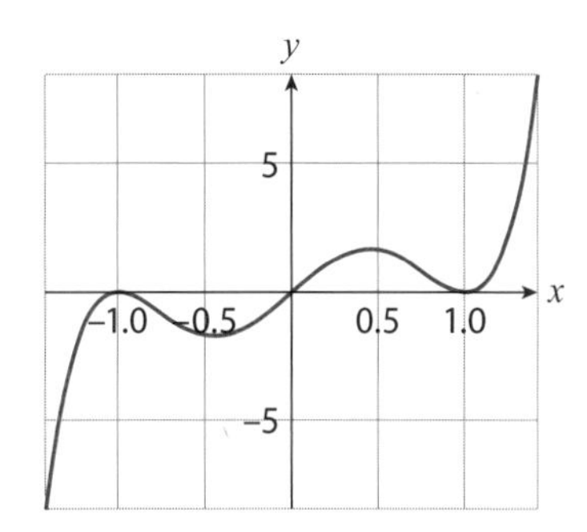

19. $f'(x)=0$ **20.** $g'(x)=50x^{49}$

21. $f'(t)=\frac{8}{\sqrt{t}}$

22. $h'(s)=7s^6-6s^2$ **23.** $f'(x)=\frac{2}{\sqrt{x}}-\frac{10}{3\sqrt[3]{x^2}}$

24. $h'(s)=\frac{\sqrt{s}}{2}(3+5s)$

25. $g'(x)=-\frac{1}{(x+1)^2}$ **26.** $h'(t)=-\frac{1}{2\sqrt{1-t}}$

27. $g'(x)=2x(\sqrt{x}-14x)+(x^2+7)\left(\frac{1}{2\sqrt{x}}-14\right)$

28. $f'(x)=\frac{1}{x^2}$ **29.** $h'(x)=\frac{2x(1+x^2)}{\sqrt{(1+x^2)^2+1}}$

30. $g'(t)=\pi t^{\pi-1}$ **31.** $h'(x)=\frac{1-x}{2\sqrt{x}(x+1)^2}$

32. $f'(x)=3\left(x^3+\frac{2}{x}\right)^2\left(3x^2-\frac{2}{x^2}\right)$

33. $f'(s)=-\frac{6}{(3s-7)^3}$

34. $g'(t)=12t^{-1/5}-(3t^2+1)$

35. (a) $f'(1)=\frac{3}{2}=1.5$

(b) Since $f'(2)=1+\frac{1}{2\sqrt{2}}\approx 1.35$, the instantaneous rate of change of f is greater at $x=1$ than at $x=2$.

(c) $f(2)-f(1)=1.414\ldots$, which is less than 0.1 away from the estimate of 1.35 from part (b)

(d) The slope of the line tangent to the graph of $f(x)=\sqrt{x}+x$ at $(1,2)$ is 1.5

(e) $y=1.5x+0.5$

36. $f''(x)=12x-6$ **37.** $f''(x)=-6x$

38. $f''(x)=-(1/4)(x+3)^{-3/2}$

39. $f''(x)=\frac{3(x+4)}{4(x+3)^{3/2}}$

40. Yes: $f'(x)=(4/3)x^{1/3}$, so $f'(0)=0$. No: $f''(x)=(4/9)x^{-2/3}$ and so $f''(0)$ DNE.

41. If $f'(x)=0$ then all tangent lines to the graph of f are horizontal. Moreover, $f'(x)=0$ for all x implies f is differentiable for all x, and therefore continuous for all x. Conclusion: f is a constant function, $f(x)=c$, c a real number. If $f''(x)=0$, the previous argument implies $f'(x)=c$, c a real number. This says that all tangent lines to the graph of f have slope c. If $c=0$ then we're back to f a constant function. If $c\neq 0$ then $f(x)=cx+d$, d a real number. Conclusion: f is a linear function.

42. $j(t)=6$ mi/hr^3

43. The (instantaneous) rate of decrease is measured by $U'(t)$. If $U'(t)$ is increasing, then tangent lines to its graph slope upward, meaning $U''(t)>0$.

44. (a) The total cost of repaying a 5% interest rate student loan is $10,000.

(b) Units: $ per percent interest; interpretation: When the interest rate of the student loan is 5%, the total cost of repayment is increasing at the instantaneous rate of $1,000 per percent interest.

(c) Positive, because as r increases so does the repayment cost

45. (a) $g(0)\approx 9.8$ m/s^2 (b) $g'(h)=-\frac{2GM}{(R+h)^3}$, $g'(0)=-\frac{2GM}{R^3}\approx -3.08\times 10^{-6}$ (m/s^2)/m

46. (a) $T(9.81) \approx 2.006$ seconds

(b) $g(T) = \frac{4\pi^2 l}{T^2}$ m/s^2; $g(2.006) \approx 9.81$ m/s^2

(c) $T(g(h)) = \frac{2\pi\sqrt{l}}{\sqrt{GM}}(R+h)$

(d) $f'(0) = \frac{2\pi}{\sqrt{GM}} \approx 3.15 \times 10^{-7}$ seconds per meter. Since $f'(0)$ is the slope of the linear function $f(h) = T(g(h))$, we can interpret it as follows: For every 1-meter increase in altitude the period of small oscillations of a 1-meter long pendulum increases by about 3.15×10^{-7} seconds.

47. (a) $h(F) = s(C(F)) = 20.05\sqrt{\frac{5}{9}F + 255.37\overline{2}}$ (the bar above the 2 indicates a repeating decimal)

(b) $h(68) \approx 343.29$ m/s; $\frac{c}{h(68)} \approx 873,900$, indicating that light is almost 874,000 times faster than sound

(c) $h'(68) = s'(C(68))C'(68) = \frac{5}{9}s'(20) = \frac{5}{9}\left[\frac{20.05}{2\sqrt{20+273.15}}\right] \approx 0.32$ meters per second per degree Fahrenheit

48.

$$f'(x) = \begin{cases} -1 & \text{if } x < 0 \\ 1 & \text{if } x > 0, \end{cases}$$

and $f'(0)$ does not exist.

49. $f'(x) = 0$ for all $x \neq 0$, and $f'(0)$ does not exist.

51. $y = -2x, \quad y = 2x$

52. $f'(x) = g(x^2) + 2x^2 g'(x^2)$

53. $f'(x) = 4e^{4x}$ **54.** $f'(x) = -(2x)2^{-x^2}$

55. $g'(t) = 2(t^2 + t + 1)e^{2t}$ **56.** $h'(z) = \frac{e^z - e^{-z}}{2}$

57. $f'(x) = \frac{2x}{x^2+5}$ **58.** $f'(z) = \frac{(1 - z\ln(3z))e^{-z}}{z}$

59. $h'(t) = \frac{1-t^2}{t^3+t}$ **60.** $g'(t) = -\frac{2e^t}{e^{2t}-1}$

63. (b) $T'(0) = -27.03$; interpretation: when the coffee mug is taken off the warming plate, its temperature is decreasing at the instantaneous rate of 27.03° F per minute.

(c) $T'(t) = -27.03e^{-0.318t}$ (d) $T(t)$ has the horizontal asymptote $y = 75°$ F; interpretation: eventually the coffee will cool down to $y = 75°$ F (the ambient temperature).

64. (a) a; interpretation: in the long run, only $100a$% of the information learned initially is retained.

(b) $R(t) = e^{(\ln 0.7)t} = (0.7)^t$

(c) $R'(1) = 0.7\ln 0.7 \approx -0.25$; interpretation: one day after learning something new (and assuming no review in the interim), your retention of the information is decreasing at the instantaneous rate of 25% per day.

65. (b) The instantaneous rate of change of the probability of wind gusts of speed 0 mph occurring near the turbine is a.

67. $f'(x) = 12x^2 - 3\cos x$

68. $f'(x) = \frac{\cos x - 2x\sin x}{2\sqrt{x}}$

69. $f'(x) = \frac{1 - \tan x + x\sec^2 x}{(1-\tan x)^2}$

70. $f'(z) = \cos z - 1$

71. $g'(x) = -\sin x - 2\cot x \csc^2 x$

72. $h'(t) = \frac{t\cos t - \sin t}{t^2}$ **73.** $g'(t) = -\frac{1}{1+\sin t}$

74. $h'(z) = 2z^3 \sin z(2\sin z + z\cos z)$

78. $a = 0$: $\theta = 0$; $a = \pm 1$: $\theta \approx 71.6°$; interpretation: the tangent line is horizontal at $x = 0$ and is inclined about 71.6° from the x-axis at $x = \pm 1$.

79. (b) 0; as the number of triangles in Exercise 60 of Appendix B inscribed in a circle of radius r grows arbitrarily large, the instantaneous rate of change of the sum of areas of those triangles approaches zero (in a nutshell: "at $n = \infty$" we stop adding area to $A(n)$).

80. (a) Amplitude: θ_0; period: $2\pi\sqrt{\frac{l}{g}}$
(b) $\theta(t) = \frac{\pi}{60}\cos(\sqrt{9.81}t)$
(c) $T = \frac{2\pi}{\sqrt{9.81}} \approx 2.00607$
(d) $T\left(\frac{\pi}{60}\right) = \frac{2\pi}{\sqrt{9.81}}\left[1 + \frac{1}{16}\left(\frac{\pi}{60}\right)^2\right] \approx 2.00641$
(e) $T'\left(\frac{\pi}{60}\right) = \frac{\pi}{4\sqrt{9.81}}\left(\frac{\pi}{60}\right) \approx 0.01$; interpretation: when the initial amplitude of a 1-meter-long pendulum is 3°, the period is increasing at the instantaneous rate of about 0.01 seconds per degree.

Chapter 4 Exercises

1. $L(x) = 0$ **2.** $L(x) = 1 + \frac{1}{2}(x-1)$

3. $L(x) = 1 - (x-1)$ **4.** $L(x) = 8 + 12(x-2)$

5. The actual value is $\sqrt{10} = 3.162\ldots$; using $f(x) = \sqrt{x}$ with $a = 9$: $\sqrt{10} \approx 3 + \frac{1}{6}(10-9) \approx 3.167$.

6. The actual value is $(1.01)^6 = 1.0615\ldots$; using $f(x) = x^6$ with $a = 1$: $(1.01)^6 \approx 1 + 6(1.01 - 1) = 1.06$.

7. The actual value is $\frac{1}{\sqrt{3}} = 0.57\ldots$; using $f(x) = x^{-1/2}$ with $a = 4$: $\frac{1}{\sqrt{3}} \approx \frac{1}{2} - \frac{1}{16}(3-4) = 0.4375$.

8. The actual value is $\sqrt[3]{2} = 1.25\ldots$; using $f(x) = x^{1/3}$ with $a = 1$: $\sqrt[3]{2} \approx 1 + \frac{1}{3}(2-1) \approx 1.33$.

9. (a) $(-\infty, -3)$ and $(2, \infty)$ (b) $(-3, 2)$
(c) $x = -3$ and $x = 2$ (d) local max. at $(-3, 81)$, local min at $(2, -44)$

10. (a) $(-\infty, -1)$ and $(1, \infty)$
(b) $(-1, 0)$ and $(0, 1)$ (f is not defined at $x = 0$)
(c) $x = -1$, $x = 0$, and $x = 1$
(d) Local max. at $(-1, -2)$, local min. at $(1, 2)$

11. (a) $\left(\frac{1}{2} - \frac{\sqrt{5}}{2}, \frac{1}{2}\right)$ and $\left(\frac{1}{2} + \frac{\sqrt{5}}{2}, \infty\right)$
(b) $\left(-\infty, \frac{1}{2} - \frac{\sqrt{5}}{2}\right)$ and $\left(\frac{1}{2}, \frac{1}{2} + \frac{\sqrt{5}}{2}\right)$
(c) $x = 0.5$, $x = \frac{1}{2} \pm \frac{\sqrt{5}}{2}$
(d) Local max. at $\left(\frac{1}{2}, \frac{9}{16}\right)$, local mins. at $\left(\frac{1}{2} - \frac{\sqrt{5}}{2}, -1\right)$ and $\left(\frac{1}{2} + \frac{\sqrt{5}}{2}, -1\right)$

12. (a) $(-\infty, -6)$ and $(0, \infty)$
(b) $(-6, -3)$ and $(-3, 0)$ (f is not defined at $x = -3$)
(c) $x = -6$, $x = -3$, and $x = 0$
(d) Local max. at $(-6, -12)$, local min. at $(0, 0)$

13. (a) $x = 3$ (b) $x = 1$

14. (a) $x = 3$ (b) $x = \pm 1$

15. (a) $x = 1$ (b) $x = 0$

16. (a) $x = 1$ (b) $x = 0$

17. (a) $(0.5, \infty)$ (b) $(-\infty, 0.5)$; infl. pt. at $x = 0.5$

18. (a) $(-\infty, 0)$ (b) $(0, \infty)$; infl. pt. at $x = 0$

19. (a) $(-\sqrt{3}/3, \sqrt{3}/3)$
(b) $(-\infty, -\sqrt{3}/3)$ and $(\sqrt{3}/3, \infty)$; infl. pts. at $x = \pm\sqrt{3}/3$

20. (a) $(-\infty, \infty)$ (b) none; no infl. pts.

21. 2/3 in./min.

22. $\frac{1}{16\pi}$ cm/s **23.** $\frac{3}{1000\pi}$ liters/s

24. $3\sqrt{5}$ ft/s **25.** $\frac{20}{3\pi}$ cm/s **26.** $\frac{5\sqrt{10}}{2}$ m/s

27. 50 mph **28.** 2.5 ft/sec

29. The point $\sqrt{2}/2 \approx 0.7$ miles directly east of her starting point

30. (a) $p(x) = 350 - \frac{x}{100}$ (b) $p(17,500) = \$175$

31. (a) $\overline{R}'(x) = \frac{xR'(x) - R(x)}{x^2}$
(b) $\overline{R}'(x)$ is undefined when $x = 0$; $\overline{R}'(x) = 0$ when $R'(x) = \frac{R(x)}{x} = \overline{R}(x)$; x-values satisfying this last equation are such that the average revenue generated by selling x units is equal to the revenue generated by selling x units.

33. When standing at ground level, an increase in elevation of 1 meter decreases the acceleration of gravity by approximately 3.08×10^{-6} m/s^2.

34. The absolute max is 2,500; the absolute minimum doesn't exist because the product is $x(100-x) = -x^2 + 100x$, which has no minimum value.

36. $\frac{25}{11}(3\sqrt{3}-4)$

39. (a) Increasing on $(-\infty, 1)$; decreasing on $(1, \infty)$ (b) 1 (c) local maximum at $x=1$ (d) absolute minimum at $x=2$; absolute maximum at $x=1$ (e) concave down on $(1,2)$; no inflection point inside the interval

40. (a) Increasing on $(0, \infty)$; decreasing on $(-\infty, 0)$ (b) 0 (c) local minimum at $x=0$ (d) absolute minimum at $x=1$; absolute maximum at $x=2$ (e) concave up on $(1,2)$; no inflection points inside the interval

41. (a) Increasing on $(2, \infty)$; decreasing on $(0,2)$ (b) 2 (c) local minimum at $x=2$ (d) absolute minimum at $x=2$; absolute maximum at $x=1$ (e) concave up on $(1,2)$; no inflection points inside the interval

42. (a) Increasing on $(0, \sqrt{e})$; decreasing on $(\sqrt{e}, \infty)$ (b) $\sqrt{e}$ (c) local maximum at $x=\sqrt{e}$ (d) absolute minimum at $x=1$; absolute maximum at $x=\sqrt{e}$ (e) concave down on $(1,2)$; no inflection points inside the interval

43. (a) When $b>1$, $\ln b>0$, and so $f'(x)>0$ for all x. This implies that f is increasing (Theorem 4.1) for all x, and thus has no local extrema. When $0<b<1$, $\ln b<0$, and so $f'(x)<0$ for all x. This implies that f is decreasing (Theorem 4.1) for all x, and thus has no local extrema.

(b) Since $f''(x)>0$ for all x, Theorem 4.7 implies f is concave up for all x. Therefore, there are no changes in concavity (i.e., no inflection points).

44. (a) For $0<b<1$: decreasing on $(0,\infty)$. For $b>1$: increasing on $(0,\infty)$. Since g' never changes sign, g has no local extrema.

(b) (a) For $0<b<1$: concave up on $(0,\infty)$. For $b>1$: concave down on $(0,\infty)$. Since g'' never changes sign, g has no inflection points.

46. (c) Here's the graph:

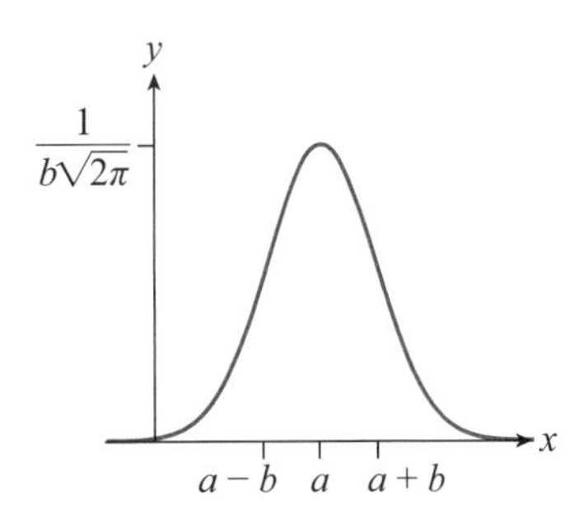

47. (a) $R'(\lambda) = \frac{e^{2/\lambda}(2-5\lambda)+5\lambda}{\lambda^7(e^{2/\lambda}-1)^2}$ (b) Since $R'(\lambda) > 0$ for $0<\lambda<0.4$ (approximately), $R'(\lambda)<0$ for $\lambda>0.4$ (approximately), and R is continuous, it follows from Theorem 4.4 that R has an absolute maximum at $x \approx 0.4$

48. (a) $G(t) = e^{1-e^{0.085t}}$; $G(0)=1$, which tells us that the probability of surviving to age 0 after a successful birth is 100%.

(b) 0; this tells us that Gompertz survival curves predict that the probability of surviving approaches zero as the survival age gets arbitrarily large.

(c) Interpretation: The probability of surviving to age t after a successful birth decreases as age increases.

(d) Interpretation: The probability of surviving to age t after a successful birth is decreasing at an increasing rate as age increases.

49. (c) $v = \frac{1}{\sqrt{e}} \approx 0.6$ m/s **50.** $\frac{2\pi}{3}$ ft/s.

51. $\frac{40\pi}{3}$ miles/min.

53. (a) None (b) decreasing on $(0,\pi)$ (c) local maximum at $f(0)$; local minimum at $f(\pi)$ (d) absolute maximum at $x=0$; absolute minimum at $x=\pi$ (e) concave down on $(0,2\pi/3)$; concave up on $(2\pi/3,\pi)$; inflection point at $x=2\pi/3$

54. (a) Increasing on $(-\pi/3,\pi/3)$ (b) none (c) local minimum at $g(-\pi/3)$; local maximum at $g(\pi/3)$ (d) absolute maximum at $x=\pi/3$; absolute minimum at $x=-\pi/3$ (e) concave up on $(-\pi/3,0)$; concave down on $(0,\pi/3)$; inflection point at $x=0$

55. (a) Increasing on $(0,\pi/6)$; decreasing on $(\pi/6,\pi/2)$ (b) $\pi/6$ (c) local minima at $h(0)$ and $h(\pi/2)$; local maximum at $h(\pi/6)$ (d) absolute maximum at $t=\pi/6$; absolute minimum at $t=\pi/2$ (e) concave down on $(0,\pi/2)$; no inflection points in the interval of interest

56. (a) Increasing on $(\pi/2,3\pi/2)$; decreasing on $(\pi/4,\pi/2)$ and $(3\pi/2,7\pi/4)$ (b) $\pi/2$ and $3\pi/2$ (c) local maxima at $g(\pi/4)$ and $g(3\pi/2)$; local minima at $g(\pi/2)$ and $g(7\pi/4)$ (d) absolute maximum at $x=7\pi/4$; absolute minimum at $x=\pi/4$ (e) concave up on $(\pi/4,\pi)$; concave down on $(\pi,7\pi/4)$; inflection point at $s=\pi$

58. (c) Since $0\le\mu\le 1$, then $0\le\mu^2\le 1$ and $1\le 1+\mu^2\le 2$. It follows that $1\le\sqrt{1+\mu^2}\le\sqrt{2}$, and so $\frac{1}{\sqrt{1+\mu^2}}\le 1$. Multiplying by the nonnegative number μmg yields $\frac{\mu mg}{\sqrt{1+\mu^2}}\le\mu mg$. Finally, $\mu mg\le mg$ follows from $\mu\le 1$.

59. (a) $r(0)=a(1-e)$, $r(\pi)=a(1+e)$. Note that $r(\pi)=r(0)+2ae$, so that $r(\pi)-r(0)=2ae>0$

(c) closest: $r(0)\approx 9.14\times 10^7$ miles; farthest: $r(\pi)\approx 9.46\times 10^7$ miles

61. (c) $t''(x)>0$ on $[0,L]$

63. (a) When n is large, $\frac{2\pi}{n}$ is near zero. Letting $x=\frac{2\pi}{n}$, the result then follows from $\sin x\approx x$ (recall (4.14)).

Chapter 5 Exercises

1. $A(t)=10t$ **2.** $A(t)=t-t^2/2$

3.

$$A(t)=\begin{cases}\frac{t^2}{2} & 0\le t\le\frac{1}{2}\\ \frac{1}{4}+\left(t-\frac{1}{2}\right) & \frac{1}{2}\le t\le\frac{3}{2}\\ \frac{5}{4}+\frac{1}{2}\left(t-\frac{3}{2}\right)(4-2t) & \frac{3}{2}\le t\le 2\end{cases}$$

4. (a) 0.25 (b) 0.75 (c) 1.25

5. (a) Left: (2, 3); right: (0, 2) (b) $\int_0^1 v(x)dx=\frac{3}{4}$ is the distance traveled in the first unit of time; $\int_0^3 v(x)\,dx=\frac{3}{4}$ is the displacement during the first three units of time.

6. (a) Here's the graph for a random x-value in [0, 1]:

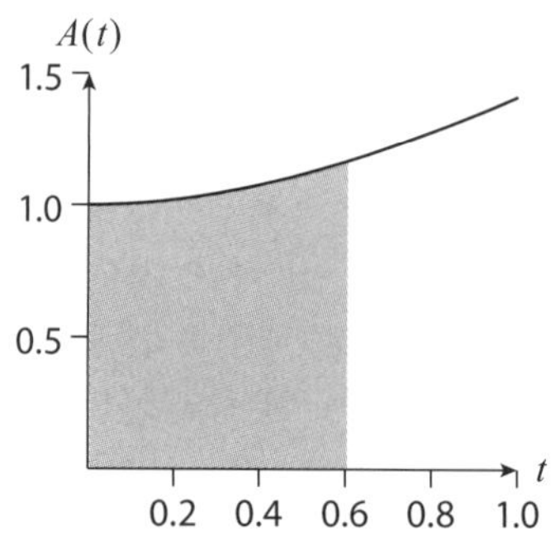

(b) $A'(t)=\sqrt{1+t^2}$; since $A'(t)>0$, it follows that $A(t)$ is increasing everywhere, including on [0, 1].

(c) $A''(t)=\frac{t}{\sqrt{1+t^2}}$; since $A''(t)>0$ for $t>0$, it follows that $A(t)$ is concave up on the subinterval (0, 1] of [0, 1].

7. (a) $A'(t)=t$ (b) $g'(t)=2t^3$ **8.** $\pi/2$

9. $f(x)=0, f(x)=\frac{1}{2}x+C$ **10.** 3

11. 1/2 **12.** 2

13. (a) F.T.C., Theorem 5.1 (b) the Sum Rule, Theorem 3.3

(c) Setting $t=a$ yields $A_{f+g}(a)=A_f(a)=A_g(a)=0$ (since the area bounded by $x=a$ and $x=a$ of f, g, and $f+g$ is zero). This forces $C=0$, which then yields the (integral) Sum Rule.

14. (a) The slopes of the tangent lines to the graphs of $A(t)$ and $d(t)$ are the same at every t-value. (b) $g'(t)=0$, because $g'(t)$ is the difference of the slopes of the tangent lines to the graphs of $A(t)$ and $d(t)$ at the same t-value, and by (a) this difference is zero. (c) $g'(t)=0$ says that every tangent line to the graph of $g(t)$ has slope zero. This implies that $g(t)$ must be a constant function. And since $g(t)=d(t)-A(t)$, it follows that $d(t)-A(t)$ is constant.

15. (a) $L(0)$ is the percentage of the nation's income earned by the bottom 0% of households, which is zero. $L(1)$ is percentage of the nation's income earned by the bottom 100% of households, which is 100%. Thus, $L(1)=1$ (recall that the percentages x and $L(x)$ are converted to decimal form by dividing them by 100). Since x and $L(x)$ are percentages, they range from 0% to 100%, or 0 to 1 in decimal form. Thus, x and $L(x)$ are both numbers between 0 and 1. (b) If every household has the same income, then the bottom x% of households will earn x% of the nation's income. This implies that $L(x)=x$. Using this in (5.26) yields an area of zero, and therefore $G=0$. (c) The percentage of the nation's income earned by the bottom x% of households is less than x%. (d) When $L(x)<x$, we have $2x-2L(x)>0$. And since G is the area under the graph of this function between $x=0$ and $x=1$ (recall (5.26)), $G>0$.

16. There are roughly 24.5 blue-colored boxes under the curve. Each such box has area 0.05, so an estimate for $\int_0^T c(t)\,dt$ is $(24.5)(0.05)=1.225$. Therefore, $F\approx\frac{A}{1.225}\approx 0.82A$.

17. 8 **18.** 2/3 **19.** $\frac{3}{\sqrt[3]{4}}$ **20.** $\frac{7}{10}$

21. $\frac{2}{3}x^{3/2}-4\sqrt{x}+C$ **22.** $\frac{y^3}{3}-\frac{3}{2}y^2+2y+C$

23. 7 **24.** $\frac{16}{15}$

25. $-\frac{2}{15}\sqrt{1-t}(3t^2+4t+8)+C$

26. $\frac{(2\sqrt{2}-1)a^3}{3}$

27. Theorem 5.2 is being used, but $f(x)=x^{-2}$ is not continuous on the interval $[-1,1]$.

28. 7 **29.** 8 **30.** 2 **31.** $\frac{1}{2}$

32. Assuming $r(t)$ is continuous, Theorem 5.2 tells us that the integral represents the net change in oil consumption between 2017 and 2027 (measured in barrels of oil). Since we expect the world's oil consumption to continue increasing over the decade starting in 2017, we expect that net change to be positive (meaning that we expect greater world oil consumption in 2027 compared to 2017).

33. (a) v_x is constant and there are no forces being considered in the horizontal direction, so $x(t)=v_xt$.
(b) $A=-\frac{g}{2v_x^2}$, $B=\frac{v_y}{v_x}$, $C=d$; B is the ratio of the vertical and horizontal velocities of the object, and C is the initial height of the object.

34. (b) For 1 hour, ≈ 1.73 feet; for 2 hours, ≈ 1.49 feet

35. (a) $P'(100)=-23(100)^{-1.23}$; interpretation: when 100 units have been produced, the production cost is decreasing at the instantaneous rate of change of $23(100)^{-1.23}$ \$/unit (about 8 cents per unit).
(b) $P(n)=100n^{-0.23}$ ($P(1)=100$ was used to show that the arbitrary constant $C=0$)

36. (a) Theorem 4.5

(b) By (a), at every x-value in $[t, t+\Delta t]$ $f(x) \geq f(m)$. Therefore, the area under the graph of $f(x)$ and between $x=t$ and $x=t+\Delta t$ will be greater than or equal to the area under the graph of the constant function $f(m)$ and between $x=t$ and $x=t+\Delta t$. Similar reasoning, starting from $f(x) \leq f(M)$ at every x-value in $[t, t+\Delta t]$, establishes the other inequality.

(c) The leftmost integral in (b) is the area of the rectangle of height $f(m)$ and width Δt; similarly, the rightmost integral is the area of the rectangle of height $f(M)$ and width Δt. Using these results and dividing the inequality in (b) by Δt yields the inequality in (c).

(d) As $\Delta t \to 0$, m and M approach t (since both m and M are in the ever-shrinking interval $[t, t+\Delta t]$). Thus, $f(m)$ and $f(M)$ approach $f(t)$ as $\Delta t \to 0$. It follows from the inequality in (c) that the middle term approaches $f(t)$ as $\Delta t \to 0$ too, which is the claim in (d).

41. $\frac{2\sqrt{2}}{3}$ **42.** $e^{3x}+C$ **43.** $\frac{5^x}{\ln 5}$

44. $\frac{2}{3}\left[(1+e)^{3/2}-2^{3/2}\right] \approx 2.9$ **45.** $\ln 2 \approx 0.7$

46. $\ln(\pi+e^x)+C$ **47.** $\frac{1}{3}$

52. (a) $p'(x)=\frac{1}{\ln x}$; interpretation: For large x, the number of primes less than or equal to x is increasing.

(b) $p''(x)=-\frac{1}{(\ln x)^2}$; interpretation: For large x, the instantaneous rate at which the number of primes less than or equal to x is increasing is decreasing. Said differently, our result indicates that although the number of primes less than or equal to x gets larger as x gets larger, the rate of increase ($p'(x)$) slows down ($p''(x)<0$) the larger x gets. Even more succinctly: primes spread out as x gets larger.

53. (a) $\Delta P \approx 2\pi x p(x)\Delta x$; interpretation: Compared to a radius x from a city center, the population living within a radius $x+\Delta x$ is larger by about $2\pi x p(x)$. Note that $p(x)>0$ since it counts people.

(b) $P(x)=600(1-e^{-\pi x^2/100})$; the limit is 600, and tells us that as we consider radii arbitrarily far away from the city, the population living within those distances of the city center approaches 600,000 people.

54. $\frac{t^4}{4}-\sin t+C$ **55.** $-\cot x+\cos x+C$

56. $-\frac{4}{3}(\cot t)^{3/2}+C$ **57.** $\frac{2+\sqrt{2}}{6} \approx 0.57$

58. $\sqrt{2}-1 \approx 0.41$ **59.** $\theta-\ln|\cos\theta|+C$

60. $\frac{2}{\pi}$ **64.** (b) $\frac{\pi}{8}$

62. (a) a liters per second (b) this question is about the period of $v(t)$; the answer is $\frac{2\pi}{b}$
(c) $\frac{2a}{b}$; interpretation: the volume (in liters) of air *inhaled* during one respiratory cycle

63. (a) $a=74, b=2$ (c) $c=\frac{\pi}{12}$

Bibliography

[1] Fernandez, O. (2014). *Everyday Calculus: Discovering the Hidden Math All Around Us.* Princeton, NJ: Princeton University Press.

[2] Gellish, R. L., Goslin, B. R., Olson, R. E., McDonald, A., Russi, G. D., and Moudgil, V. K. (2007). Longitudinal Modeling of the Relationship between Age and Maximal Heart Rate. *Medicine & Science in Sports & Exercise* 39(5), 822–829.

[3] West, G. B., and Brown, J. H. (2005). The origin of allometric scaling laws in biology from genomes to ecosystems: towards a quantitative unifying theory of biological structure and organization. *Journal of Experimental Biology* 208, 1575–1592.

[4] Speakman, J. R. (2005). Body size, energy metabolism and lifespan. *Journal of Experimental Biology* 208, 1717–1730.

[5] Savage, V. M., Gillooly, J. F., Woodruff, W. H., West, G. B., Allen, A. P., Enquist, B. J., and Brown, J. H. (2004). The predominance of quarter-power scaling in biology. *Functional Ecology* 18, 257–282.

[6] "What Is Inflation and How Does the Federal Reserve Evaluate Changes in the Rate of Inflation?" Board of Governors of the Federal Reserve System, January 26, 2015. http://www.federalreserve.gov/faqs/economy_14419.htm (Accessed May 15, 2017).

[7] Fernandez, O. (2017). *The Calculus of Happiness: How a Mathematical Approach to Life Adds Up to Health, Wealth, and Love.* Princeton, NJ: Princeton University Press.

[8] Wilson, Chris. "This One Simple Tool Could Save Your Relationship." TIME.com, 24 July 2017. http://time.com/4867730/relationships-money-tool/ (Accessed August 15, 2017).

[9] Frankenfield, D., Routh-Yousey, L., and Compher C. (2005). Comparison of Predictive Equations for Resting Metabolic Rate in Healthy Nonobese and Obese Adults: A Systematic Review. *Journal of the Academy of Nutrition and Dietetics* 105(5), 775–789.

Comment. In the study the authors note that "the Mifflin-St. Jeor equation is more likely than the other equations tested to estimate RMR to within 10% of that measured, but noteworthy errors and limitations exist when it is applied to individuals and possibly when it is generalized to certain age and ethnic groups."

[10] "Wind Chill Chart." *National Weather Service.* National Oceanic and Atmospheric Administration, n.d. http://www.nws.noaa.gov/om/cold/wind_chill.shtml (Accessed May 15, 2017).

[11] World Bank. "Poverty and Prosperity 2016: Taking on Inequality" (PDF). Figure O.10 Global Inequality, 1988–2013.

[12] Hubble, E. (1929). A Relation Between Distance and Radial Velocity Among Extra-Galactic Nebulae. *Proceedings of the National Academy of Sciences of the USA 15(3),* 168–173.

[13] Child, J. M. (1920). *The Early Mathematical Manuscripts of Leibniz.* Chicago: Open Court.

Index of Applications

Physical Sciences

Life Sciences

Business and Economics

Sports

Social and Behavioral Sciences

Other

Subject Index